Dugald Furguson

Vicissitudes of Bush Life in Australia and New Zealand

Dugald Furguson

Vicissitudes of Bush Life in Australia and New Zealand

ISBN/EAN: 9783337313500

Printed in Europe, USA, Canada, Australia, Japan

Cover: Foto ©berggeist007 / pixelio.de

More available books at **www.hansebooks.com**

コニ

IN

AUSTRALIA AND NEW ZEALAND

BY

DUGALD FERGUSON

(AN OLD COLONIST)

LONDON
SWAN SONNENSCHEIN & CO.
PATERNOSTER SQUARE
1891

THE ABERDEEN UNIVERSITY PRESS.

CONTENTS.

CHAPTER I.,
" II.,
" III.,
" IV.,
" V.,
" VI.,
" VII.,
" VIII.,
" IX.,
" X.,
" XI.,
" XII.,
" XIII.,
" XIV.,
" XV.,
" XVI.,
" XVII.,
" XVIII.,

CONTENTS.

		PAGE
CHAPTER XIX.,		125
„ XX.,		131
„ XXI.,		137
„ XXII.,		142
„ XXIII.,		156
„ XXIV.,		167
„ XXV.,		173
„ XXVI.,		184
„ XXVII.,		196
„ XXVIII.,		202
„ XXIX.,		205
„ XXX.,		214
„ XXXI.,		218
„ XXXII.,		223
„ XXXIII.,		228
„ XXXIV.,		238
„ XXXV.,		243
„ XXXVI.,		253
„ XXXVII.,		259
„ XXXVIII.,		265
„ XXXIX.,		275
„ XL.,		284
„ XLI.,		286

		PAGE
CHAPTER XLII.,		298
,, XLIII.,		301
,, XLIV.,		304
,, XLV.,		313
,, XLVI.,		320

BUSH LIFE IN AUSTRALIA.

CHAPTER I.

WHAT a host of associations start up at the mention of Home, especially in connection with my last view of it, in the sunny month of June; with its garments of green and purple on mountain and valley; its green waving woods and bright glistening water; the evening skies fringed with golden and purple clouds; the noon-day sun, with its golden beams, lighting up hill, wood, and water, with a dazzling brightness that seems now truly the light of other days. For, in all my wanderings since, in sultrier climes, what sun has shone so brightly, or what landscape looked so truly lovely, as those scenes of my childhood amongst the hills of Argyleshire, and by the shores of Lochfyne?

Wordsworth has some beautiful, though fanciful, lines written on the scenes of his childhood, in which, adopting the Platonic theory of a previous existence, he accounts for the golden associations of childhood's years, by the idea that an infant's soul is filled with memories of its former blissful state, and that these memories grow more faint as the child's years increase. Be that as it may, I can readily imagine of those endowed with the poetic faculty that the season of their unconscious, inarticulate childhood, is the true period of their espousal by the muse. I am not now referring to those robust geniuses who, even in their tender years, show the bent of their irrepressible natures, but only to those whose gifts are of a weaker growth; and who do not discover their talent till years have matured their powers. Such I would advise to return in thought to the habit of their youthful ideas, when everything appeared earnest; when the melody of birds, the humming of bees, the very shape of the summer clouds, the daisies and primroses, had a meaning, a purpose, an existence to them, and were fraught with an exquisite sense of harmony, which they implicitly accepted, and never dreamt of analysing. It is as we grow older, and have to deal with stern realities, when the ideal faculty becomes hampered in its scope by the encroachment of practical necessities, that we learn to smile at these undefined crudities that were once sources of such intense enjoyment to us.

Yet so strangely constituted a being is man, to whom always "Distance lends enchantment to the view," that, as the child grows up to manhood, his ideas of happiness become simply inverted. For, as now we look back to the season of our youth as the time of our purest happiness, then, though richly enjoying that season, we ever looked forward to manhood's golden future—when, freed from the shackles of school, the foreground was teeming with wealth.

Such, at least, were my ideas, when, a raw youth of eighteen, I prepared to start for Australia to join a wealthy, "squatocratic" relative, to whose friendly offices, with all the confidence of youth, I doubted not that I might leave the sole responsibility of my future advancement in life. With such golden prospects before me, it was with a light heart that I shook hands with my aged mother at Granton pier, and stepped on board the steamer that was to take me to London, *en route* for Australia. My noble, Roman spirited mother, how little did we then think, either of us, that we had taken our last look at each other on earth, and that I, who so lightly and carelessly bade thee adieu, was then being shot forth by the strong bow of destiny, across the far rolling ocean, thenceforth to be joined to an order of things, that the decrees of fate had determined should be ever separated from thine! May peace be with thy spirit, for this world now knows thee no more, though surely thou art at rest, and hast thy reward.

In the latter end of June in the year 1850, the same that witnessed the first discovery of gold in Australia, I sailed from the East India Docks in the good ship " Sussex ". After a prosperous voyage, and what was then regarded as a moderately quick one, of something over a hundred days, we sighted land, and shortly after dropped anchor inside the Philip Heads. How bright still seem all those days in retrospect. I remember with what exultation I hailed the sight of that land, that to me seemed stored with so many golden prospects. It was a cloudless, sultry October morning, and after my first delightful gaze at the blue headland, my overflowing feelings suddenly found vent in the words of a song that had been popular during the passage—

" Hurrah for the fields of Australia," etc.

I enthusiastically led the chorus, energetically marking the time with my feet as I paced along the deck.

Everything seemed bright and wonderful to me then—even the crew of the quarantine doctor's boat that presently came alongside excited my admiration, as I remarked their bronzed faces, which gave them a kind of foreign appearance in my eyes.

Melbourne was at that time only just emerging out of its chrysalis state. The habitations still chiefly consisted of huts and tents run up at the time when this land of promise was only known in connection with its pastoral interests. But, with the sudden discovery of untold treasures, large and more pretentious buildings, chiefly hotels and stores, were everywhere seen starting into existence. All was commotion and confusion, and diggers, maddened with excitement, scattered their easily-won money around them in handfuls, till the stories told of their frenzied extravagance made one marvel.

But I waited in Melbourne only until the first steamer started for Geelong, for to a merchant of that town I had a letter of introduction. Furnished by this gentleman with directions for my guidance, I started next day on foot across a vast plain, the timber land on the other side of which I entered late on the same afternoon. Mistaking my road, I there made my first acquaintance with Australian bush life by passing the night in the forest and being almost devoured by mosquitoes, until in sheer desperation I, long before dawn, started up and resumed my journey. After blundering a while amid the dark forest, I fortunately made up to the homestead I was in search of.

The men's cook, who had but lately risen, and had a fire blazing in the hut on my entrance, made me heartily welcome to a substantial breakfast. Directed by him anew, I once more started off through the forest, and after contriving to lose myself only twice again, I came across some bullock drivers who were driving their teams in the same direction, and who safely deposited me at night at the hospitable abode of a neighbouring squatter, from whose place on the ensuing day, after a short journey of four or five miles, I finally reached the home station of my kinsman.

He received me courteously, but himself had no need for my services, and accordingly, as he was then about to despatch a bullock team with goods up to another station owned by him in the Wimmera district, some hundred and twenty miles further up country, I was after a few days sent on to this station by this means.

Mr. M'Elwain, my kinsman, was a gentleman of rather pleasing appearance. He was of a slight but wiry build, well formed, and with a rather dark complexion, and hair and whiskers of a dark brown colour. Courteous and affable in manner, he was nevertheless keen in business, and a man who looked to a full return for all his payments. Owing his own considerable fortune solely to his habits of perseverance and frugality, in his ideas of the proper regimen for young men in husbanding their resources, he combined almost meanness with abstemiousness, so that, although he could prove himself a fast friend to those who secured his confidence, yet he was not a man likely to run much risk by too rashly furthering the interests of anyone of whose prudence and thrift he was not perfectly satisfied. With him the great end of life was to make money, not indeed to hoard it, but as a necessary means of making life enjoyable, and to those who, by habits of carelessness or wantonness, dissipated means which might have been husbanded for this end, he could give no sympathy.

I mention these particulars now, as they may help to explain the reasons for the sudden collapse of my expectations from the good offices of a man for whom I nevertheless continued to entertain sentiments of sincere respect. He was then over forty-five years of age, though only recently married.

My journey with the bullock dray occupied about a fortnight. During that time the only incident worth recording was a slight fracas betwixt myself and the bullock driver. He was a young man rather older than myself, who, in view of his four years' residence, laid claim to all the privileges of colonialism, the which implied, in his opinion, a decided superiority over me. He proceeded accordingly, under this belief, to hector and domineer over me, until at last, in a fit of exasperation, I vindicated my Highland blood, and "New Chum" spirit, by heartily cuffing him. With the exception of this trifling matter, the journey to me afforded a source of continual pleasure. It was bright summer weather, though extremely sultry. For a considerable distance, the road wound chiefly through the primeval forest, whose deep glades fairly echoed to the loud cracks of the bullock whip, as, with its cruel lash often applied in mere wantonness, the driver urged on his slow team, whilst the flocks of green and crimson plumaged parroquets fluttering from tree to tree, or clamorous cockatoos filling the air with their discordant cries, were things that, together with the novelty of my situation, were to me a continuous feast of pleasure. But travelling over the vast, treeless plains that, later on, spread for miles along our route, was not quite so enjoyable, save in the evening camp by some cool watering place. It was then that the "billy" brewed tea, the ash baked damper, the fried chops, were discussed with that keen relish which only a youth of healthy appetite and bounding spirits could fully appreciate.

At length, situated in a forest of box, gum, and oak, through which we had been journeying for some hours previously, I saw a paddock fence, and knew we had reached our destination, and that I should travel no further, for some time, at least. I mean, of course, that we had reached Mr. M'Elwain's upper station.

The buildings of the homestead were few and simple. The chief one was a moderate-sized wooden building, containing about four rooms, erected by Mr. M'Elwain with his own hands, for he had formerly resided there himself. Its walls inside were done up with lath and plaster. In front was a verandah, into which one room opened, whilst to the rear its dimensions were extended by what, in New Zealand, is known as a "lean-to," the whole being roofed in with wooden shingles. Adjacent to it was another hut that served as a kitchen, and, irregularly scattered in front of these two, were several more wooden huts, some built of split slabs, others merely of logs, answering respectively as stores and shearers' hut, whilst one was occupied by the weekly station hands. Outside the paddock fence, about a quarter of a mile off in the open forest, was an immense building, whose rafters appeared to fairly bend beneath the superincumbent weight of bark and logs, by which it was roofed in, and the roof secured. This was the woolshed, in front of which were erected high posts to support the lever that then formed the primitive method of compressing the wool in the frames that were fixed inside the building. Besides these, the woolshed was almost surrounded by necessary yards for the management of the sheep in driving them into the shed, branding, etc. This brief summary completes the description of the buildings of a station whose wool returns represented ten thousand sheep, and which had, in addition to that stock, a few hundred head of cattle. I, immediately on my arrival, introduced myself to Mr. M'Lean, the overseer, and presented my credentials; the result was that I found myself forthwith attached to the regular staff in the capacity of store-keeper. This situation I accepted contentedly enough, hoping that it would be merely the first rung of the ladder, the last of which my aspiring ambition fondly pictured, would at no distant date be the chief managership of that very station. I may as well say here that in this situation I remained for about two years, at the end of which time Mr. M'Elwain suddenly disposed of his station by sale. This circumstance occasioned a sudden collapse to all the pleasant little airy structures I was continually in fancy erecting regarding my rapid success in life. Mr. M'Elwain, who really had no further occasion for my services, now kindly, but with his usual caution as to committing himself, offered to find me a situation with a neighbouring squatter. But, from the well-known sordid disposition of the latter, my further trial of station life would have been so uncongenial, and the remuneration that he offered so slight in comparison to what I deemed my merits deserved, that I declined the offer in disgust. Upon this, my kinsman, no doubt regarding me as being altogether too high minded, after giving me some excellent advice for the regulation of my future movements, and with many expressions of good-will for my welfare, bade me farewell.

Although the daily events of my two years in this station were too flat and commonplace to interest a general reader, yet it was here that I made the acquaintance of Selim, the old horse, a sudden sight of whose grave, by flooding my mind with old memories, suddenly inspired me with the idea of recounting them all in a written narrative. But, as all the circumstances connected with that event form a somewhat lengthy episode, I will reserve the account of them for another chapter.

CHAPTER II.

"My beautiful, my beautiful, that standest meekly by,
With thy proudly arched and glossy neck, and dark and shining eye."
—*Mrs. Norton.*

THE time had verged upon twelve months since my arrival at Mr. M'Elwain's station. It was September, the month that ushers in the wool season in that locality, and preparations for beginning the shearing were in a forward state. The shearer's hut, generally unoccupied for the major part of the year, now presented a somewhat bee-hive-like appearance, with all its tenants crowding about the door, or sitting down on a form alongside. The course of a fine spring day was waning towards its close, when the sudden clamorous barking of the dogs signalled the approach of a stranger, and looking down the track, a single horseman was seen approaching at a quick walking pace towards the home-station: his appearance excited considerable comment from the shearers and other hands who were assembled round the door of the hut.

After several remarks had been ventured regarding the probable identity of the stranger, one of the men suddenly exclaimed:—

"It's George Laycock, I know his horse; there isn't a nag in the Wimmera can walk with him."

"Is it George Laycock, Driscoll?" said another. "So it is. I wonder what he is after now. I thought he was hunting after wild horses; by-the-bye, has he still any hope of getting Major Firebrace's mare?"

"Why, the old fool will never get her," said a third. "He has been five years now trying to catch her, and he has as little chance of getting her now as ever he had."

"Well, now, I believe different," here chimed in another, an elderly-looking man. "I knew George in the Sydney side, and a better stock-rider you won't find in the colony, and I know that he wouldn't keep on after that mare if he didn't think as how he'd yet get hold of her. But George is a cove as will stick to a thing he goes after—he'll never cave in as long as he believes there's a chance to get it."

"Well, but," replied the former speaker, "see how long he's been after that mare now, surely he ought to have knowed long ago if he had a chance or not."

"That's all very well," replied the elderly man, "but George has told me hisself that he never yet seed a proper chance to get a good hunt after the mare; there is always so much scrub where she is; and besides it has been only now and then he's been after her in these five years. See how long he's been putting up the stockyards and wing fences to run her into!"

"And has he been to all this trouble for all this time, and got nothing at all to show for it?" chimed in another voice.

"No fear," replied the elderly man. "Leave George alone for that; I believe George makes a good thing out of these horses, for I s'pose he has never been out after them yet but what he brings home one, and sometimes two and three with him."

"Has any one of you ever seen this mare?" some one now demanded. "Is it certain she's as good as people make her out to be?"

"Seen her," replied the man who was addressed as Driscoll, "I seen her often afore she got away. She was a splendid looking animal to be sure—a dark chestnut. I believe the major paid 200 guineas for her, and he refused 100 guineas afterwards on his bargain. She was a pure blood. She got out of the paddock when there was a mob of wild horses about at the time, so she just went away with them; and since then she's got as wild as a deer, and gallops straight for the scrub when any one comes in sight of her, and she can just go like the wind."

"I suppose," one of the interlocutors here remarked, "she'll have some foals by this time, but then they'll not be much worth, as they'll be got by some scrubber of an entire."

"No," replied Driscoll; "in the mob with which she runs there's a splendid looking entire. He's not a blood, of course, but the foals out of that mare, if they are got by that horse, will be something worth having, I can tell you."

By this time the subject of their remarks had dismounted to take down the slip panels, and leading his horse forward, and saluting the men, he ungirthed his saddle and turned his horse out to graze. The animal first, however, proceeded to enjoy the luxury of a good roll, then getting up and shaking himself soundly, went quietly munching the herbage around him. He was rather a fine-looking animal, high in the withers, with shapely legs and long pasterns; altogether an animal well calculated to support fatigue, and with considerable capacity for speed.

His rider was a man evidently about fifty years of age, rather under-sized, but powerfully built. His features, encircled with a bushy grey beard, were open and honest, though they expressed marks of impatient irritation. His eyes were restless and seemed filled with a rather unwonted light, as if something exciting was just then occupying his mind. That such was the case was soon evident, when, after the first general salutation to the assembled groups of, "Good evening, lads," and a more particular greeting to the elderly man who had lately expressed such a decided opinion of his abilities, whose hand he cordially grasped with the remark, "Well, Caleb, how are you getting on?" he at once delivered himself of the matter that was exercising his mind, with the abrupt remark, "Well, lads, are any of you fancy riders inclined for a gallop, there is a splendid chance now for any one who wants to get hold of a good nag?"

"Where is the mob, George?" inquired Driscoll. "I wouldn't mind having a ride, if they weren't too far away; is the major's mare among them?"

"Yes, she's there, and all within ten miles of us now," replied Laycock quickly. "Do you know," he eagerly continued, "I never did have such a chance of getting that mare as I had this afternoon. I had been shepherding them for days, and to-day the whole mob came to drink at the Bonyup water hole; if I'd had two or three chaps along with me I could have ridden them at once into the yard; so I came away here to see if I could get two or three of you chaps to go along with me. We can ride out and camp at the water hole, and then on the following morning we can easily get them in the bush, they won't be far away. I know where we shall be able to drop upon them."

Laycock's information produced a lively sensation amongst the shearers, most of them young men and bold riders, and a half dozen or so instantly volunteered their services, but expressed surprise at Laycock letting his horse loose if he intended riding forth that evening. "Oh, he's all right; I thought he would be the better for a roll, I can catch him anywhere."

Meanwhile, to all this conversation I had been a silent, but highly interested listener, and, charmed with the novelty of the adventure, the idea instantly possessed me to make one of the party. Signifying this determination forthwith to the others, my proposal was received with ready approbation.

Among the station horses was a bright bay gelding with dark points, named Latrobe, that had often excited my admiration. He was a magnificent looking animal, above sixteen hands in height, with an eye that under excitement literally blazed like a live coal; yet, he was entirely without vice, but from one slight blemish consequent on his excitability, by which he was so inflamed when galloping in pursuit of horses, that he became almost unmanageable, he had been slighted for "mustering" by the overseer, who was rather a timid rider. Yet, precisely for this reason, I now resolved to make him my charger for the occasion, with the reflection that as hard galloping was to be the chief means towards the success of our enterprise, Latrobe could hardly do too much of it. After a hearty supper had been done ample justice to we all started off in high spirits, prepared with blankets for camping out for the night.

The twilight had faded, although it was still as bright as day from the light of the moon, that now at its full, shone calm and broad over the trees, when we reached the water-hole where we all dismounted, and tethering our horses securely, prepared cheerfully for passing the night there. But Laycock, with the foresight of a prudent commander, instead of seeking repose, determined if possible to ascertain the whereabouts of the horses, so as not to lose time needlessly in the morning, and at once started off on foot on the track of the troop. In about two hours he returned in high glee, stating that he had heard them neighing in the forest, just about where he had expected they would be.

I may as well here describe the position we were placed in. The water-hole at which we were camped lay through open forest country about ten miles due north from the station. At this water-hole, however, there was a small plain two miles or so in circumference, which on the other side was fringed with a belt of the mallee scrub that is spread so widely through parts of Australia. This belt of scrub was about three miles in breadth; then the country again expanded for a space into open forest land that was again enveloped on all sides by the interminable scrub.

It was in the open space of forest land that Laycock had discovered by the sound of their neighing the whereabouts of the horses. So our plan of operations for the morrow was carefully laid down that night, viz., to start by four o'clock next morning, and quietly outflank the horses before attempting to come within sight of them, and then, as the dawn appeared, to charge suddenly from different points so as to head them directly for the water-hole, where there was a yard ready for their reception, and into which we doubted not that by proper management, we should be able to drive them.

Exactly by four o'clock the next morning accordingly, we were all in the saddle, indeed, I question if Laycock that night closed an eye, so excited was he at the near prospect of securing the prize after which he had toiled for so many years. Under his guidance, after entering the mallee, and proceeding quietly for fear of alarming the quick ears of the animals we were in quest of, we made a wide detour, never entering the open forest till we were able to do so on the side opposite to where we wanted to drive the horses to. That none of the wild troupe might have an opportunity of eluding us among the dark shades of the timber, we made no show of discovering ourselves till the dawn was fully in, when, first spreading ourselves out in the form of a semicircle, the forest suddenly resounded with the loud lashings of our stockwhips, as we all instantly galloped from our different stations towards the astonished herd. Then occurred one of the most magnificent spectacles it has ever been my good fortune to witness. In instant reply to our demonstration we heard the loud neighing of the entires—of which there appeared to be several in the troop—while one, a stately looking animal with flowing hair, and of a dark brown colour, that appeared to be the leader of the troop (which might have numbered about fifty head), without thinking of his own safety, first dashed completely round the mob collecting them all into a close body, then, placing himself at their head, and laying back his ears, dashed straight at George Laycock, in the centre of the advancing riders, with the apparent purpose of overthrowing his horse, and so breaking through the cordon that was intercepting the retreat of himself and his troop to the scrub. But Laycock, who was well accustomed to such encounters, was provided against the emergency. In such a case the lash of a stockwhip would be of small use in preserving a horseman from the attack of an infuriated stallion. But just as this one, open mouthed, was preparing to leap upon Laycock's horse, the sudden detonation of a revolver (blank charge), with the flash and smoke in his face, sent him flying in panic in an opposite direction. Then all the riders, with whoops and yells, plied every nerve to lash the terror stricken troop before them. For a short time it appeared as if they were going to have everything their own way, until the mob began to enter the mallee towards the little plain where the water-hole and yard were ; when the leading stallion, that appeared to be an animal of great spirit and courage, again galloped along in front of his companions, and wheeling sharply round, made another desperate attempt to break through the line of riders and making a wild launch of his heels at one of the horses, whose rider, unlike Laycock, unprovided with a pistol, lashed vainly at him with his whip—he fairly dashed past him. Then ensued a scene of indescribable confusion, as the whole troop following their leader, poured through the gap in the line of riders like water through a sieve, despite the furious efforts of the horsemen who, with their long whips lashed frantically in the horses' faces, and, as the line contracted in the effort to secure the parts through which these breaches were made, galloping round either end of it, and so making off in the direction whence they had come. Above all the tumult, Laycock's voice stern and loud could be heard, shouting to the men to keep back from the horses, but such was the excitement and confusion, that intelligent co-operation was hopeless ; in a few minutes the whole mob was irretrievably dispersed.

While these events were passing, I, whose station had been on the northern extremity of the line of riders, found myself, when the horses made their furious stampede, suddenly confronted by a small group that had detached itself from the main body and which evidently intended outflanking me.

It consisted of a superb looking chestnut mare, and I at once surmised it must be the same that Laycock had toiled so long and so arduously to secure. With her were three other younger looking animals, colts or fillies, that by their close attendance appeared to be her progeny.

How, until that moment, I had conducted myself I hardly remember. Being fresh to such work, during the chase I experienced the wildest exhilaration of spirits, and in my excitement I almost forgot what I was doing. But on being thus confronted with the mare, whose identity I immediately divined, in instant anticipation of her intention I gave Latrobe the spur, and thus caused her to head in a more north-westerly direction. Even then, I fear, I was less animated by an intelligent wish to wear her back to the water-hole, than by the mere youthful pleasure of galloping after something, still I had the idea of trying to gain on her opposite flank, and thus to prevent her escape in the direction of her own haunts.

Latrobe behaved gloriously. His mettle required no quickening with either whip or spur, so gallantly did he stretch out to lay himself level with the animal he was chasing.

Between him and the mare, at the start, there might have been a couple of lengths; and I noticed, as we flew along, that although Latrobe gained nothing on the mare, neither did she leave Latrobe behind. This might, however, have been accounted for by the fact that she was heavy in foal, and hence running at a great disadvantage. Among the progeny that followed her there was a three-year-old colt that next to the mare herself excited my admiration by the easy, springy gallop that enabled him apparently at will to lead the chase, as he occasionally shot ahead, kicking up his heels in mere wantonness. The other two, that seemed to be one and two-years-old respectively, found it more difficult to keep pace with the others and gradually dropped behind. For half an hour the chase continued, during which time we must have traversed ten or twelve miles at least, sometimes through open forest, but principally through scrub. How I saved my head from the stems of the latter as they flew back from the pressure of the racers ahead I cannot tell. I seemed so transported with my situation that I felt no fear, and experienced no difficulty in accommodating myself to its dangers, whilst, as the supple body of my horse doubled beneath me in his bounds over fallen timber, or in crashing through resisting saplings, I felt as if I were but part of him, so thoroughly had my enthusiasm and spirit warmed up all my nerves.

At length through the mallee, whose dewy tops were now illumined by the bright beams of the risen sun, I descried the openings of a plain, which we almost immediately after entered. From a piece of rising ground, however, on the edge of it, I cast a glance forward, and saw the interminable scrub apparently spreading again for miles beyond.

On entering the plain, that was of an oval form, not over two miles in length or one in width, the mare suddenly wheeled to the right towards the upper end, about a half mile further on. Latrobe was now white with foam, and as the mare appeared to be gradually extending the distance between us, I was thinking, as we were rapidly approaching the termination of the plain, with all the scrub ahead of us, that further pursuit would be hopeless, when I was suddenly made aware of a violent crashing among the bushes to my right, and just as the mare was within a hundred yards of the scrub, George Laycock, his horse literally a sheet of foam, dashed out ahead of us, and, crossing the mare's path again, with a shot fired almost in her face, so startled her that she wheeled round in her tracks. Then immediately crack, crack, crack went the lash of the stock-whip, resounding like so many reports of a rifle, as, giving the animals no time to consider, he flogged them right on before him. "Keep 'em on, keep 'em on, young fellow," he shouted, wild with excitement, still energetically plying his whip, which he seemed able to manage as he pleased whilst at the top of his speed, pressing hard upon the mare's heels, or galloping from side to side, as she turned her head in either direction, his horse the while with scarcely a motion of the rein wheeling as short as if on a pivot, in obedience to each impulse of his will.

As the horses wheeled back on their former course, they were joined by the two younger beasts, which had fallen considerably to the rear. They were now all urged towards the opposite extremity of the plain, and as we approached it I noticed two lines of rail fencing along the edge of the timber on either side of me, and converging gradually, and I instantly divined that they were wings leading to some adjacent stockyard.

As we approached to within the reaches of these fences, the old man's excitement seemed to become uncontrollable, as, furiously lashing with his whip, he shouted to me to use mine. Indeed his objurgations at times were hardly complimentary to me, but these, supported by the important service I had been able to render, and buoyed up by the excitement of the chase, I hardly heeded. My whip, in fact, I carried more for form's sake than for use, as, not being yet expert with a stockwhip, in the rapidity of my morning's gallop I had scarcely attempted even to flourish it.

At length we found ourselves fairly within the protecting wing fences, and after another half mile through forest country, during which the fences rapidly converged and led into a substantial stockyard, whose slip panels were down as if in expectation of our present emergency, the whole troop were fairly driven within its secure enclosure. Instantly springing from his foaming horse, and without uttering a single word, Laycock banged the panels into their sockets, and fixing them there with wooden pegs, lashed them securely with a strong rope; nor even then did he appear satisfied, until he had carefully walked round the yard and critically examined each panel as he passed. Not till this was done did he give utterance to his satisfaction in words too forcible for insertion here. His fevered impatience now wholly subsided, and the lines of his face relaxed into the open, frank expression that seemed to be natural to them. Cordially shaking my hand, he greatly complimented me on my conduct in the chase, which indeed was more owing to the mettle of my horse than to any judgment that I had been conscious of.

"Look here, young fellow," he remarked, "I reckon there's not a rider in the county would have done better than you have done this morning, although you are a new chum. Now I'll tell you what I'll do with you: you see that bay colt, the three-year-old? I've long had my eye on that beast, as I reckon him to be near as good an animal as his mother. I think he must be got by that stallion that charged me that time. Well, I'll give you that colt, and what is more, I'll take him in hand for six months for you, and if you can find a man in the country as can beat old George Laycock at breaking-in a horse, why, I'd like to see him, that's all! And after I have done with him, I reckon £100 would be too little a price for him."

To this condition I gladly enough agreed, and considered I had performed a very fortunate day's work. Laycock appropriated to himself the lion's share of the booty in the mare and the two remaining young ones, a colt and filly. This was only his due, considering the labour he had previously expended to enable him to at last secure his prize.

We now proceeded to examine the various points of the captured horses. The mare was indeed a most superb-looking animal of a rich, dark chestnut colour. The broad forehead, expansive nostrils, almost square-cut nozzle, large prominent eyes, still aflame from her late excitement, proudly arched neck, sloping shoulders, and springy, taper legs, on whose pastern joints not one long hair was visible, were all models of beauty that it would have delighted the heart of a jockey to view.

But the colt that was destined to be my prize naturally absorbed the major part of my attention. His bright bay colour, that was not marred by a single white hair, contrasted strongly with the deep black of his mane, and the lower part of his legs. Whilst the well-formed head, large expressive eyes, and dilated nostrils were inherited from his mother, in the greater luxuriance of hair in his mane and tail, that also showed slightly in his pastern joints, he betrayed the coarser strain of his bush sire. This, as it added strength, was in my eyes rather an improvement than a defect in view of the use for which I should require him, while his noble withers and swelling ribs, his short back, and that part that to horse dealers is technically known as the coupling, which in him was so short, that the last rib almost seemed to lie against the hindquarters, all promised a strength and endurance that made Laycock predict that no ordinary amount of fatigue would affect him.

I asked him how it came that he had been able to arrive so opportunely to prevent the mare from reaching the scrub. He replied that, on the sudden dispersion of the horses, he had noticed the mare, for which alone he had eyes, in the midst of the troop, but losing sight of her immediately after in the scrub, he rode on for a few minutes, and then, chancing to look round, he just caught sight of her, with me in full chase, disappearing through the forest. Instantly divining the final direction in which she would head—an old haunt of her's—he rode straight for the head of the plain, and, having the advantage of reaching it in a straight line, whilst we had made a slight curve, he was thus enabled to arrive at the critical moment.

We now unsaddled our horses, which, on our dismounting, had stood where they were, panting heavily from the strain of their late exertions.

Laycock, after critically examining Latrobe, praised him highly, and indeed the brave animal deserved it, for without his spirit and ability our present captives would by then have been miles away in the pathless mallee scrub. We rested ourselves for some time in a hut that, together with the yard, had been built by Laycock long before, in view of the present contingency, and, by his foresight, there was also an ample supply of flour, tea, and sugar, so that we were provided with all necessaries for quickly allaying the cravings of keen appetites that our morning gallop had considerably quickened. Brewing a dish of tea and baking some *Johnnie cakes* on the coals, we soon fell to, and heartily enjoyed our simple breakfast.

After a delay of some hours, Laycock took his bridle in his hand and went into the yard alone (he would not allow me to come with him) to try and secure the mare. At first she seemed frantic when he attempted to get near her, and, snorting loudly, impetuously charged almost over him. Steadily keeping his eye upon her, however, until in time he seemed to master her by its influence, she at last allowed the bridle to be slipped over her head. Feeling the bit between her teeth once more, all this noble animal's instincts of obedience seemed to return, and quietly allowing the saddle to be girthed tightly round her, after a few restless motions such as might almost have been made by a spirited stable-fed horse, her captor placed his foot in the stirrup and quickly vaulted into the saddle, when the mare walked quietly away as if she had already forgotten, or now resigned, all further thoughts of her late freedom.

Laycock next attempted what seemed a still more improbable feat, and still alone. For after dismounting from the mare he returned to the hut, and choosing from some breaking tackle that lay there, a strong halter of plaited hide, to which was attached a rope of the same material, he, to my utter astonishment, went into the yard with the evident intention of securing the bay colt with this halter.

At first the animal seemed to grow frantic at the sight of him, but he quietly persevered for hours in walking after him with his eye fixed upon him, and eventually succeeded, with the assistance of his own horse, which he led up beside the colt, in getting the head of the latter fairly into the halter, and then by degrees he soothed and coaxed him until by nightfall he could handle and caress him all over.

The result of all this was, that the next morning I returned to the home-station to fetch some provisions and blankets, and then stayed for two more days with Laycock, who refused to allow the colts and filly to leave the yard until he had made them all so tame that he could lead them away with halters; feeding them meanwhile with oak branches and long grass that he cut by the edge of the water-hole.

The colt, which from its noble appearance I had named Selim, was delivered into my hands on the expiration of the period during which Laycock had promised to keep him in hand for me. His qualities I have already hinted at. In the course of the following pages there will be related some incidents in connection with his endurance and fidelity that will show how far George Laycock's estimate of his own knowledge of horse-flesh and capabilities as a trainer were justified.

CHAPTER III.

ON returning me my colt, thoroughly docile and educated in all his proper paces, Laycock volunteered a short characteristic piece of advice regarding the treatment that I was to pursue with him.

"Look here, Mr. Farquharson, a horse like this is what a man can only hope to get once in a lifetime. There are indeed very few men who ever get hold of a real good horse, and there are still fewer who, after they have got a good horse, are able to keep him good. Most people when they do get hold of one are bound to get showing off what he can do, and the more willing a horse is, the faster they will take what is in him out of him; for a horse's wind and limbs can only stand a certain amount of wear, and the rougher that wear is, the sooner he breaks down. But you use this horse properly, and only force him when you have need to do it, and above all never lend him to anyone, and he may serve you well for twenty years to come. See this old horse of mine, you would not think from the way he galloped that day after the mare that he was twelve years old; but he is. I broke him in myself, and I always looked after him properly. When you are going on a journey, and want to canter or trot (it is all one to Selim, he can go the one pace as well as the other), just keep on at that pace till you see his hair getting wet, then check him, and walk him till he is dry again; and you will always, in this way, get over the longest journey with the least stress to your horse. Remember that galloping up a hill is hard on a horse's wind; but galloping down a hill is just as bad on his limbs. So, unless you are pushed, never go up or down a hill, even if only a little one, at anything but a walk. When using the horse for hard work for days together, mustering, never run him down; when you see him begin to look fagged, get another horse if you can, so as to give him a spell. Look after him, feed him well—keep him always in good working order, and you will never be sorry for it." And with this shrewd advice the friendly old fellow shook hands with me and rode off.

As I never met George Laycock again, I may as well state here what little I heard of his subsequent career, and of his celebrated mare. Almost immediately after taking the mare home, he was offered by a neighbouring squatter of sporting proclivities three hundred guineas for her; this offer George refused, as, having a taste for horse-racing himself, he determined to use her as a stud mare for the turf. As it may seem strange that George should have been able to retain such a valuable animal unchallenged by her former owner, it is as well to mention here, that Major Firebrace had long before left the country, and had, before he left, resigned the right of the mare, in the event of her capture, to Laycock, who had formerly been for a long time in his employment, and for whom, as a man, he had some regard.

The two younger offspring of the mare, I was afterwards informed, George had disposed of for sixty guineas apiece, but his hope of success on the turf he was fated never to realise. In little more than twelve months after the mare came into his possession, poor Laycock died at Melbourne, and the mare then passed into other hands.

But to return to my own affairs. The period of my employment on Mr. M'Elwain's station, that, as I have already stated, covered about two years, terminated in a few months after Laycock delivered Selim into my keeping.

The excitement due to the extraordinary gold discoveries

was now at its height, and as I was casting about in my mind for some future employment, it may cause surprise to some of my more adventurous readers that I did not at once embrace this opportunity of securing a ready-made fortune in some of the rich mines of Bendigo or Ballarat, etc. But, apart from my natural bias in favour of country pursuits, I never had any kind of speculative turn in my nature or love for games of chance. Besides, never having been accustomed to manual labour, I felt repelled by the idea of the severe exactions of pick-and-shovel work that such a style of life necessarily demanded. So, being now so well provided with horse-flesh, and hearing that such work was well paid, I resolved, in preference, to try and get employment at either sheep or cattle driving from some of the numerous dealers who were constantly on the roads taking stock to the markets; and work of this kind I soon found.

I must now ask my readers to pass with me in silence over the eight uneventful years which followed upon my leaving my kinsman's station.

A chief part of this interval was employed in driving cattle and sheep to the city and digging markets, or store stock from distant inland localities. In this way I traversed extensive tracts of country, both in Victoria and the neighbouring Colonies, until, finding myself in possession of no inconsiderable capital, I, in concert with another man, took up a small run in the Upper Murray.

This investment proved unfortunate. In the first place, my partner contracted dissipated habits, and squandered large sums in racing; and, finally, a fell disease (catarrh) made such havoc among our sheep—thousands dying in the course of a few weeks—that, with the heavy loss caused by my partner's debts in addition, matters came to such a pass, in two years, that I was glad to come out of the concern, with little more than £100 in my pocket. And thus, at one fell sweep, vanished almost all my savings, the fruits of years of toilsome driving on the roads in all kinds of weather. All my hopes of affluence and comfort were, for the time being, dashed to the ground, and I was now no nearer the object of my ambition than when I left Mr. M'Elwain's station eight long years before, with the added disadvantage of being that much older, too!

Yet, in all the vicissitudes of my calling, I never forgot George Laycock's advice with reference to the usage of my horse.

The result of this was, that after eight years' continual service, saving, of course, intervals of necessary rest—old Selim was still as hale a horse, and as free from splints and strains, as on the day when he careered before me and Latrobe with his dam.

After the disposal of my station affairs, and pending my decisions as to my next move, I had ridden down to Melbourne. That city was now making prodigious strides commercially, and was extending on all sides.

A few days after my arrival there, I was sitting in the reading room of Mack's hotel, when I suddenly attracted the attention of an elderly gentleman of respectable appearance, who, with gold-rimmed spectacles on nose, had been engaged in reading one of the daily papers, on my entrance. Happening to look up from his paper, this gentleman gazed earnestly at me for a moment, then, rising from his seat, he advanced straight towards me, just as a flash of memory revealed to me the features of my kinsman and quondam employer, Mr. M'Elwain, now slightly worn by Time, that had silvered his whiskers, and thinned his hair.

Shaking my hand heartily, he enquired after my welfare, and the nature of my pursuits since I had last parted with him, expressing surprise that I had not thought fit to communicate with him in the interval. I accordingly gave him a brief account of my experiences that, promising and flourishing, for a few years, had had such a disastrous consummation, in the loss of my sheep, and ruin of my credit, through the wasteful extravagance of my partner.

Mr. M'Elwain listened very attentively to all I had to tell.

"This is a bad business," he remarked, "after so many years of careful industry on your part. It is a pity," he continued, "that you should have been so careless in allowing Wilson so much scope with your affairs, without keeping a watchful check upon him."

"I had no idea of anything being amiss," I replied. "He always appeared to be so plausible, and in the details of banking accounts, he had so much more business knowledge than I had, that I left it all to him, until it was too late to avert the ruin that had overtaken us, so that, with the loss of the sheep, I found there was scarcely anything left to meet the overdrafts on our funds."

"Yes," replied Mr. M'Elwain, speaking with his wonted plainness of manner, "but that was where you were very wrong; you should not trust to any man, no, not even to your own brother, the oversight of your private business. Every man should always know for himself how his bank account stands."

With this evident truism I freely concurred, adding rather

bitterly, that the error I had committed was now made sufficiently apparent to myself by the penalty I had paid.

This experience, indeed, exposes a weakness in my character that almost unfitted me for purely business details. Fond of physical exercise to a degree, when indoors my inveterate habits of reading yet rendered the duties attendant on book-keeping a bore that I was always only too willing to relegate to any one on whose ability, I presumed, I could depend. Contented with a general knowledge of the condition of the stock before my eye, and with the rough estimate of value that I formed thereon, the congenial companionship of Scott, Byron, Campbell, Moore, Goldsmith, Burns, Macaulay, and Addison were to me a source of delight, to which those attendant upon casting up bank accounts or studying the possibilities of overdrafts were as nothing.

It was to this well-known habit of mine that Mr. M'Elwain's strictures mainly pointed. After a pause of calm reflection he again remarked:—

"I have just been thinking of a situation that should suit you. There is a gentleman, at present residing in this hotel, a squatter on the Lower Murray, who this morning told me that he wanted to find a trustworthy man who could undertake the management of a considerable station he owns on the Darling River. If you think that this situation would suit you for the present I make no doubt but Mr. Rolleston would take you upon my recommendation. That recommendation, I think," he added with a smile, "I may safely venture to give; perhaps in a year or so you may be able to start on a run of your own again."

To this very kind proposal I, of course, gladly acceded, thanking Mr. M'Elwain for his thoughtfulness on my behalf.

Shortly after this conversation Mr. Rolleston himself entered the room, and to him I was accordingly introduced by Mr. M'Elwain. He was a respectable, elderly looking person, with fair or almost yellow hair and whiskers, now slightly grizzled, and blue eyes, and rather well formed features. He was beneath middle height, with a quiet expression of countenance, combined, however, with a strong interest in worldly matters which had left its mark in the shrewd lines about his eyes.

On Mr. M'Elwain mentioning my capabilities to him as being sufficient for the post at his disposal, his conversation with me instantly assumed a business turn. He informed me that the station he wished me to take charge of carried about 20,000 sheep. It was situated, as I had already been told, on the bank of the Darling, about 180 miles from the confluence of that river with the Murray. Besides this sheep-run he also informed me he had another run on the opposite side of the river solely devoted to cattle purposes, on which there might be somewhat more than 1,000 head depastured. This run was under the immediate control of a stock keeper, who was sufficient for all the practical management of the place, but, as his education was very limited, I was, with regard to the pure business department, to have the supreme charge of both runs.

"But in dealing with this man I fear you will have need of some tact to avoid a difficulty," remarked Mr. Rolleston, who, unlike most people of a speculative turn, seemed of a contented mind and unwilling to make innovations on any established order of things. "You see," he continued, "he is a most trustworthy and excellent man, and as true as steel, and in all matters connected with the management and working of cattle a most efficient servant; but, like most old hands, he is very impatient of control, and will brook no interference from any one, not even from myself, in his own arrangements with the stock. With my last manager he was continually at loggerheads, and on several occasions I had great difficulty in persuading him not to leave me altogether. This would grieve me, and I am persuaded that by a little conciliation and leaving him to his own measures with the stock (which I have hitherto found to be uniformly excellent) a manager would find a pleasant man and a useful ally in Benjamin Lilly in working both the runs."

To this I made answer:—

"With such a man as you describe Lilly to be it only requires a little sense and judgment to get along very well. With such men the rule should always be, to deal with them in a common-sense manner; and, as their jealousy at the interference of others is usually associated with ideas of incompetence on the part of the meddling party, by showing them that for what you do you have a good reason, and by sometimes deferring to them they will come to respect you and then—unless they are really capricious or jealous-minded fellows—once the confidence of such men is gained you may reckon on them as on those on whose sympathy and aid you may rely."

After some more preliminaries had been satisfactorily discussed, it was finally agreed that I should at once proceed to take charge of Mr. Rolleston's station, at the liberal salary of £300 per annum; terms, that after my late gloomy prospects, seemed by no means an unpropitious commencement for a fresh start in life.

Next morning, accordingly, after again shaking hands with Mr. M'Elwain, and bidding farewell to my new employer, I started off on my journey to the Murray. Traversing a country rich in plain, and magnificent forest land, and passing through Bendigo and Castlemaine (whose gutters and tunnelled gullies, still resounding with the rocking of the diggers' cradles, bore witness that eight years of incessant toil had left unexhausted their golden treasures), and many a minor township and exploded rush, I at length struck the Lodden, a tributary of the Murray, and journeying along its banks for some hours, first sighted the larger stream at Swan Hill; I had travelled a good two hundred miles, but with Selim's splendid walk and sweeping canter, I easily accomplished it in three days.

At Swan Hill, the Murray, one of the most navigable of Australian rivers (none are remarkable for their size), is about eighty or more yards wide. It is a noble looking stream, flowing steadily in one deep, stately current, in a westerly direction, and marks the boundary betwixt Victoria and New South Wales. Turning my course down the river, the country that had hitherto presented a rich, grassy appearance, now assumed a more barren look, being covered with timber and scrub, and showing a sandy soil. In about another hundred miles, I made Mr. Rolleston's station. Riding into the yard, I dismounted, and leading my horse towards the stable (an unwonted luxury at a station), I delivered him over to the charge of the man who fulfilled the duties of groom, and finding my way to the house, I was duly ushered in by a tall, blooming, and rather nice looking damsel, with fair curling hair, into the interior of Mr. Rolleston's house.

CHAPTER IV.

THE appearance of the station buildings I will forego describing. Like the homesteads of most squatters in those days, they were chiefly constructed out of the primitive materials of split slabs or round logs and bark. There was the inevitable woolshed seen in the distance with its surrounding yards and fences. Even the house of the owner was of the most unpretending appearance, with about half a dozen rooms or so, and chiefly distinguished in structure from the other buildings, not so much by its size, as by the fact that the slabs with which it was built were sawn instead of split, and that it was shingle roofed, all the other buildings being simply covered in with bark. Almost attached to this was a smaller building, in which, as I subsequently learned, the overseer, Mr. Campbell, resided, whose wife indeed also acted as housekeeper to Mr. Rolleston, Mr. Rolleston himself being a widower.

On entering the house, the interior of which I discovered to be much more pretentiously furnished than the exterior gave promise of, I found myself in the presence of three ladies in a rather tasteful looking sitting-room, arranged with almost fashionable fastidiousness.

The elder of these ladies might have been a little over forty years of age; plainly, but neatly dressed, there was that look of calm and easy self-possession in her manner, that always denotes the presence of a superior mind. Her complexion was pale, and she had dark hair, and in person she was tall and graceful. This was Mrs. Campbell, the overseer's wife. Beside her was a girl, evidently about seventeen, her daughter, as I was given to understand. Unlike her mother, her hair resting in curls on her shoulders, were of a rich golden colour. Her face was at once sweet and attractive. Her eyes were blue, her cheeks dimpled as she smiled. Her figure, though well formed, yet scarcely promised to reach her mother's stately height. Altogether her expression was at once engaging and innocent.

But my attention was chiefly engaged by the appearance of the third lady, who more nearly approached my ideal standard of beauty than anyone whom I had hitherto met.

Of middle height and slight figure, with dark brown hair, crimped and waving on either side of her brow, over which it was evenly parted and allowed to fall down in soft, glossy ringlets behind her neck, a slight though well-formed forehead, with brows as neatly outlined as if drawn there with a pencil; brown eyes, and a rich, yet softly toned complexion; a small and straight nose, and lips that, parting with the arch humour that constantly animated her intelligent face, seemed formed (as I mentally noted in a glow of poetic enthusiasm) for loving and kissing; while two rows of dazzling teeth completed the outlines of this very pleasant picture.

In the presence of this attractive company I at first experienced considerable embarrassment.

After my long residence in the bush, where I was deprived of the advantages of intercourse with ladies during the early period of manhood, when one is more amenable to their refining influence, I felt that in the polite usages of society I was deficient and rustic.

But rallying my energies for the occasion, I at once introduced myself to them, mentioning my name, and stating my business.

Thereupon the elder lady rose, and with quiet self-possession at once mentioned who she was, and introduced me to the other ladies, the dark beauty being Miss Rolleston, the only daughter, and, indeed, child, of my employer; and the fair one her own daughter, Mary. Then, expressing a kind concern for my fatigue, she showed me into the room I was to occupy for the night, and on my return to the parlour gave me a cordial invitation to draw up my chair near to the fire; to which, as the evening was a chilly one, I immediately responded.

Making an effort to overcome my rather clownish diffidence, which I felt that a bashful silence would only render more conspicuous, I endeavoured to put myself at my ease by some commonplace remarks in the way of opening a conversation with my fair companions. My efforts would have been crowned with but little success, however, but for the vivacious manners and natural frankness of Miss Rolleston, that soon relieved me of any responsibility in supporting the conversation between us. Miss Mary occasionally chimed in, but with a quiet grace, in which, however, was mingled a vein of such clever humour, that I felt myself quite affectionately impelled towards her.

After all the inevitable topics of the varied phases of the weather, both past, present, and to come, had been discussed with all the deep interest due to matters fraught with most important consequences to us all, the scope of our remarks gradually widened, as, like an inexperienced swimmer, who, after the first few spasmodic strokes, gains courage, and learns to strike out more steadily, so I gained self-assurance in the course of these commonplaces, and was buoyed up in venturing into the deeper waters of a more interesting conversation. After the weather, we discussed surrounding objects: the garden flowers, visible from the window, and the books in the library bookcase. When it came to the turn of the books, Miss Rolleston revealed the bent of her literary taste, that chiefly inclined to works of poetry and romance, Sir Walter Scott being her idol in both departments.

At that time photography was still in its infancy, and the photograph album was not so conspicuous in every drawing and sitting-room as in these days, when it proves such a stereotyped theme of interest for the recreation of visitors, and by whose means even a bashful man can sustain his part by simply showing some sign of interest as the various photographs are being turned over and descanted upon, with the names and histories of their originals. There were, indeed, a few daguerrotypes, as such works were then termed, framed in morocco cases, of a few of the town friends of the Rollestons, besides some of Miss Rolleston and her father, that were handed to me for inspection ; but to me both of these seemed to be tawdry, washed-out presentments.

By this time I felt so perfectly at home, that I ventured the remark, " Do you, Miss Rolleston, prefer staying in the bush here to living in the town ? I should think that to a young lady so much isolation from society and the constant sameness and tameness of everyday life in a wilderness like this, must be very irksome in comparison to the pleasure and gaiety of life in town."

"Psha," replied the laughing girl, "I am glad to get away from all those tiresome amusements at times ; when in town there is nothing else, night after night, but some party or play to go to, till I am glad to get a spell in the bush again, for I am a child of nature, and it is amongst nature's charms that I enjoy myself most."

" You appear to be romantic," I remarked, with a smile.

"Oh, quite so. Mary and I, during our walks down the glades of the forest, as we listen to the magpies' songs and watch the bright-winged parroquets fluttering from bush to bush, and tree to tree, often declare that we could stay here all our lives."

" That is as much as to confess that you both intend to marry squatters ? " Mary at this simpered a little, but her livelier friend instantly retorted, " Not a bit of it ; I love living in the bush, but a squatter whose ideas never soar above the contemplation of his flocks and birds, or the returns of his wool, would never win my heart, nor Mary's either, I hope ".

" In fact, like all young ladies, the idea of marriage is the very last thing that either of you ever contemplates ? "

" Just so, for my part if ever I do marry any one, it will be some dashing bushranger, with whom I can roam through the bush and live in some romantic cave like Maid Marian with Robin Hood."

" A strange idea for a wealthy squatter's daughter," I replied, and jestingly added, " What a pity Captain Melville's career was so soon stopped, for as far as bravery and dash went, I should think, Miss Rolleston, he might have suited you ? "

" Indeed ! " she replied, " I fairly dote upon the character of Captain Melville, he was just my ideal of a bushranger—handsome, brave, and chivalrous—and I think it was a cruel shame he was dealt with as he was at last."

" He certainly extorted people's admiration," I answered, " by the almost heroic attempt he made, at the head of the other prisoners, to break away from the warders in the hulks—nor was this admiration a bit lessened by the spirited defence of himself that he subsequently made at his trial, but, Miss Rolleston, dashing highwaymen are very well to read of in romances, but in actual life the best of them are but dangerous pests, and the more quickly their careers are stopped the better for all honest folk."

"But you must allow, Mr. Farquharson, in view of Captain Melville's many redeeming points, that he was very harshly dealt with when condemned to that hopeless term of thirty-two years—a sentence, which, it is supposed, caused that brave man, in a fit of desperation, to put an end to his own life."

I quietly replied, "You simply regard this man from a purely romantic, and hence sympathetic point of view, Miss Rolleston; but the real pith of the matter is, does a handsome figure, and dash, and bravery, entitle one man to take away the property of another? I have indeed myself but little sympathy with lawlessness in any form, and a good deal of the daring and courage generally associated with the actions of these gentry, I greatly suspect, if the truth were known would resolve themselves into little more than bluster and bravado; for I always consider that there is far more true manhood in facing life's legitimate hardships, with their generally slow rewards, than by shunning the former in order to expedite the latter by lawlessness."

"I am glad to hear you speak so sensibly to the madcap," here chimed in Mrs. Campbell with a smile. "I am sure with all the high-flown, romantic ideas that she gets from the silly novels that she will persist in reading, she is getting into the habit of regarding everything through such exaggerated colouring, that I really begin to fear that her notions of social propriety are getting quite perverted."

At this grave fear Miss Rolleston simply laughed merrily.

"And you, Miss Campbell," I remarked, "are you such a lover of the bush that you too would prefer to marry some dashing bushranger?"

At this remark she simply smiled at first, and then made reply quietly; "I love the country, but I like to get down to town occasionally, especially when mother can come too".

"Oh, Mary is but an old fashioned little goose," exclaimed her lively companion. "She would be content to go poring all day over some book; unless I were here to rouse her up I don't know what would become of her; but I am going to take her down to Adelaide with me next month, when I intend introducing her to all the dashing young gentlemen of my acquaintance, whose flattering attentions will cause her to assume such a conceited air, that you will not know her to be the same girl when she comes back."

"That is to say if you can spare any of their attentions from yourself," replied Mary slyly, her cheeks dimpling.

"Oh, you sly little puss," exclaimed Miss Rolleston laughing, "who would think that there was so much mischief in you. Your sister Jessie could hardly be more teasing!"

"Miss Mary's sister is in town then now, I presume?" I here remarked.

"Yes, she has been in Adelaide for about two months with a friend. When you meet with her, Mr. Farquharson, you will not be dull, I can tell you, but I warn you to see to your heart; Miss Campbell makes conquests wherever she goes."

"Thank you, Miss Rolleston," I replied laughing, "I will not forget your cautions if ever I have the pleasure of meeting Miss Campbell, although if she bears any sort of likeness to the family here I should certainly consider myself flattered by any notice from her."

At this designed compliment, Mary bashfully simpered, whilst her mother acknowledged her sense of it with a pleasant smile.

In the meantime supper had been prepared, and I hardly required the kind and hospitable attentions of Mrs. Campbell to do ample justice to her homely but substantial fare. While discussing this meal, Mr. Campbell, who till then had been absent taking a ride through the run, suddenly joined us at table. He seemed a person of a quiet, practical appearance, low in stature, but with rather well-formed features, and hair and whiskers of a reddish brown, slightly sprinkled with grey.

Supper over, Mr. Campbell and I had a long conversation on various topics connected with station management; stock, their proper usage, breeding and feeding coming in for a chief share in the discussion. In this conversation, coming as it did within my own proper provinces, I could of course hold my own. Yet, I confess that I was not sorry, when, after a while, on Mr. Campbell's being obliged to go out to give some directions to the men concerning the morrow's duties, I was again left to sustain a more diffident part in conversation with the ladies. But now, so much had the frankness of their manners put me on a footing of confidence with them and dissipated my natural backwardness in female society, I found the task to be a comparatively trifling one. Thus the conversation, chiefly carried on in a tone of light pleasantry, was sustained with great spirit for some time, until, to my great contentment, Miss Rolleston seated herself at the piano, and, almost unsolicited, began to play.

I am naturally fond of music and especially devoted to the plaintive strains of the ballad minstrelsy, with which my country is so eminently associated. It was therefore with something like rapture that I now listened to Miss Rolleston's exquisite rendering of songs that she herself accompanied on the piano with great sweetness and art; and as she next rattled off with spirit some stirring strathspey I felt my feet involuntarily moving an active accompaniment to the tune.

Miss Rolleston's mother had been Scotch and had also been her daughter's sole musical teacher.

The evening at last came to an end, and my interest in the sociable household, whose acquaintance I had now made, was nothing lessened, when at its close family worship was conducted by Mr. Campbell—a custom that I now witnessed for the first time since my departure from Scotland.

After solacing myself, and refreshing my horse with another day's rest at this pleasant establishment, and after receiving many expressions of hope for my speedy return, accompanied by warm pressure of hands at parting, I at length started for my new home, situated nearly two hundred miles up the Darling.

I will refrain from any further description of my journey, simply remarking that bush and plain alternated throughout the way, the chief feature in the landscape however being the salt bush that served in lieu of grass, that, save on the sand hills in the spring, is here scarcely to be seen. This salt bush is found to be very nutritious for both sheep and cattle. The leaves have a sharp, salt flavour, but otherwise are not disagreeable to the taste, and can be easily masticated without becoming reduced to that stringy pulp that necessitates the spitting out of the masticated parts of other shrubs. Indeed, I have known instances where, on the failure of the supply of ordinary salt, shepherds have been prevented from suffering any great inconvenience by the use of this plant, simply eating the leaves of the salt bush with their meat. But a stranger, accustomed only to grass country and unacquainted with the nutritious properties of the salt bush, would be puzzled to know how the sheep contrive to live at all, on gazing around him upon nothing in the way of herbage but these shrubs, in size and shape like heather, and, in the summer, by the road side usually bare of leaves.

Accommodating my stages to the situation of the various home-stations along my route that offered opportunities for convenient breaks in my journey, I reached the end of it on the fourth eve after my departure from the Murray, when I rode into the home-station where I was for a time to be ruler supreme.

CHAPTER V.

THE buildings of the homestead were of the usual description of station residences—the huts simple and rude, with woolshed conterminous. Dismounting to let myself through the gateway panels, I rode up to the house that was to be my future abode. It was constructed of sawn slabs of short lengths laid horizontally in grooves into strong squared posts, also sawn, and contained about four rooms, with a kitchen attached.

As I approached the house, a tall, thin, but muscular looking man, came to the door in answer to the loud barking of the dogs that had instantly saluted my appearance.

He was dressed in a tartan jumper, with tight breeches and knee-boots. His brawny, sunburnt throat was bare, and a cabbage-tree hat completed the attire of a man whom it needed no second glance to assure one was an old colonial. He might have been about thirty years of age. His eyes were keen and searching; but when he smiled an expression of dry humour lurked about the deep lines, or crow's feet, that radiated from their corners. His face was deeply bronzed, as if from long exposure to the Australian sun, and slightly fringed with a beard and moustache that, like his hair, were coal black; his eyes, too, were of a dark brown colour. In a word, I at once knew that I stood in the presence of that same Mr. Benjamin Lilly whose good graces I had been so strictly cautioned by Mr. Rolleston to endeavour to secure.

His appearance altogether was bold and independent. His regularly formed features wore more of a sarcastic than any other habitual expression. His independent bearing was not belied by his manner and the bluntness with which he addressed me in return to my salutation as I rode up to the door. "I s'pose you're the boss," he briefly demanded. I replied that I had come up with the purpose of taking charge of the station. "Good job," he replied shortly, "I'll go back now to my own quarters." And after coolly directing me to where I could put up my horse, which was simply to let him loose where he was, and without another word my rather unceremonious friend stalked down to the river's edge, and there jumping into a canoe paddled across the water.

By this time the man who acted as cook had come out of the kitchen-hut, and understanding the quality of the stranger to be that of his future " boss," relieved me of the task of ungirthing Selim and letting him loose to graze, for a stable here was an unheard of luxury in those days.

Entering my new home, I first glanced around and took note of its appearance and contents. I found myself in a comfortable enough sitting-room, or rather, I should say, a room that answered to the several requirements of several rooms in a fashionable house, being parlour, dining-room, and library all in one. There were besides, three other rooms used as sleeping apartments, with one or two small closets for odds and ends.

All were plainly but conveniently furnished, and a small stock of useful and entertaining books promised me the means of mental recreation during my hours of leisure.

On the station several men were employed throughout the year—a bullock driver, two bushmen, and the home-station shepherd, besides a cook attendant upon them—occupied a hut that I had already passed about two hundred yards from my own house.

These details I then took no immediate steps to acquaint myself with; of course my attention was at first more agreeably engaged in the contemplation of the supper which the cook had set on the table shortly after my arrival. This I discussed with such appetite after my long day's ride that I never noticed the entire absence of vegetables from the meal.

Supper over, after a little time spent in examining the contents of the books on the shelves, that comprised, as I was delighted to find, the works of several of my own favourite authors, and that were of an imaginative, discursive, and historical character, I retired to rest, and so soundly did I sleep that it was eight o'clock on the following day ere I awakened.

After breakfast I at once set out to make a formal inspection of the various features and requirements of my charge and the immediate occupations of the station employées, etc.

This indeed did not occupy much time, for the hands were but few, and their duties of the simplest description, mending fences or splitting timber for new ones, etc. After speaking a few words to the men whom I met leaving the hut on their way to their work, I went into the shed and spent some time there in making a particular survey of its general qualifications for the various purposes for which it was designed.

Like almost all inland buildings of a similar class erected at that period these qualifications were mediocre enough. Having finished my observations there, I next entered the men's hut to see what arrangements were there provided for the comfort of the station hands. This, as a bush hut, where the chief conditions looked for were, security from rain and wind, I found on the whole to be fairly satisfactory. The simple furniture of table and forms and the goodly array of bunks in view of the vastly increased demand for accommodation at the sheep-shearing season were, if rudely, yet substantially constructed.

The cook had by this time completed the clearing up of the breakfast dishes, and had set his hut in order for the day, and I saw at a glance, from the general look of order and cleanliness in the place where the pannikins on the wall fairly shone with polish, that he was master of his profession. He was a thin, under-sized man, with flaxen hair, very upright in his carriage, high in his opinion of himself, and exceedingly talkative.

His manner was amusing, nor was he in the least embarrassed at the presence of his new " Boss," but being evidently of the opinion that Jack is as good as his master, especially when Jack knew his work, he at once bluntly engaged me in a conversation that quickly merged into a mere narration of his own personal experiences. Judging by results as detailed by himself at least, Charles Knight (such was his name) was evidently a person who always rose superior to every situation. The interest which he strove to arouse in one's mind, by a minute account of all the details of some circumstance that was sure to have a triumphant issue in his own favour, was not a little increased by his singular manner of telling how, in every dilemma, he debated the case with " Self," as though " Self " were some second party standing beside him.

This habit, which I soon understood to be an inveterate one, very naturally gave rise to his receiving the nickname of " Self "—a thing, however, that did not disconcert Knight in the least, so powerfully was he sustained by a knowledge of " Self's " intrinsic merits. Thus, whenever he had occasion to go outside for anything, and he happened to pass along where some of the men were at work, their general style of salutation would be—not, however, in any sneering tone, for Charley's abilities as a cook were too much respected for him to be made a laughing-stock—" Oh, here comes Self, and pray how is Mr. Self?" To which Charley, passing with his upright carriage and stately walk, would, with the most placid equanimity, reply, " Self is well. Self is very much at your service."

On the present occasion, the description he gave me of his difficulty in having to make shift in baking bread without a camp oven, on his first entering upon his duties as cook in the time of my predecessor, is a fair sample of the general style of his conversation.

" When I came on to the station first, sir, there was no camp oven for the men's hut. The cook had broke it, and there was not another one on the station, except the one in the kitchen, and, of course, the cook there could not do without it ;

anyway, he wouldn't. So that the men's cook just had to muddle away with dampers in the ashes, for the men's meals. Now, a damper is all very well once in a way, but, as a general thing, I don't believe in dampers myself—the bread is too close grained, and aint so easy digested as light bread. Not but what I can bake a damper, with here and there a man, myself, but I always go in for light bread myself, which, I think, is the most wholesome bread for men. However, when I came here, I saw there was no light bread, and no oven to bake it in neither, and I was told that Old John in the kitchen would only growl if I were to ask a loan of his. Now, here was a fix, but fixed, I was resolved, I would not be. 'Now, Self, old man,' I says, 'here's a go for you; are you going to let your invention get bested now, Self? You won't give in without a struggle, my boy, will you? No, by the hokey, Self, you won't.' With that I just looks round the hut, and I very soon spies one of the pots they boil the meat and duffs in; and, at once, an idea flashes through my mind—'Bravo, Self, my boy,' I said, 'I thought you would not be easily bested, as far as ingenuity could get you out of a difficulty'. Now, sir, I says nothing to the men, but I goes straight away up to Old John, and asks him for some hops. 'What do you want with hops?' says he; 'you have no camp oven, and I can't let you have mine, you know.' 'Never you mind,' I says, 'what I want with them; I won't trouble you about your camp oven, any way.'

"So I makes yeast and says nothing to the chaps, and they didn't ask me what I was doing either. But when my dough was ready I made up as much of it as would fill the bottom of one of these tin plates, and I put that on the hot coals and turns the pot over it, and put an iron hoop round the feet of the pot so that it could hold a good supply of coals to bake the bread with, and it did bake most beautiful. Well, sir, you'd have laughed to see how astonished the men looked when they came home to their supper that night and saw my pot loaves. They were so tired of the dampers the last cook had been baking them —and sods of dampers some of them were I believe—that they were quite delighted, and, by Jove, they polished off my three loaves I had ready for them that night before they went to bed, and I had to keep baking on till twelve o'clock that night for a fresh start in the morning, when I worked away with the pot, and kept the men in light bread till the train went down to the Junction and brought me up this camp oven. Now, of course I can do the baking with more ease to myself, but the bread is no better than what I baked under the pot for all that. Here, sir, just taste that bread. That's the kind of bread I always bake for the men; how do you like it, sir?"

"The bread is very good, and I am very glad to see the men so well provided with a cook as they seem to be in you, and I have no doubt while you do as well as you seem to be doing now we shall all get on very well together. As long as I see that a cook is not wasteful with the stores, I never care to limit the working hands to regulation rations, so if you only let me know when you run short of anything in the way of common stores or supplies for puddings I will always see that you have a fresh supply." With these words I left the hut.

This preliminary survey being so far satisfactorily completed, I next, with the hope of ascertaining by a study of his manner in conversation what prospects of co-operation in the duties of my new position I might count upon with my rather cavalier acquaintance of the night before, directed my steps to the river's bank, and cooing on one of the black fellows—who were camped on the further side—I was by him quickly paddled across in a bark canoe.

Lilly's hut was cosily situated close by, in a bend of the river, thickly sheltered with a dense belt of shady gum trees. On my way to the hut I passed by a substantial stockyard, fitted up with every improvement for the management of cattle. I took notice particularly of the neat manner in which all the buildings about this yard were completed. They were so altogether different to the rough joinery incidental to mere bush carpentry, as to almost merit the character of ornateness. I saw that all the tenons were exactly fitted into their corresponding mortices, and everything bore the impress of no ordinary workman.

I found Lilly outside the hut door plaiting a stock whip. I may mention here that Lilly had a cook for himself, and he was the only other person with him in the hut.

Evidently Mr. Lilly regarded my movements with a jealous eye. He doubtless thought that I was there with some interfering purpose on hand, which he was determined to resent forthwith as an encroachment on his proper domain. His plan of campaign evidently was to simply preserve a respectful distance between us. In this spirit his reply to my morning's salutation, though civil, was brief.

But what a strange creature man is, and what little things will at times instantly revolutionise his most steadfast purpose!

Here was Benjamin Lilly, for instance, firmly resolved on keeping me at a jealous distance, as a man from whom he apprehended only annoyance to himself. Yet, by a few simple and wholly unpreconcerted remarks on my part, not only was that feeling of suspicion removed, but in less than half-an-hour Lilly and I were on a footing of most amicable friendship with each other.

As I spoke, I inadvertently glanced at the hut and noticed the same superior workmanship there as I had already admired in the yard. It was a simple affair: built like my own house of short sawn slabs, laid horizontally in grooved posts. Yet, in this slight matter, there was displayed a compactness, a finish, I had almost said a taste, not often displayed in the construction of a stock-keeper's hut.

"You appear to have had a rare workman here, Mr. Lilly," I remarked. "Was it the same carpenter who built the house across there who built your hut?"

"No, I guess not," replied Lilly, with a slightly supercilious smile.

"No," I said. "For the matter of that I might have known it, for the house by no means shows the marks of good workmanship that this hut does. He must have been a rare workman who built this. I suppose it was the same man who built your stockyard? I took particular notice of it as I came along. Now, the fact is, I have had some considerable experience with stockyard requirements, and I believe I never have seen a yard more handily arranged or showing such good work in its construction as this one does."

I noticed that Lilly's features at this relaxed into a rather pleased expression. "I suppose that you, at least, know the man who built them. Is he still about here?"

"Well, I believe he aint very far off," he quietly replied.

"Mister," here interrupted the cook who had been passing in and out of the hut during this dialogue, "Lilly built the yard and the hut hisself."

"Indeed," I answered in genuine surprise. "Are you a carpenter then, Mr. Lilly?"

"I never spent a day in a carpenter's shop in my life," he replied.

I now took notice of the whip on which he was engaged. "You seem to be as good at plaiting whips as you are at carpentry," I said. And in truth the plaiting and general form of the whip, so far as he had gone, seemed to be as well executed as if it had been done by a professional whip-maker.

As a further proof of this man's ingenuity, the cook now presented for my inspection a stockwhip handle, beautifully carved and embossed with pin heads, that filled me with admiration as I gazed on its curious workmanship.

"Well, Lilly, without the least bit of flattery, I must tell you that this article, for beauty and finish, exceeds anything of the kind that I ever came across in my life; it must have cost you immense pains and patience to have put so much work into it."

Lilly's reserve, that had been gradually thawing under the compliments that I really could not help paying him, and the genuineness of whose quality the simple earnestness of my manner must have at once attested, now completely gave way, and he became quite communicative. For Lilly was not only an ingenious workman, but it was a matter of pride to him to be thought so; and if there was one weak point on which, in colonial parlance, Lilly could be got at, it was his vanity over his own ingenuity.

But I soon found out that what I had already seen were only a few samples of Lilly's mechanical abilities. Indeed, it would have been hard to state what thing connected with the work on a station or anything connected with the necessaries of life Lilly did not do equally well, and was not almost equally proud of.

His cabbage-tree hat was solely his own manufacture, save in the preparation of the material; when he chose he could make his own trousers and mend his own shoes, whilst in his own particular sphere as a stockrider or horseman he was acknowledged to be without his equal on the Darling. He also, when the shearing season came round, and there was no particular call for his services with the cattle, went on to the shearing-board, and here his superior skill was again completely manifested, for not only did his tallies of 170, 180, and even 190 place him beyond the reach of the keenest competitor, but the quality of his work was far above that of any other shearer in the shed, and, indeed, it was apparent to anyone that were Lilly pushed by a rival of more equal power, or were he to put on a slightly rougher cut, even 200 itself would not always measure the quantity of his day's tally.

I have not yet exhausted the category of all of this ingenious man's talents, for on emergency Lilly could also act the part of cook, when he would turn out a "feed" that would make the mouth of an epicure water. No pastry-cook could beat him in that special department, and, above all other things, he excelled in baking a damper.

I have said that Lilly was vain on the point of his ingenuity. But were I to name the particular accomplishment of which he felt positively proud, I should certainly say it was in baking a damper. Next to the damper came his shearing; this afforded him a constant source of triumph, as he was always able to keep the lead in the shed against all comers, and for it, too, his name was known throughout the colony. As regarded the dampers, he was not at all inclined to make himself cheap, for it was no small difficulty to get Lilly to undertake to bake one at all, and it was, indeed, only under the pressure of very special circumstances that he could be persuaded to do so. When he did consent, it was with such elaborate preparations, such minute attention to details, such certain anticipations of triumphant success, that—but I shall weary my reader. One other characteristic trait and accomplishment I must, however, mention, and that is his penchant for bullock-driving. Like all other work that he turned his hand to, in this also he excelled. But having ideas of his own as to how these animals should be driven (he contended that not one in a hundred calling themselves bullock-drivers knew their work) he had, by careful selection from the stock on the run during several years, gradually acquired a noble team that he himself had broken into yoke, and would suffer no man but himself to drive. So determined was he on this point, that he had been more than once on the point of leaving the station because of my predecessor's wish to have these bullocks used for the general work of the station, at his pleasure.

In finding names for his bullocks, too, Lilly was original, always asserting that the names of bullocks should correspond with their places in the team. For instance, he said that as leaders were adapted by mental rather than physical qualities for their positions, so for them he chose names corresponding to the moral positions that they both held. Thus the strikingly appropriate names of "Dauntless" and "Fearless" respectively distinguished the near and off leader of this model team; both of these were red-spotted, spreading-horned, fine young bullocks. "Dainty" and "Davy," the next pair, were both as remarkable as their leaders for their comeliness and activity: "Dainty," black spotted, with curling horns and white forehead; "Davy," strawberry coloured. But as the team strengthened towards the pole, the more portentous names of "Roderick" and "Bauldy" represented two wide-girthed, strong-ribbed animals, both sure and unfailing in the yoke. "Roderick" was black, and Bauldy darkly brindled. In the pole bullocks the perfection of strength was fairly represented. Two noble-looking animals they were; both vast in size, and well proportioned as they were vast; deep chested and straight chined, no common names could be borne by these impersonations of strength.

It seemed that among the few books that Lilly had found patience to peruse (for he was no great reader) Miss Jane Porter's *Scottish Chiefs* was a notable one. From the time of reading it Lilly became a most enthusiastic admirer of Scotland's mightiest champion, although he himself was an Irishman. He firmly believed Sir William Wallace to have been one of the greatest men who ever lived; and of his hero's strength he had an almost fabulous conception.

In the truth of the book Lilly firmly believed, and anyone who ventured to speak disparagingly of its facts, or hinted at their being embellished with fiction, he regarded with the utmost contempt. For his favourite pole bullock then, a noble animal, with large red spots and upturned horns of proportionate size, Lilly found the appropriate name of "Wallace," but for the off pole bullock, an animal even more massive, if possible, than the other, Lilly could find only one name—a scriptural one. Of course, all these particulars of Lilly's idiosyncrasies and characteristic traits I learned afterwards. I am now stating them here by way of anticipation, to present Lilly's character at once to the reader, to avoid having to continually recur to it in the course of our further acquaintance with each other. Meanwhile I had suddenly advanced so much in Lilly's estimation, that seeing me still admiring the stockwhip handle (it was of the beautiful variegated myalwood, one side being like mahogany and the other of an orange colour), he volunteered to make another like it for me. Of course I was right well pleased at the offer, and thanked him accordingly.

We then talked about station matters in general, and after some conversation, it was agreed that on the following day he should accompany me on a tour through both runs. I then left him, not a little pleased at the unexpected turn matters had taken, and the sudden breaking down of the barriers of jealous suspicion that at first sight appeared to bar the way in any approaches I might venture to make towards the confidence of my new acquaintance. Nor was my satisfaction lessened on account of the sterling character that lay behind this rough independence of manner. All strained relations were thus obviated between us. I spent the remainder of the day in a general survey of matters around the homestead, which for a station, appeared to have been fairly well attended to.

CHAPTER VI.

IT was a beautiful morning in the mildest season in Australia —the fall of the year—when Lilly and I sallied forth on our tour of inspection over all the run. The sun shone brightly without the excessive heat of summer. Altogether it was a day that made it a pleasure to live, and the feeling of exhilaration inspired by the prospect of a ride over wide plains, through pathless woods, and dense scrubs, was indescribable. This feeling is greatly enhanced by the sensations of being borne along by a noble animal that seems to share in its rider's pleasure, needing neither whip nor spur, and snorting and tossing its head as if in proud consciousness of its power, a word or even a motion of the bridle being enough to increase its pace to a trot, canter, or a gallop at its rider's pleasure.

It was in this mood that I gazed around in admiration at the fine pastoral country through which we rode—large plains interspersed with belts of timber or sand-hills, now green with verdure, whilst in the back-ground the far off mountain ranges loomed blue in the distance.

My companion evidently had a taste for horseflesh, for sometime after starting I noticed that he keenly scanned Selim's various points, appearances, and paces. He made no remark however. He himself was well mounted on a dapple grey mare that went along at a quick amble that Selim's longer stride nevertheless kept even pace with.

Our tour over the sheep run, with its frontage of fifty miles of the river upwards, counting from the home-station that was situated ten miles within the lower boundary of the run, was a greater distance than could be satisfactorily covered in one day's ride. Consequently, after calling at the huts and viewing the shepherds' flocks on the run, as night was coming on we put up on our way back at an out-station that was in charge of an under-overseer named Bellamy, whose duty it was to keep an eye on the sheep at that end, besides seeing that the huts were kept duly supplied with monthly rations.

Here I had an opportunity of verifying the wonderful truthfulness to nature of Charles Dickens' seemingly improbable characters.

In one particular of Mrs. Bellamy's manner I saw a living embodiment of the widely-famed Sairey Gamp in that worthy's inveterate custom of exalting her importance and respectability in the eyes of her acquaintances by continual references to her bosom friend, the immaculate Mrs. Harris. In a certain Mrs. Williams, Mrs. Bellamy also had a bosom friend, to whom, in all recitations of her past experiences, and these were constant, she continually referred as to some oracle. Indeed the amount of moral obligation she seemed to be under to this estimable person must have been simply incalculable, as in all the changing complications of her life—whether in grief or pleasure, in times of trial or times of triumph—this same infallible Mrs. Williams seemed to have been ever at hand to sustain or to sympathise with her, to advise or to rejoice.

The resemblance to Mrs. Gamp, however, held good only in this single point, for fresh, tidy, and by no means ill-looking, Mrs. Bellamy was otherwise a very estimable person. But she was one of the most incessant talkers to whom I ever listened, and whilst listening to her continual references to her friend Mrs. Williams, I could not help wondering whether she had ever come across the great humourist's famous character, or whether she would really recognise her own failing if she were to read it.

However, in blankets provided by the station for such emergencies, I was made exceedingly comfortable by Mrs. Bellamy for the night, whilst Lilly in a possum rug that he had carried strapped on to his saddle, passed the night with equal satisfaction and comfort in one of the empty bunks.

The following day we returned home, deferring the inspection of some back country to another time.

On the morning following we rode out to see the cattle run. The country there I found to be somewhat similar in appearance to that set aside for the sheep, only that here there appeared to be more grass country than on my run, which was chiefly covered with the salt bush. Here, too, there was a fine lake of water, around which the grass seemed as luxuriant as in an English meadow.

The cattle appeared to be quiet and in fine condition, and at the crack of Lilly's stockwhip rounded up as we passed along.

Crossing over one of the plains studded over with several small mobs of cattle, Selim, checking himself in his usual steady pace, suddenly snorted, and lowering his head with an impatient jerk, as if to free himself from the restraint of a bridle, betrayed signs that long acquaintance with his habits enabled me at once to interpret as a desire to pursue some object that had then just caught his eye. It was not the cattle, for, unless urged on, of them he took no heed.

"My horse must see something," I remarked to my companion. "I can tell by his manner there is something he doesn't like."

"It may be a snake," said Lilly, glancing keenly round. "Horse are all skeered at these varmint; but see there! that dingo ahead there! he is sitting down in front of that cow and calf, beside the bush yonder." Being an old colonist, like too many of his class Lilly's expressions had frequently a sanguinary flavour that I cannot venture to transcribe here, but looking in the direction indicated, I descried a yellow dingo, or native dog, about three hundred yards ahead of us, seated as described by Lilly. As the wind was blowing towards us, he, absorbed in the contemplation of a newly dropped calf, evidently had not as yet noticed our approach.

This at once accounted for Selim's sudden emotion, as many a chase had I had on his back after similar quarry. Advancing quietly, although with difficulty restraining my curveting steed, we managed to approach within about a hundred and fifty yards ere the dingo detected us, when it instantly started off at a surprising pace across the plain. It only needed the slackening of the bridle and a "hist on, Selim," to send my brave horse off like a shot on his track, while Lilly's, with equal mettle, darted along for a while with even pace.

For several miles we kept on thus, Selim gaining slowly but surely on his canine quarry, whilst Lilly's plucky, but less bottomed mare, was left a length or so astern. At length Selim came like to trample on the dingo's quarters. Knowing what would occur, I accordingly strengthened myself in the saddle for the event. In a chase such as this, the management of which might be freely committed to Selim's own discretion, I had only to concern myself with the preservation of my own seat in the saddle. Almost simultaneously with my precaution came the need for it, as the dog, to avoid his impending doom, suddenly wheeled round beneath Selim's belly. Arresting his onward course by planting his fore legs with a shock on the ground, that would most certainly have sent an unguarded rider over his head, Selim, seeming as if he had thrown himself round on his hind legs, wheeled almost as suddenly as the dog had done.

Lilly now checked the dog, that again dashed past Selim, and again, as if on a pivot, Selim wheeled after him. But it would be impossible to give a detailed account of this exciting chase for the next half-hour, during which our active quarry kept us in pursuit; doubling back under the belly of either horse alternately as each one overtook him. Suffice it to say that Lilly at length, coming close upon him, as he was exhausted with his exertions, and sweeping his whip, handle foremost, round his head, caught the dog with it over the skull and felled him, and instantly dismounting, dispatched him with his stirrup iron, that he had for that purpose dragged off the saddle as he threw himself from his horse.

"By ——, Mr. Farquharson," he exclaimed, as we both stood looking at the dead dingo, our horses panting heavily the while, "that's a sweeping horse of yours; I used to think that as a stock horse there were not many like Coleena in the colony, but now I fairly give your horse best over her."

I smiled at the well merited compliment to Selim's qualities, but remarked :—

"You need not, however, Lilly, lose conceit of your mare in being beaten by my horse; knowing what he is, I think it no small thing in her favour that she was able to stick to him so close as she did in the chase."

"Where did you get your horse, might I ask? where was he bred?" he asked.

"In the scrub," I replied, "and out of the scrub, when he was little more than a three year old colt I helped to run him." I then detailed to him the circumstances that the reader is already acquainted with, regarding the manner of Selim's capture.

"George Laycock," mused Lilly, "I think I ought to know that name. Yes, I remember. When I was a bit of a stripling a man of that name was with me on a station on the Darling Downs; he was a grand stockman."

"I believe that must have been the same, as I heard that he had come from that part of the country," and I then told him the report I had heard of Laycock's death.

I should here remark by way of apology for Lilly's inveterate habit of swearing, that not only had he spent all his manhood and a great part of his youth in the colony at a time when such language as seems simply shocking to refined ears was the general custom then, and deemed to be actually harmless, but also that he had been sent out to the colony a convict.

To lessen the effect of such a stain upon his character in the reader's eyes it must be added that it was for no breach of the moral law that he had been so punished, unless the seemingly venial one of transgressing on the preserves of some wealthy landowner may be regarded as such. At a time when offences of this kind were most stringently dealt with, this had constituted the crime for which at the age of fifteen he had been condemned to take his place among social criminals by a Christian legislature.

Fortunately for Lilly, however, he was not long left to the contaminating influences of mere felons. After the manner of those days he, shortly after his arrival in the colony, was hired out by a neighbouring squatter, a man of integrity, though of blunt and unpolished manners, and in his service he remained until his seven years of penal probation had expired. Consequently, although Lilly's manners and sense of propriety had been considerably blunted by the unavoidable influences of his surroundings, still his moral principles remained uncorrupted, chiefly through the example of his worthy employer. Therefore, though a careless enough manager with his own affairs, yet in situations of trust he had been proved to be so reliable and faithful that the stain of his former transgression was now rarely, if ever, remembered against him.

Though occasionally he would allow a coarse enough expression to fall from his lips, yet the mere habitual use of foul language he eschewed, while sheer blasphemy he strongly reprobated. Of his views on this point a story fits in here of how he once chastised a bullying sort of fellow whose language disgusted him. The story is worth telling if only as an indication of what poor Lilly's semi-enlightened mind deemed the utmost limit of immorality in language. This man was working the bullocks, a task he was manifestly unfitted for. Laying the cause of the slow progress he was making to the account of the poor animals instead of to his own want of skill in guiding them, he was, whilst shouting to them in strains that were horrible for their filth and blasphemy, flogging them in an equally brutal and altogether needless manner—a proceeding which always stirred up Lilly's anger. Watching the fellow for a short time, until he at last fairly paused for breath in the midst of a torrent of foul expressions, Lilly remarked quietly :—

"Well, old man, feel any better after that ?"

"What! You go to h—l" was the savage rejoinder.

"Oh, never mind me" (in the same tone of quiet sarcasm as before) "but don't you think you'll sweat yourself too much if you use the whip as you are doing ?"

"I'll d—— soon lay the d—— whip over you, you etc., etc., etc., if you don't clear out of my way."

"Or, suppose you lay down the whip altogether," was the now stern reply—as Lilly suddenly stepped back and slipped the jumper over his head, which while speaking, he had been quietly unbuttoning ; a signal that the bullock-driver, who was a sort of fighting man in his way, quickly acknowledged by dropping his whip and squaring up to Lilly. But he soon found himself in the hands of a man who was as much his superior in the science of fisticuffs as he was with the bullock-whip. After three or four rounds, in the course of which he found himself—in the phraseology of the ring—each time sent to grass in a very summary fashion, he was glad to slink off to the hut, carrying with him a bloody nose and bunged-up eye, as some token of Lilly's prowess. While by way of enunciating to his discomfited opponent the moral principle whose infraction had cost him such a condign castigation, Lilly, while replacing his jumper, emphatically remarked, "I'll teach you to swear like another d—— Christian". Equally characteristic was his sarcastic remark to another ruffian who was actually cursing the Deity, and whom Lilly recommended to challenge his Maker down to fight.

In Lilly's eyes it was evident that swearing could be only deemed offensive when it assumed the character of blasphemy.

Language that included only such words as damn he reckoned was simply harmless, nay, on occasions even manly, with the exception of that one word whose use he looked upon as foolish. But I am sorry to say that among the list of harmless expressions, he freely admitted that of his Maker's name. Yet, with singular inconsistency, the name of both the second and third persons in the Trinity could not be used without blasphemy.

Lilly's reverence for the Deity manifested itself also in another peculiar fashion of his own, in his sense of obligation for the mercies of food. Most people, even those of a semi-religious profession seem to be horrified at the idea, when in a company of strangers, of returning audible thanks to their bountiful Maker in providing so liberally and continuously for their wants. To such people, poor unenlightened Benjamin Lilly, who otherwise on religious subjects was almost entirely ignorant, would have acted as a fine rebuke by his blunt mode of acknowledgment for such mercies on these occasions. Not that Lilly believed in, much less attempted to speak, a formal prayer. With him in this particular, brevity was certainly the soul of wit. Yet if brief, I question if it was any less heartfelt than the most solemn utterance of Christian professor or Rev. Dr. of theology. Lilly's manner of acknowledging his sense of God's mercies to him then was simply thus. On the conclusion of every meal he first placed both hands on the table, and actually looking upward, emphatically said, " Thank God for a good feed ".

When it is borne in mind that Lilly was a man of strong force of character, and therefore of strong influence over most of the men with whom he came in contact, it may be well questioned if the repetition in such a manner of this simple, and even uncouth form of praise (for praise in a sense it really was), had not some moral effect on his rough associates. " For why," as Lilly remarked in reference to this practice of his, " if a man ain't ashamed to thank another man for giving him a five pound note—and far less than that, too—should he be ashamed to thank God for giving him both food and appetite to eat it, without which all the five pounds in the world would be of no value to him ? "

There is some philosophy there, reader, and that too of a kind that might well tend to shame men much more enlightened and educated than was strong, common-sense Benjamin Lilly : men who, while professing to believe in a God, yet by their irreligious and thankless lives practically ignore him.

CHAPTER VII.

THE time at the station glided smoothly by, and soon overseer and men got into the easy relations of crank and eccentric towards each other; that is, we were at home with one another's ways, and I knew what each one's capabilities were, and what reliance I could place on them, and they on their side understood what measure of work I required from them. As I had taken over the work at the close of the shearing season, and matters there had subsided into their wonted groove, life on the sheep station was monotonous enough, only occasionally enlivened by some stir on Lilly's side of the river, as the requirements of branding calves or other work incidental to the management of cattle necessitated a muster among the stock.

It might have been a little over two months since my arrival on the station, when, more for the sake of enjoying the exercise than from any demand of duty, I mounted Selim, and spent the major part of a sultry day in riding over Lilly's run, and observing how his cattle were looking. Tired with my ride, on returning to the river, instead of sending Selim across, and getting a black fellow to paddle me over, or paddling myself over in Lilly's canoe, I dismounted and entered Lilly's hut. The apartment into which I first entered was that set aside for his cook. Although there were voices in Lilly's end of the building, feeling languid and weary, I simply stretched myself on a bench that was fitted up with scoured sheepskins as a sort of primitive couch. The cook evidently had gone across the river to enjoy a confab with one of his professional brethren of either the men's hut or the house kitchen. The voices that I heard at the other end of the house, and whose utterances I could easily follow, came from Lilly and a poor traveller whom he had been lately entertaining: in this particular, Lilly was left solely to his own discretion, and not only was his hospitality at all times liberally extended to all passing wayfarers, but, by choice, he either entertained a friend, or a chance traveller whose necessity interested him, or to whom he was attracted, for days together. But as Lily at no time encouraged waste, idleness, or as the colonial term expresses it—"loafing"—it was well understood in these occasional instances of extended hospitality that either the peculiar necessities (such as sickness or weakness) of the person, or some other circumstance, justified his action. Anyhow, Mr. Rolleston's expenditure was in no way augmented by Lilly's profuseness with the station's rations, beyond the usual proportions entailed by the prescriptive exercise of station hospitality. And, indeed, his carefulness in deprecating anything that savoured of waste was a prominent trait in this intelligent man's character. Be it observed, however, that in those days, almost on every station, the claims of hospitality were so freely acknowledged that the consumption of rations by passing travellers was just reckoned among the necessary expenses of the station. Yet how much such a tax must have added to the expenses of the station's rations may be judged, when it is remembered that scarcely a night passed without two or more callers, and, towards the shearing time, this number was frequently increased to about a dozen every evening.

The stranger in question, whose voice I was now listening to as he spoke to Lilly, was a poor traveller whom I had observed about the latter's hut for some days before, and who had excited his consideration by his utterly forlorn appearance when he called at his place one evening and asked for a job, and, failing that, a night's shelter. The poor fellow seemed to be utterly dispirited and done up with that particular day's tramp. He had travelled a distance of three and twenty miles, a great part of it without water, and the day had been extremely sultry.

Together with the natural fatigue from such a journey, his difficulty had been not a little increased by the fact that his boots were almost falling from his feet, and when he arrived at the hut he was limping painfully. Altogether, from his ragged appearance and utter exhaustion, he looked a most utterly woe-begone object.

He was a tall young man of a rather simple yet interesting appearance, with blue eyes and fair hair. He had only left home a few months before and had come to try and make his fortune in the diggings, but was like to have found a famine there instead. To make matters worse, he belonged to that class of pretentious respectability that looks upon the idea of self-support by mechanical, much more by mere manual labour, with feelings of repugnance. But ruin of his people's financial position at home had compelled him to attempt doing something for himself, and now, by his entire loss at the diggings of what little means he had brought out with him, he seemed to be as wholly prostrated in spirit as he was exhausted in body. Not being quick and energetic, or, as it is more happily expressed in the Scotch phrase—" Guid at the uptak," in his efforts to secure some sort of employment his green, unhandy appearance had hitherto either procured him a prompt rebuff where some work might have been found for a more workmanlike applicant, or, if taken on at hazard, a contemptuous dismissal after a few days' experience of his utter inability to handle any tool that he was set to work with. Yet, as it afterwards turned out, he was by no means unwilling to be taught, or above learning any kind of work he was set to do.

Thus, penniless, and almost bootless, and seedy-looking in every way, the poor fellow continued to wander on until, after many weeks of weary travelling, he had penetrated thus far into the interior of Australia.

This was, indeed, a case for the exercise of Lilly's compassion, for, underneath a rough, sarcastic exterior, that worthy fellow had no small stock of that precious commodity at hand. I have already mentioned that amongst his other accomplishments, Lilly could mend his own boots. But the task of repairing this young man's boots, which, the day after his arrival, he actually attempted, was beyond even Lilly's mechanical ingenuity. Throwing aside this hopeless task, he did something infinitely better, for he procured him a new pair from the store slops. This gift he afterwards followed up with the presentation of an entire new rig-out; not even excepting a pair of comfortable blankets. This, however, was done on the understanding that if the young man found employment he was afterwards to recoup Lilly for this outlay. Such employment, he told him, he thought he could induce me to find for him until the lambing season came in, when there would be plenty of work for all. However, he was to stay where he was until fully recovered from his fatigue; and if there could be no work found for him on the station, he might at some future time repay Lilly for his clothes. So far this arrangement had been satisfactorily agreed to by the traveller.

Dressed up in his new rig, and fully recovered from his fatigue and depression, William Lampiere—for such was the young man's name—had rather a pleasant look, and Lilly seemed to conceive quite a friendship for him. He was evidently greatly amused at his frank simplicity, to which was apparently joined a perfect truthfulness of manner. If anyone delighted in a spirit of truth, it was Benjamin Lilly, and the habit of exaggeration, or in colonial parlance "blowing," was his special aversion. "What is a man if you can't believe what he says?" was his usual strong expression of contempt with reference to the slightest indications in any one of this habit—a habit that is, indeed, only too common in the colonies.

But to return to Lilly and his new friend, whose voices I have already stated were audible to me where I was then languidly reclining. From the tenor of their remarks, it was not long ere I was aware that Mr. Lilly was engaged in his favourite occupation of "stuffing"—a habit that seems inconsistent with Lilly's love of truth, but which was notwithstanding an inveterate one with him whenever he found a favourable opportunity for indulging in it. In the present instance, in the confiding simplicity and charming inexperience of his new friend, he had an opportunity of indulging this propensity to the very "top of his bent," and I must certainly allow that he seemed to be then making the most of his opportunity.

I have already referred to the light in which Lilly viewed the almost incessant use in colonial phraseology of the adjective "damn".

With this knowledge of Lilly's true view of the use of this word the reader may be able to appreciate from the following conversation the exquisite, though strange, humour he was then enjoying at the expense of his guileless companion.

When their conversation first became intelligible to my ear Lilly was remarking: "Do you know how to cook? but of course how can a new chum like you know anything about such work!"

"No, I know nothing except what I saw at the diggings. My mates swore at me very much when I tried my hand at a damper once—and indeed no wonder, for I just spoiled it. Then they would not let me do any more of the baking, and what meat was to be roasted one of them did at night, for he seemed to be handy at cooking. All I did was to attend to any boiling meat that was being done, besides always boiling the billy."

"Ah! I thought so; well, never mind, while you are here I will get Jack to put you up to the proper way of cooking all that needs to be done on a station. Now you watch him with all your eyes, for cooking is a handy thing to know at all times, and I'll show you how to make a damper properly, and if you take up my lessons anything like at all, you will be able to bake a cake of bread that will be fit to be set before the governor. If there is no work for you after I have done with you at a small job I have on hand, that we will start at in a few days, when you are well spelled after your long tramp, I may get the Coni* to give you a job at hut keeping. There is not much wages at it, but it will be a fine and easy job for you, and will do anyhow till you see a chance of shepherding: there will be better wages at that job. But mind," continued Lilly impressively, "when you start cooking, whatever you do, keep yourself clean, and always wash your hands before you handle meat, or start baking, for if there is anything that I can't bear more than another, it is a dirty, lazy cook." (Lilly had a most

* Bush term for the employer.

religious faith in the maxim, that cleanliness is next to godliness.) He continued in the same impressive tone, "Another thing I want you always to remember is, to be good to travellers".

"Oh, certainly," assented Lampiere readily, "I hope I shall always be careful on that point, when I remember for how much I have been indebted as a traveller myself."

"Yes," remarked Lilly bluntly, "but it is very often the case that men who have loafed over half the country when out of a job, when they have got a billet as a cook themselves have been the very first to kick against cooking for poor swaggers, and it is often such dogs as these that have given stations a hungry name, for when the squatters themselves have been willing enough to allow rations, these wretches have kicked up a row about cooking them. That is how the pannikin of flour system started in most places."

"What kind of system is that?" enquired Lampiere.

"Why," replied Lilly, "on stations like these that have been spoilt with dogs of cooks who refuse to cook for swagsmen, the Coni just gives them a pannikin of flour, and some tea and sugar, and sends them up to an empty hut to cook for themselves— a hungry, slovenly and wasteful fashion it is. But you, when you see a traveller in sight, mind, whatever you're doing, just drop it, and fly about setting your tucker ready for him at once. If the billy happens to be empty, pick it up and run to the creek, break your d—— neck, but what you will let the traveller see, be he common swagger, or 'sujee swell,'* that you are glad to make him welcome. What is there to blow about in giving a poor traveller a feed? There is in my eyes nothing so degrading as one man letting another feel under a sense of obligation to him for a meal. It seems to me to be much the same as one man feeling under an obligation to another in being permitted to live so much longer on the earth. As you said you have done some clerking at home," he here abruptly asked, "why didn't you fall back on that trade again, instead of trying to get a job at hard work?"

"Well, you see," his companion replied reasonably enough, "I wasn't known in the town, or indeed in the country for that matter, and although I did try and get a situation on Ararat, and called at several offices where they were advertising for clerks, I couldn't get on. Then I had no money left, so I couldn't stop there without money, and I didn't know what to do. At last I thought I would just do the same as I saw others do, who went travelling through the place with

* A bush expression of contempt for a person of more flashy appearance.

their blankets strapped up and hung over their shoulders, and once started I just kept on travelling ahead: that is now six weeks ago."

"Well, sure enough, that was better than loafing about town, and shows you have some spirit of independence about you, anyhow. How much money had you when you started digging?"

"A little over £50—all that remained of what I had when I left town, after paying passage money and expenses till I reached Ararat, which I did from Melbourne by coach. Don't the money melt away quick from a fellow out in these Colonies! I found it did anyhow, and I didn't spend any in foolishness either."

"How came you to fall in with your two mates?"

"Well, you see, on the day after I reached Ararat, I was walking out through the diggings, and came up to where two men were having dinner, and as I stopped to look at them one asked of me if I had any tobacco. I said no, that I did not use it, on which he remarked, "What! a swell like you and can't afford to smoke!" I just laughed a little and said nothing. Then, as both of them kept passing some jeering remarks on the swell who could not afford to buy tobacco, I was about to turn away and leave them, when one of them said, "Well, swell, don't you think that we diggers are very vulgar, eh? how would you like to be a digger?" I answered very quietly, "That is just what I want to be, and I am now looking about to find any digger who would like to take in a partner".

"Why," the same man answered, "have you any 'hoot' to start with?" "Any what?" I said, for I did not understand what he meant by "hoot". They both laughed at this and said I was precious green, and then asked me if I had any money to start digging with, on which I told them I had about £50. Then after some whispering together, the first man said to me, if I liked to take my luck with them I could. They were running short of money themselves, and although £50 was not much of a capital, still it might do as things were, and I should have a third share when the claim was bottomed; that they expected would be in a fortnight more from then, when they were sure there would be a small "pile" for each of us.

"And what then?" enquired Lilly.

"Well, I didn't like these men at all; they wanted me to play cards for money on the first night that I joined them. This, however, I refused to do. Then they told me that I should have to give them what money I had, to pay for what expenses they had already incurred in working the claim, and

4

also for rations to keep them going. And then they often swore at me because I was so unhandy with the tools, although I wrought till my hands were full of blisters and did everything they told me. They even made me wash their clothes for them, for they said it was the rule in the diggings that greenhorns had always to do that for the experienced hands. That had to be done, too, on Sundays. But as I would rather not break the Sabbath, I sat up late at night when washing time came round. Well, I believe I was very glad when the claim was bottomed, although after washing-up my mates told me it was a 'duffer,' and didn't yield enough to pay their salt."

"But were you there to see for yourself when they were washing up?" said Lilly.

"No. I was kept at other work. I went over and saw them sometimes at the cradle, but I could never see anything like gold in it. But I do believe they cheated me for all that, for, while waiting about the township for several days looking after a situation, I could see them drinking and shouting all the time, and quarrelling and fighting most brutally, while I had not as much money as would pay for a loaf of bread, and I was two days at last with only one meal, till I started on the road and got a good one at the first station I came to."

"Ah, my lad. the d—— wretches had you properly; they must have been sods of men to take advantage of you like that. Why, they just swindled you right and left. They saw you were a greenhorn, and they were no men anyhow. I guessed as much so soon as I heard you tell about them trying to cadge tobacco and then joking you about it as they did. Fancy diggers cadging tobacco! Of course there are such men, but they are only low-lived curs who would do such a thing as that, though there are no more straightforward independent men in the country than diggers as a rule; but you fell into the hands of sharks that time, my lad."

"Won't the overseer be angry if he sees me stopping about here so long?" Lampiere enquired.

"The overseer? no; why should he? He never interferes with what I do," replied Lilly.

"What sort of a man is he? is he a nice person to speak to?"

(The reply expressing Lilly's opinion of me, to which I was now an involuntary eavesdropper, was too interesting to me for me at this point to offer any interruption to it, as I was curious to hear Lilly's frank opinion of me behind my back. Perhaps I might not have been quite so curious, knowing Lilly's bluntness of manner, had I not been fully convinced that his opinion of my merits was not an unfavourable one.)

"He is a very nice sort of a man," replied Lilly decidedly.

(You see I was not disappointed about his opinion.) "A man who understands his work, and although he is very quiet he'll take no nonsense from any one. While a man is able for his work he'll get on very well with Mr. Farquharson."

"What sort of a man is he in appearance?"

"Why, didn't you see him over at the stockyard with me the day after you came here?"

"I saw somebody; but I am rather near-sighted, and indeed I did not look much, as I was reading at the time."

"Faith, he saw you quick enough then, for he asked who you were. What sort of a looking man is he? Well, he's a tallish, smart made, and not at all a bad-looking man, a little browned with the sun, with blue eyes and darkish hair, but very quiet and thoughtful in his manner. He doesn't speak much, but what he does say is to the point; just such a man as I like to deal with."

"But if he is sharp in noticing about a man's work, I am afraid I shall never be able to please him, as I have not learnt to do any kind of work, and it seems to come so awkward to me when I try; although I should be most willing to learn if any one would just have a little patience with me at first."

"Never you fear for Mr. Farquharson, if he is able to give you a job: he always knows when a man is trying to do what he can, and when he sees that, he never says anything to him. But I am going to start a little bit of post and railing about the place and build a cow shed, and you'll be my mate, and I will soon put you in the way of using tools in a handy way, so that if you keep your eyes open you may be soon able to use an axe or adze, saw or spade, with some of the flash bushmen. Of course you'll not be up to doing much work of any kind at first, but after a few days, when you begin to get seasoned, I'll make you keep out of my way, and teach you to look spry."

"I am sure I shall be most willing to learn, and will thankfully, for a month or more, do what I can, just for my food."

"Well, never mind about that at present; it will be a week or more before I am ready to begin. You can just cut firewood for the cook in the meanwhile. I will show you how to sling the axe about—you can always be learning."

"I will indeed, and I am very much obliged to you for such kind thoughtfulness on my account."

"Oh, bother obligations. You must learn to be a little more rough and ready in this country. Just take as much as you can get, and instead of being very thankful for what you get, try and be bounceable for more."

"Yes?" (apparently uttered in surprise).

"You see, when you fell across these digging swindlers, if

instead of always thanking them, which I suppose you did, when they were only codding you, you had at once told them to go to h—— when they first began taking a rise out of you about the tobacco, they would have looked upon you as a man of spirit, instead of which they simply looked upon you as a fool, and so just treated you like one."

"I would rather not swear."

"Well, I must allow that *useless* swearing is a very silly thing, but then sometimes it is necessary to make people keep their distance. Why, there are plenty of such chaps as them digging friends of yours who, if they thought a man had no spirit in him, would just jump on him. But just give them a good swearing and they see at once they have to deal with a man who will stand no nonsense from them."

"Does the overseer swear?"

"Well, no, I can't say as I ever heard him slip an oath since he came here. I have heard him speak pretty sternly too on one or two occasions to men who were not shaping well at their work. In fact, they were right down loafing, but he cleared them out pretty quick. But as you don't like to swear, and you must say something you know or you won't have the life of a dog with some men, I'll tell you how to manage, and that will do just as well. For instance, there's the word 'damn': there is no harm saying that word any more than when you cut your finger to speak about your bloody finger. Well then, although hell is a swearing word, yet there's no harm in saying 'blazes,' is there?"

In response to Lilly's question Lampiere admitted that in these single words he could see no immorality.

"Well," continued Lilly, pressing his point with all the ardour of a Jesuitical casuist in making as simple an admission open the way to an equally doubtful conclusion. "If you don't want to say 'What the d—— h—l?' which is certainly swearing, and good Colonial swearing too, there can be no harm in saying 'What the d—— blazes?'. That simply means nothing, and yet sounds as strong as the other."

To this plausible induction again Lampiere gave his assent as to the seeming innocuousness of the phrase. The question being thus settled, Lilly next proceeded to impress on his friend some useful advice for his future guidance.

"Now then, you be always sure and remember when you see any one attempting to bounce or in any way to crow over you to say, 'What the d—— blazes do you mean?'—will you?"

"Well, yes, if I see any real occasion for it I will."

"Is that all the attention you intend to pay to my advice and me trying to colonise and make a man of you?" asked Lilly in a reproachful tone; "do it on all occasions, I tell you, if it is only in a common argument. If you don't, everyone will look upon you as just a milksop or common Johnnie Raw and bounce you as they like. But if you do what I am telling you, any person who might be setting you down as a greenhorn will very soon draw in his horns and say to the other chaps, 'My word, I thought I could do as I liked with that young fellow, but I was near putting my foot in it! By the way he expresses himself I see he has some spirit in him. Still waters run deep. I'll have to mind my p's and q's and not rouse him up. My word, I'll keep civil to him.'

"If it is only in a common argument, I say, and you see the other man is inclined to lift up his voice authoritatively as if giving you to understand that his way of putting it is to be the only way, just get up on your feet (Lilly apparently suiting the action to the word) look straight at him, and pull your hat over your brows in this fashion—it gives one a more determined look—and then say distinctly, 'What the d—— blazes do you know about it?' He will have to be a pretty bounceable customer if that won't put him down."

At this stage I regret to say that the moral effect of this rather singular lesson on the mind of this simple-minded pupil, with its accompanying illustration, was entirely destroyed by my involuntary interruption of it; for my sense of humour was so exquisitely tickled while listening to this singularly original advice, heightened by my mental picture of Lilly's appearance while giving it—the whole being, I felt sure, from my now familiar experience of his habits, a piece of wilful burlesquing on his part—that it now fairly mastered all my endeavours to restrain it, and first finding vent in a silent but continuous giggle, it soon broadened in volume into a loud, side-aching guffaw, which at once apprised Lilly that it was not the cook that he had as an audience to his lecture. As to the cook indeed, he was too well accustomed to similar freaks of manner for it to have concerned Lilly to know that he had heard every word he had spoken.

But with the immediate consciousness of the ridiculousness of his position in my eyes, when he caught my eye as he threw open the door to see who it was who had laughed, for the first and only time I saw Benjamin Lilly's countenance fairly put to the worse. But it was only for a moment.

Recovering instantly from his confusion, the more quickly as he seemed to read from my manner my perfect sense of the burlesque he had been enacting, he remarked:—

"You see, this young man is such a green new chum that I have been trying to put him up to a wrinkle or two in colonial ways".

"Quite right, Lilly," I made ready answer, "and I make no doubt but that if he is attentive to your instructions he will be so thoroughly initiated in these ways that he will be able to hold his own in any company. Well, young man, as Lilly has been giving you his idea of my character, that, I think, upon the whole is pretty near the mark, and for which I am certainly obliged to him, you will know what to expect from me as an employer. As Lilly proposes to take you in hand for a short time, I advise you to pay attention to all he tells you, for you may consider yourself lucky in getting such a teacher as Lilly to instruct you as he proposes to do. After he is done with you I daresay I shall be able to find you some kind of job; that is, of course, if Lilly, after trial of you, is able to speak promisingly of your work."

For this Lampiere gratefully thanked me in an earnest way that convinced me that there was something really genuine about him.

CHAPTER VIII.

THE result of this was more satisfactory to me than I ever imagined it could be. After the first few initiatory lessons, Bill Lampiere displayed such an interest in and consequent aptitude for his work as to transcend even Lilly's expectations, though the latter had felt sanguine, at the outset, of being able to make something out of him. It was not, indeed, that Lampiere's abilities turned out to be better than Lilly had imagined them to be: indeed these he found to be dull enough to have taxed the patience and temper of any ordinary instructor. Lilly, however, proved himself to be no ordinary instructor by the manner in which he soon made his directions clear to the confused mind of his apprentice.

Indeed it was soon self-evident that, although lacking in what a phrenologist would term "concentration," a want which prevented his bringing his ideas to bear rapidly on any new problem with which he was confronted, and which, therefore, hindered his gaining a general idea of a subject, yet, when once supplied with the reason why such an operation should lead to such a result, his natural understanding could grasp the full bearings of the case at once.

Observing this, I doubted much if his mind, clearly intelligent enough, was not of that nature that from natural nervousness simply intimidated its possessor with purely imaginative difficulties. But Lilly with his usual quickness of insight soon understood what was the cause of Lampiere's apparent slowness of intellect; and patiently accommodating his instructions to this mental difficulty, he was in a short time so gratified at the progress that his pupil was making in a knowledge of his work, that he predicted that he would soon make him so proficient in the use of his tools as to be able to put some of the " fancy " bushmen in the shade.

But if Lampiere had a mental defect that seemed to militate against his ready attainment of handiness with tools, he had another, a physical one, that without Lilly's patient instructions would have seemed a more insuperable difficulty in that direction.

Apart from the difficulties of manipulation common to all novices, there are some people who have much greater difficulties in training their hands to the use of unaccustomed tools than others have. This too, when a considerable experience of the work enables them to have a perfectly correct mental idea as to how the work should be done : yet what the mind directs shall be done, the hand finds itself unable to perform. I suppose a more relaxed state of the nervous system preventing all the manual muscles from being kept well in hand—if I may be allowed to repeat myself—is the real cause of this weakness.

This, however, was a thing that, left to himself, would have simply rendered Bill Lampiere a most handless workman at everything he attempted.

Against this natural defect, Lilly's patient instructions, however, proved to be a most powerful remedy. Continually watching him, and at the same time keeping his patience and temper in a most wonderful manner (for Lilly was naturally a quick-tempered man and with others easily provoked to irritable expression, if work was unsatisfactorily performed) he would take the tool out of Lampiere's hands, and would direct him to watch how he held it, and to hold it so himself, until, in a few weeks' time, with an axe or adze, cross-cut saw or augur, spade or pick, Lampiere showed a familiarity and handiness equal to any other man in the station—Lilly alone excepted.

Even allowing for what I know of the genuine kindness of Benjamin Lilly's heart, all this patient attention and consideration for this young man to me appeared to be simply inexplicable, until I was put in possession of the key to the riddle.

This key, that was furnished by Lilly himself, gave me a fresh insight into this worthy man's many-sided nature.

I have already intimated that Lilly's education was very limited. A bare knowledge of reading and writing and a few rules in arithmetic comprised it all; neither had he naturally any great taste for reading. Yet all this did not prevent an excessive admiration on his part for the knowledge in others that he lacked so much himself, and, among the various attainments of literary men, he looked upon poetry as the very cream of all.

So then the whole secret of Lilly's attention and consideration for Lampiere, that had first begun in simple commiseration for his forlorn condition, and was further attracted by the man's simple integrity, was now strengthened by a genuine admiration. It transpired that Bill Lampiere was a bit of a poet, and this discovery had raised him immeasurably in Lilly's estimation. It was this apparent feeling of respect for Lampiere on Lilly's part that I had lately observed and that had begun to puzzle me so much.

It might have been about a fortnight after his arrival that Lampiere's poetical accomplishments discovered themselves in a couple of verses that he had scrawled with a pin on a tin pannikin at the conclusion of a meal. These verses caught the cook's eyes, and he showed and read them aloud to Lilly, who was so delighted with them that he got Jack ("cabbage-tree Jack," as he was termed, from his constant occupation of weaving this kind of hat in his leisure hours, from the proceeds of which he considerably augmented his salary), who was a much better penman than himself, to transcribe them for him.

A few days after this important discovery, Lilly having come over to the store for some nails he was in want of for his work, after some conversation on various matters, suddenly remarked to me:—

"Mr. Farquharson, that young fellow is a poet".

"A what? Whom do you mean?" I replied in some surprise at this unwonted communication.

"That young fellow who is working with me. Now you're a man of education; don't you think that this is a first-rate production?—at least read it and tell me if you don't think it is."

I took the paper from Lilly's hand, not a little amused at the eagerness of his manner, as he appeared to consider that what he was handing to me for my inspection was a perfect gem of poetry. Well, I read it over attentively, and then reperused it with deeper interest; for although it was only a comic thing, yet throughout there was a spirit of decided humour and a kind of epigrammatic force in its conclusion, that in my eyes certainly could lay claim to some sort of merit of its own, while such pith and vigour seemed astonishing as coming from such an apparently mild young man as its author. However, I will let the reader judge of the composition for himself.

BENJAMIN LILLY'S WELCOME.

Be you traveller, or overseer,
  Or working man who shall come,
Here's bread and meat, sit down and eat,
  And have a feed and welcome.

If gentle, snobs, or men in mobs
  Of twos and threes together,
Here's lots of tea, then dine with me,
  And fill yourselves and slither.

Certainly delicacy is not the chief characteristic of the production, and the last line, that seems to offend in this point, Lilly regarded as the very essence of the wit of the verses, and roundly avowed his opinion that it was fully equal to Burn's famous verses on Inverary hospitality. Indeed, this it might well be; for, apart from the pointed bitterness of the latter, I am not aware that it has any particular claims to literary merit.

I will here again digress to remark that Robert Burns was another of Lilly's ideal men. His devoted admiration for the Ayrshire bard, and, as I have already stated, for Sir William Wallace, and his enthusiasm about the great Scottish victory of Bannockburn—than which themes none more frequently occupied his mind—seemed more befitting a Scot than a native of the Emerald Isle. Of the history of his own country he seldom spoke, and when he did mention it, it was slightingly, as though he considered that there had been some mistake or mismanagement about it.

Yet that he was none the less a patriotic Irishman, his pride in reference to the Irish soldiers in the Peninsular War, and at Waterloo, proved, and any one would have found also who had ventured to make his country the subject of sneering comments in his presence. Yet even on this subject he was extremely impartial, and he freely admitted faults and excesses in the Irish character, and blots in Irish history, made more conspicuous by the natural impulsiveness of the people.

But to return to our subject. "Now, Mr. Farquharson, don't you consider that there is real genius in that piece?" Lilly enquired, when I had finished reading it.

"Well, Lilly, there is enough merit in it to lead me to think that if Lampiere were to seriously give his mind to the task, he could produce something respectable."

"I think that is very good as it is."

"Well, for a piece of broad humour it will do very well; but do you know if he has written anything besides this? I daresay he has, as the easy manner in which these lines run, show me that it has not been his first attempt."

" Faith, you may say that. When I spoke to him about it, he showed me a manuscript book very nearly full of poetry."

" Indeed ? " I replied.

" Yes, indeed ; I think that young fellow will make his name yet. He showed me a poem of three or four hundred lines about his leaving home and voyage to Australia. I tell you it is just capital !"

" Well, well, he may come to something yet ; after all, it may be merely rhyme with nothing in it."

Lilly, however, vigorously protesting that there was something in it, shortly afterwards left me.

I will not say but that, after that, I looked with something like a kindly interest on Bill Lampiere.

Personally I always had a kindly leaning to the reading of poetry, especially of an objective and heroic nature ; such as Pope's glorious Iliad, and Scott's stirring lays, with Burns, Byron, Campbell, and such inspiring writers. But the entirely metaphysical, or subjective writings of such men as Browning, Swinburne, etc., and even those of the Poet Laureate himself —" In Memoriam " and a few others excepted—I never felt myself equal to.

In a word, I had by nature that amount of love for poetry which would make me readily believe that, had the circumstances of my younger days been more favourable to its culture than the life of action which I had led, this love would have in time developed itself into a chronic malady of rhyme in myself. I presume, however, that the world has been no loser by the fact that these symptoms were nipped in the bud by the force of circumstances.

CHAPTER IX.

THE lambing season was past and our percentage had been favourable, and, with some new ideas on management on which I had resolved, I hoped that the following year would be even more satisfactory.

At this juncture I received a letter from Mr. Rolleston desiring my presence for a few days, and requesting that if practicable I would take Lilly with me, as he wished his judgment on a small herd of cattle, consisting of a bull and some heifers, that were then under offer to him.

" That is," ran his letter, " if both of you can arrange to leave together. As the lambing is now over, and it will be some time yet ere the shearing begins, I think you both should be able to spare this much leisure. I would wish you could do so, as I should like to hear both your opinions about the cattle, that I intend sending up to Lilly's run if I purchase them. There will be men sufficient here to drive them up with you.

"Yours faithfully,

"WILLIAM ROLLESTON."

The contents of this letter interested me not a little, for, as the reader may guess, in all these months that had intervened since I had left there, I had never forgotten the pleasant society at the Murray Station. And often, during my solitary rides over the run, or on my couch at night, had the sweet vision of Miss Rolleston, with her elastic figure, her sprightly manners, and fascinating appearance, brightened my thoughts, whilst, though in a more measured degree, the sweet face of Mary Campbell, and that of her kind mother, also moved me. They formed a charming framework and background to the pleasant picture that now it seemed I should so soon behold again.

This coming pleasure seemed the greater for being so wholly unexpected, for, chained to my post, with miles between us, I had hardly dreamt of again coming into contact with these ladies for two years to come, at least, when I might deem my length of service sufficient to warrant my applying for a furlough. But what might not occur during the long interval of two years! As I thought thereon I sighed!

Yes, sighed! for confess it I must : the image of the wealthy squatter's daughter had gained a complete mastery over my thoughts—nay, if you will, my heart. Ah, what foolish air castles will people—especially love-sick people—construct under the most disadvantageous circumstances! Mr. Rolleston's reputed wealth, of a quarter of a million at least, was surely enough to have daunted any less aspiring man than a poet or a lover.

For how could any man in his senses imagine, that a parent, with the advantageous prospects that such wealth commands, would allow his only daughter to mate with his overseer?

But what will not a lover dare to hope for? what ways will he not devise for combating the most tremendous obstacles? Thus it was with me. My concern was only for the maiden herself; as to connecting her with ideas of wealth and self-aggrandisement, I believe I was honestly guiltless, and as a set-off to the father's wealth I counted largely on my own natural resources of energy and capacity, which, with the vague hope of Mr. M'Elwain's promise with reference to some day giving me a start on a station of my own, I looked upon as a capital with which I might yet rear an independent fortune. Was not Mr. Rolleston himself an overseer too at one time? I soliloquised.

And thus I hoped, and thus I hailed the near prospect of again meeting with the object of all my castles in the air.

About the propriety of getting away I saw no difficulty, as I could leave things as they were, in charge of Bellamy the working overseer, till our return, with directions for getting the yards, and matters generally, that required attending to in connection with the sheep-shearing, in a forward state. As for Lilly, his cattle were all quiet on their run, and would need neither attendance nor interference for some time to come.

On Monday morning we left the station, and on Wednesday evening our jaded beasts entered into what to them was the unwonted luxury of a stable, when they at once set to vigourously allaying the keen appetites that a toilsome journey had generated, by the almost as rare treat of a corn feed.

That my heart was beating a little more quickly than its wont, I will not attempt to deny, but, determined to reduce my nerves to order ere confronting the disturbing cause, I took an almost malicious pleasure in tormenting my own patience with fiddling about trifles in connection with my harness, chatting to the groom, etc., until even Lilly's forbearance began to give out. "Had we better not go and see about getting a feed?" he at length bluntly ejaculated.

Evidently Lilly was but little concerned with sentimental thoughts. "Well," I answered carelessly, as if in reality extremely indifferent about the matter, "suppose we do; it will do us no harm by this time."

Going forward to Mrs. Campbell's house, instead of allowing Lilly to knock there, and going on myself to Mr. Rolleston's, I again hesitated, designing to see who would open the door in response to my companion's summons. To my great pleasure Miss Mary Campbell appeared. On seeing me she joyfully exclaimed: "Oh, mother, here is Mr. Farquharson!" "And how is Miss Mary?" I asked, shaking her warmly by the hand. Mrs. Campbell here appeared. "Dear me, Mr. Farquharson, and how have you been getting on all this long time? I am so glad to see you again; but Mr. Rolleston is inside; you had better step into the house at once. I am sure you must be sorely tired after your long journey; we shall all be in presently. But who is this with you?" she continued; "what! Mr. Lilly! dear me, is it you? you are quite a stranger; it has been such a long time since I last saw you."

"Well," answered Lilly, "I reckon, Mrs. Campbell, it has been nigh on two years since I was down here last."

"Mr. Farquharson, how are you? I am indeed glad to see you," was the instant remark (whose genuineness was confirmed by the frank look that accompanied it) of the sprightly Miss Rolleston, on the door opening, as I was ushered into the sitting-room by the pretty kitchen-maid.

Miss Rolleston was dressed attractively in dark silk, relieved with a white lace collar, that was secured by a jewelled brooch at her pure white throat; her dark curling hair and lively dark eyes, and glowing cheeks, looked glorious in my eyes. "O that I could claim that soft beauty as my own," was my next daring thought, as my dazzled eyes fell upon her. As it was, I felt my blood fairly thrill at my heart, when her soft white little hand, in salutation, was placed within mine.

Miss Rolleston was then, as I afterwards learned, in her twentieth year.

Mr. Rolleston, hearing my name from a neighbouring room, now entered, and shaking hands with me warmly, professed his pleasure at seeing me, enquiring at the same time if Lilly had come down with me.

"I am glad of that," he said, "as I should like to have his opinion on those cattle. What percentage of lambs did you realise?" he continued.

"Ninety-seven," I replied.

"Oh, that is very satisfactory," he said, smiling and rubbing his hands expressively.

"Yes," I replied carelessly; "but, unless I am much mistaken, next year's tally, if all goes well, will pass the hundred."

"Oh, come, Mr. Farquharson, that is a rather sanguine calculation on your part. Pray, what do you intend to do to secure such an increase as that?"

"Well, I will not trouble you at present with details. You will be better able to judge of the merits of my system when you see the results," I answered.

"I see you manage to work in harmony with my old friend Lilly; I am pleased at that, as I should be sorry that anything should occur that might deprive me of that man's services. He has been in my service for the last five years, and in all that time he has proved himself a most conscientious and trustworthy person, with reference to all matters that have been placed in his charge."

"He is indeed a most valuable man," I made reply, "and I am not at all surprised, from my observation of him, about your solicitude on his behalf; but happily I chanced to fall into tune with him at the very start, and he and I have been on the most excellent terms ever since."

During the course of this business dialogue, Miss Rolleston

sat quietly engaged with some ornamental crochet work. Mrs. Campbell and Mary now entered, and the tea arrangements being then completed, we all drew our chairs, at Mrs. Campbell's request, to the table. At the same moment Mr. Campbell, who had just returned from one of his daily excursions over the run, joined us, and between him and Mr. Rolleston and myself, during the time occupied in the discussion of the meal, the conversation was chiefly taken up with business matters, save whenever and anon my watchful attention to the wants of the ladies was rewarded by a bright smile and graceful acknowledgment from Miss Rolleston, or sweet-voiced thanks from Mary.

It was only, however, after emerging from the office, to which I had retired with Mr. Rolleston and Mr. Campbell, for the more particular discussing of the business for which Lilly and I had been summoned down to the Murray, that I was enabled to indulge myself to the full in the more congenial conversation of the ladies.

Mr. Rolleston, his mind set at rest by our late satisfactory business chat, had now, as a sort of necessary relaxation, relinquished himself to a calm and uninterrupted digest of the contents of the *Argus* that Mr. Campbell had procured for him when riding on his way home past the neighbouring Post Office of Euston. Mr. Rolleston's usual interest in his paper was on that evening still more heightened by the publication of a speech of Mr. Gavin Duffy's, on the land question; for his alarming views of reform on this subject were at that time greatly exercising the minds of all squatters. Shortly after this, Mr. Duffy found reason to alter his views on this question, however, and, consequently, became a warm supporter of the very party who were at this time regarding his democratic measures with considerable apprehension, for they seemed to be aimed at the very roots of their most cherished privileges.

I have before referred to my habitually awkward and diffident manner when in the society of ladies. It must therefore have surely been with the intention of at once dissipating this silly feeling, that, in reply to my rather conventional remark on seating myself beside them, "Well, ladies, might I ask how you have been enjoying yourselves since I last had the pleasure of meeting you?" Miss Rolleston answered in a tone of pleasant raillery, whose utter brusqueness, I had almost said coarseness, at once scattered all ideas of conventional formality to the winds.

"Oh, brawly, brawly, thank you for speering, hoo hae ye been getting on yoursel'?" cleverly mimicking the Scotch accent, that indeed she might well have been at home in seeing it was her mother's. For Mrs. Rolleston, although a lady of education I was given to understand, was fond of expressing herself at times in her native Doric tongue. I was certainly greatly taken aback at this rather free and easy speech from a refined young lady; but it had the effect that was doubtlessly intended by the clever girl, of at once setting me at my ease with the company in general, and especially with her; a result all the more difficult to accomplish from the strong passion with which I viewed her, and that communicated even more nervousness to my manner than was natural to it.

"You seemed," she continued, "so much engrossed about your lambs and sheep that I scarce hoped that we poor ladies would be honoured by your notice at all."

"Honoured by my notice, Miss Rolleston! nay, surely you are now speaking but sarcastically! 'taking a rise out of me,' in fact, as the colonial phrase is; take care," I said with a smile; "if you will flatter my conceit in this way, by attributing such consequence to my notice, you may yet have reasons for feeling yourself bored by too much of it."

"Never mind," she replied, laughing, "Mary is here to help me, and between us both we shall be able to bear the threatened penalty cheerfully. What do you say, Mary?"

"Oh," said Mary with her quiet dimpling smile, from which at times a quiet spirit of roguish humour was not very far removed, "I shall be only too delighted to get as much of Mr. Farquharson's company as he will condescend to bestow upon us. Do you think, mother, that Mr. Farquharson will bore us much?"

"Nay," replied Mrs. Campbell in her usual quietly-pleasant tones (honest, quiet, sleepy Mr. Campbell had long ago retired to rest), "I should be sorry to think so of Mr. Farquharson; but I do think there is a serious danger of his being bored by two such rattlepates as you and Rachel."

"Now, Mrs. Campbell," cried Miss Rolleston, "it is too bad of you to be always lecturing us, and calling us rattlepates. Now I am sure you will not meet anywhere two more orderly and demure, staid and precise girls than we are. Mr. Farquharson, I hope you will take our part and not believe this charge against us."

"Well, if quick wit and lively manners are necessary adjuncts to such characters, I should believe so far in Mrs. Campbell's charge, that a dull person must soon feel out of harmony with your society."

"Thank you for your excellent discrimination, Mr. Farquharson," cried Miss Rolleston. "Neither Mary nor I have any patience with dull people, for between mere dulness and gravity of manner there is a great difference."

I smiled in acknowledgment of the compliment to my usual demeanour, that I felt was implied in these words. I presumed that the ladies had been in town since I last saw them. "Oh, for months," said Miss Rolleston gaily; "and we enjoyed ourselves immensely during the Christmas festivities. Don't you think that Mary has much improved since you saw her last?"

I had already remarked that she had, especially as regarded the more womanly development of her form; but the quiet repose of her manner that conferred upon her a grace peculiarly her own, was still the same. At her friend's remark she only smiled but said nothing.

"You must feel lonely upon the Darling Station, so far from congenial society; for I believe it is a very lonely and dreary looking country up there," Mrs. Campbell remarked.

"Well, yes, Mrs. Campbell, in a measure it is so, if a person went by mere appearances only. Yet it is wonderful how soon people learn to accustom themselves to all sorts of unlovely surroundings. Yet apart from the necessary interest that attention to one's duty requires, and that of itself soon tends to familiarise and accustom one's minds to these, there, as elsewhere, the absence of one charm is compensated for by the presence of some other. If as regards scenery for instance, the Darling country is flat and uninteresting, as is indeed the case with most of the Australian interior—yet to me this seems amply compensated for by the unconfined freedom of space on every side. There is something akin to a feeling of proprietorship in the extensive tracts of bush and plain, and unlimited extent of back country, over which one can ride at will, no one venturing to check or hinder anywhere."

As I have already remarked, there was a quiet dignity and grace of manner about Mrs. Campbell that secured my respect. As I scanned her quiet orderly features, that in their calm repose appeared to me at times to show the impress of former sorrow, I often speculated as to what her history might have been previous to her marriage. Doubtless, I mentally remarked, some other person than Mr. Campbell had once coveted the possession of these superior charms; and how came it, that a plain, practical, commonplace man like Mr. Campbell had won her? About some of my surmises at least I found out shortly afterwards that I was correct. Some twenty years ago she had lost sight of a twin brother to whom she had been tenderly attached, and over whose mysterious disappearance she had bitterly mourned.

It seems that he had become involved in some pecuniary difficulties with a bank; into these difficulties he had been betrayed by an imprudent and unscrupulous partner, for whose misdeeds the bank authorities sought to make him responsible. He escaped from engagements he found himself unable to meet, and the threatened penalty of imprisonment, and disappeared no one knew whither, nor, in the long interval that had since elapsed, had any trace of him been found. Therefore, under the full conviction that some evil had befallen him, his sister had long ceased to entertain any hopes of ever seeing him again.

As I now wish to introduce a new character to my readers, I will close this chapter here.

CHAPTER X.

THE morrow dawned gloriously. It was the springtime of the year and everything looked fresh and bright. The green grass clothed the sandhills—usually so bare in summer—with tempting food for the sheep, who showed their appreciation of it by the keenness with which they ran towards it—where they could, for this dainty pasture was in some places especially reserved for the breeding flocks. The innumerable cockatoos, and flocks of variegated parroquets, as if in joyful harmony with the beauty of the morning, expressed their feeling of satisfaction in their own way, by filling the air with discordant cries.

After breakfast we all - that is, Mr. Rolleston, Lilly, and myself—started off to view the cattle, for the inspection and purchase of which we had been summoned down from the Darling, and as the weather looked so tempting, the two young ladies could not forego the pleasure of accompanying us on part of our way that lay along the banks of the Murray far on in the direction they were going, in order to pay a visit at a neighbouring station a few miles off. We were all in high spirits; who would not be, in such company, and with such surroundings? but ere the end of our journey an untimely accident was like to have changed our gay laughter into bitter mourning. It happened in this wise. The station to which the ladies were bound was on the bank of the river opposite to that on which Mr. Rolleston's station was situated. On arriving opposite the station, it was found that the punt in which travellers were usually ferried across was then high and dry, on the station side, undergoing some repairs, and the only conveyance that did duty in its place was represented by a very shaky-looking bark canoe that was paddled to and fro by a black fellow. To any ordinary ladies this would have been deemed an insurmountable barrier to the further prosecution of their journey, or rather, I should say, to the completion of it, seeing that there was but the river between them and the place for which they were bound. But Miss Rolleston, fearless by nature, and made more so by her hardy bush training, simply laughed at the idea of letting her visit be deferred on this account, and even in Mary Campbell's eyes, though constitutionally more timid than her companion, such a frail-looking means of transit did not appear very appalling.

The question at first was how they should manage about their horses; but it being suggested that they could easily be made to swim across, the next thing was to gauge the carrying capacity of the canoe. Two persons was the usual freight borne by these frail barks, but this appeared to be a large one.

By this time I had ridden my horse a little down the bank, below where our party was standing, to see where there was a convenient place to let Selim get at the water for a drink. Looking up whilst the horse was thus engaged, I called out to the ladies, warning them to go over only one at a time.

"Nonsense," cried Miss Rolleston rather petulantly, "it will be all the greater fun to go together; we can tuck our skirts closely round us, and there will be plenty of room. Come along, Mary." Mary looked a little doubtful, and even the sable Charon appeared to look grave at the prospect of such a heavy freight for his frail canoe. "Bail budgery,"* I heard him remonstrate; but remonstrance was vain—the strong-willed Miss Rolleston would have her own way, and her weaker companion succumbed to her influence. But the black fellow's words sounded so forebodingly in my ears—for I knew that they but expressed this simple child of Nature's knowledge of the canoe's capacity—that I called out still more earnestly, "Mr. Rolleston, do not let them both go together; depend on it, when the black objects, it is unsafe to do so".

"Never mind them, Mr. Farquharson," he replied, "Rachel *will* have her own way. I make no doubt but they will be safe enough." I said no more, and the canoe was pushed off, the black pilot using his paddle skilfully. The canoe, though evidently over-weighted, as its depth in the water showed, was still being rapidly paddled across to the other side, and might have reached it in safety, but for a sudden movement on the part of Miss Rolleston. Seeing a large cod-fish suddenly dart under the canoe, and uttering a loud exclamation, the thoughtless girl suddenly leant over the side to catch a further glimpse of it, and thereby communicated such lateral impetus to the canoe (already over-laden, and which, naturally, could only preserve its equilibrium under conditions of the most perfect steadiness on the part of its occupants), that, lurching into

* No good.

the water, it was instantly swamped, and all its passengers, with a wild shriek from the ladies, were left struggling near the middle of the current.

What followed I can hardly detail.

With a loud oath Lilly flung himself from his horse, and kicking off his boots, dashed into the river and swam with desperate strokes to where the black fellow was bravely endeavouring to support the fainting form of Rachel Rolleston, a matter that her riding skirt dragging in the current rendered one of considerable difficulty. But Mary Campbell, whom the black fellow had failed to grasp, or perhaps thinking he had sufficient to do in trying to save Miss Rolleston, had not tried to grasp, was left in imminent peril of going to the bottom of the Murray.

I have said already that I had gone down the bank a little way to water Selim. Be it understood that the wash of the current tended directly from the centre of the stream towards the bank where I was then watering the horse. As it happened I had remained seated in the saddle while Selim was drinking; also, through the natural leeway that the canoe made while being paddled across the stream, it was almost opposite to where I then was.

From my boyhood I have been a practised swimmer, but I did not trust to my own powers in attempting to rescue either of the endangered ladies, having much more efficient means for that purpose then under me. With Selim, at any time when it was necessary to swim a river with him, I merely needed to ride him into the deep waters, when, without the slightest hesitation, on feeling himself getting beyond his depth, he would strike out, and being a noble swimmer as a rule ensured me a dry seat in the saddle. Therefore on the present occasion, instantaneously with the canoe catastrophe, I dashed my spurs into his flanks, and with a loud shout urged him into the deep water, which he entered with a bound that caused him to sink almost over his head, the waters rising over my own shoulders. Instantly on his rising to his proper position, I urged him on to where Mary Campbell was struggling wildly in the water. She had providentially caught hold of the black fellow's paddle, at about the middle, which she held convulsively with both hands. She thus in her struggles had been unconsciously making it aid in keeping her on the surface longer than she could otherwise have managed. With Selim's powerful strokes bearing me swiftly through the water, I was soon at the poor girl's side, and at once seized her by the arm and drew her cautiously (for fear of overbalancing Selim) up half way out of the water, requesting her at the same time to let go her hold of the paddle, a request she had sense enough to comprehend, and at once complied with. Instead of returning to the side of the river I had just quitted, I now headed Selim to the opposite bank, from which a canoe was then being rapidly paddled to where Miss Rolleston, by the united efforts of Lilly and the black fellow, was also being propelled. The black fellow, intent only on bringing her to land anyhow, was endeavouring to bear his burthen to the bank she had just left, till Lilly reached him, when by his directions they faced about and made for the opposite side. This bank, though the furthest to reach, Lilly's quick sense prompted him to make for, with his almost fainting charge, because of the house and friends who were there—and the same thought had come to me—for there the drenched girls would be at once relieved of their wet clothing, and by careful nursing, protected against any possible ill effects from their ducking and fright.

While attending to Mary I had been collected enough to observe from the first that, with the black fellow's timely assistance, and Lilly's immediate co-operation, there was no occasion for anxiety on account of Miss Rolleston's safety, and now, whilst swimming shorewards with Mary, I observed that the canoe despatched by the owner of the station, who had been a horrified spectator of the catastrophe, had reached Miss Rolleston, and into it she had been lifted, by the combined efforts of the black in the canoe and Lilly and his black assistant; the two latter swimming alongside of the canoe as it returned to the bank.

Now, although I had a most genuine regard for Mary Campbell, and was sincerely grateful for having been enabled to render her the service that I did, yet I was not a little chagrined now that I had been privileged to enact so conspicuous a part in the rescue of one of these ladies, that fate had not so ordered it that the one who was the very nearest to my heart had not been also the one that had fallen the nearest to me in the water. What a noble opportunity this would have afforded, I rather disconsolately reflected, of advancing myself in her eyes, and by such a signal service giving myself some title to her affectionate gratitude! However, for this disappointment I was partially consoled by the reflection that in view of the temptation I had had for being guided by such merely selfish considerations, which might have diverted me from the path of obvious duty, my action in what I had done, had been all the more purely disinterested.

Both ladies, more frightened than hurt after their severe ducking, were at once conveyed to their friend's house, where, after about an hour, through the kind and tender nursing that they received, they, with heightened colour, were merrily chatting over the incidents of their late danger, as if they rather enjoyed the recollection of it than otherwise. After a change of clothes Lilly and I returned across the river in the canoe—Selim swimming behind—along with Mr. Rolleston, who had come across meanwhile by the same conveyance, and who, as may well be imagined, had been not a little shaken by the sight of what might have proved a tragedy enacted before his eyes. Then, remounting our horses, we resumed our business journey, for a while commenting animatedly on the scene that had just occurred; then, as the more prosaic thoughts connected with our present business reasserted themselves over our minds, gradually seeming to forget all about it.

Of the business we had in hand suffice it to say that both Lilly and I highly approved of a mixed lot of about fifty cattle, amongst which was a most lordly looking bull of superior breed, and the transaction of the sale being soon completed, we were on our way home by the afternoon.

Calling in on our way at a roadside public-house, then the only residence in what I now believe to be the considerable township of Euston, Mr. Rolleston ordered some spirits for the party, and took up the latest paper, that, as it happened, had just been left there by the mounted postman. After scanning its contents for a little while, his eye fell on the following paragraph, under the conspicuous heading of "Daring Bank Robbery," which he read aloud. "We have to record the most daring bank robbery that has ever been perpetrated in this town (Adelaide) since its foundation. Six scoundrels, all dressed in blue shirts, one of them a powerful looking ruffian with bushy black beard acting as leader, deliberately perpetrated this audacious robbery in the midst of a populous town, in broad daylight.

"Their plan of action appeared to have been as simple as its execution was prompt. Dividing their number to avoid exciting suspicion, by so many being seen together, three of these men were seen riding leisurely down the street till they approached the Bank of South Australia, when they dismounted and quietly hung their horses' bridles over the posts in front of the bank, into which the black-bearded ruffian at once entered, whilst his two companions, as if awaiting his return, stood quietly by their horses. Almost immediately after this, the other three, on the opposite side of the street and from an apparently opposite direction, rode quietly down, and also dismounted from their horses a few yards above the Bank of Australasia, and stood apparently waiting for some one, gazing meanwhile carelessly about them. This much we know of them from the testimony of a bystander, who, without having had his suspicions roused by their manner, yet chanced to have noticed them from a neighbouring corner at which he was loitering. On entering the bank, as we have said, the inmates—the usual officials and four men who were transacting business at the counter—were suddenly startled by the stern order of the bushranger, accompanied by the threat of his levelled revolver, to hold up their hands. On this order having being mechanically obeyed—for in the determined and savage lineaments of the robber leader, the alternative of death on the first symptoms of refusal or resistance, was clearly written—his two companions entered, and, first closing the outer door, they then with cords, evidently provided for the purpose, at once proceeded to bind and gag every one in the bank, first compelling the cashier, with a revolver held at his head, to disclose and unlock the safes. All the gold and silver—notes and coppers they discarded—that amounted to no less than £4000, they at once stowed away in bags they had brought with them concealed under their blue shirts. Whilst this booty was being secured by his companions, the leader went outside and across to the Bank of Australasia, when the same scene was instantly re-enacted there; and although there were seven business men here in addition to the staff, yet such was the consternation with which all were seized, that all thoughts of active resistance were quelled. Here the robbers secured £3700 worth of booty, and then, mounting their horses, they all made off in different directions. They must have been gone for fully two hours ere public attention began to be fully aroused at the unwonted appearance of closed doors at both banks during business hours. This led to an inquiry that ended in the doors being forced and the inmates of each bank found as already described. An alarm was at once raised. The rumour of this robbery spread like wildfire through the town, and the consternation at its audacity was great. Evidently, though apparently carried out in so simple a manner, the plans of the robbers had been ably matured. From their knowledge of the mysterious robbery that occurred at the south end of the town a few days ago, which the police have since been busily investigating, the scoundrels no doubt deliberately planned the present outrage, feeling secure of the absence of the police from the centre of the town for this cause—nay, who knows but that this robbery too may have been their own handiwork, deliberately executed as a diversion to draw the police thither in order to give them a better chance of effecting their more daring project with safety. As the police are, however, now in active pursuit, and Billy, the celebrated black tracker, is amongst them, it is to be hoped that such a flagrant violation of the law, committed in the very face of the guardians of its majesty, will soon be visited by condign punishment. There is every reason to believe that the bushrangers are trying to escape to the Northern Territory."

"Well, I must say that was a pretty cool proceeding," remarked Mr. Rolleston, on laying down the paper, after reading the above account.

"My faith," replied Lilly, "these fellows were up to their trade; it is easy seeing by the smart and determined way in which they did their work, that it has been an old game with them."

"The rascals! it is to be hoped the police will soon get hold of them," remarked Mr. Rolleston, who, being a man of property, had no sympathy with such instances of smartness as were hazarded at the expense of his class interests.

Our horses having been now sufficiently baited, we desired that they should be brought round to the verandah front for us, and, mounting them, we quietly rode on, reaching home at about dusk.

CHAPTER XI.

"WELL, Mrs. Campbell," remarked Mr. Rolleston, as that lady was in the act of pouring out the tea, "there was really a serious danger at one time to-day of you and I being mourners."

"Dear me, Mr. Rolleston, in what way? surely nothing has happened to the girls?" asked Mrs. Campbell, suddenly suspending her operations and looking up with a startled expression at Mr. Rolleston's words.

"Well, as it has turned out they are safe enough now, but I assure you that you have every reason to thank Mr. Farquharson there, that Mary is not now at the bottom of the Murray, where I at one time verily believed she was like to have gone, in company with that mad hussy of mine," Mr. Rolleston deliberately replied; as he slowly sharpened the carving knife, preparatory to making an incision into a juicy leg of roast mutton.

"Merciful Providence," replied the mother (almost letting the teapot drop out of her hand in her nervous agitation), "is it possible that the girls have been upset in the river?"

"Possible enough," said Mr. Rolleston. "And as I have said, but for Mr. Farquharson and his good horse, who saved Mary, and Mr. Jamieson's black fellow and Lilly who rescued Rachel, both you and I, Mrs. Campbell, should have been broken-hearted mourners at this moment." He then related the particulars of the accident.

"Truly," said Mrs. Campbell piously, "it is to our Heavenly Father, the Giver of all mercies, that our first most fervent thanks are due, and after Him, to the instruments that he was pleased to make use of; and as such, to you now, Mr. Farquharson, a mother's thanks will be always due. And may God requite your noble devotion in what you have this day done."

"Nay, Mrs. Campbell, what less than I did would any man in my place have done? while the pleasure I have experienced in being so instrumental in rendering such a service has been sufficient reward, I assure you, to me, for all I did. It would have been a pleasure to have rendered it to any one else, in a similar situation, but the pleasure is greater in that it was done to one of these ladies."

After a while Mrs. Campbell's maternal fears subsided, and she even managed to laugh at the description I gave of the ladies' improved complexions after their ducking.

"Well," remarked her husband quietly (on the affair being related to him shortly afterwards), "it may make them more heedful the next time they go into a canoe."

Next day we were mustering, and exceedingly busy till late at night, so that on entering the house I found to my chagrin that both young ladies, who had returned that day by the other side of the river (crossing over on Mr. Rolleston's boat), had retired early to bed. As I had all day been looking forward impatiently for this evening's advent, when I thought the pleasure of their society would be enhanced by the interest occasioned by the late adventure, I felt naturally disappointed; but as the next day's programme was to be a repetition of that of the day before, I was up betimes, to see how my horse was faring; this was not Selim, who was at present spelling, having had quite enough of exercise lately. Ever mindful of George Laycock's counsel, I was always careful not to keep him overtasked at any time when I could possibly avoid doing so. Here, impelled by the same motives, I was soon joined by Lilly. Lilly, I should here state, seemed quite at home in Mr. Campbell's house, and, with the blooming damsel who acted as housemaid, he was to all appearances on a footing of the most amicable friendship. I had once heard Lilly declare of that maiden, that "she was the tidiest and most civil girl" that he had ever seen—no small praise from Lilly. While standing in front of the stable, to my surprise the two ladies suddenly appeared at the door of the house, and on seeing us immediately came forward to where we were.

They were both tastefully arrayed in light-coloured morning

dresses, and Miss Rolleston, with her soft, jetty ringlets, and Mary Campbell, with locks whose golden tinge was heightened by the play of the morning sun, presented each a picture of attraction that I would fain have let my eyes linger upon, and I felt that if Miss Rolleston in beauty bore the palm from her companion, yet there was an ample margin left for independent attractions in the latter, in the sweetness and innocent grace of features that would scarcely admit of her passing unnoticed even by the side of her friend.

"Good morning, ladies; you have surely some faith in the freshening effects of the sun's earliest rays, that you are stirring so soon," was my pleased salutation as I shook them each by the hand, to which they both laughingly responded.

"And how do you find the Murray waters agree with your health by this time, young ladies?" asked Lilly sarcastically.

"Oh, charmingly, Lilly," replied Miss Rolleston. "Now really don't you think that both of our complexions have been greatly improved by our late ducking? but," she added maliciously, "I see, Mr. Lilly, it is only Tiny's face for which you have any eyes."

"Well, I dunno as it has done either of you any harm," replied Lilly, affecting not to notice the latter part of Miss Rolleston's remarks. Then looking more kindly at the fair speaker, he continued, "It would have been a cruel shame for such a purty face as that to be lying at the bottom of the Murray".

"Oh, thank you, Lilly, for the compliment; but then, I suppose this is what you would call in Ireland 'a chip of the blarney,'" replied Miss Rolleston laughingly; then added gravely, "but it was just the wish to express the thanks that we had not the opportunity of speaking before, that brought Mary and me out so early this morning, for our escape was wholly due to you and Mr. Farquharson and the poor black fellow. Mary must answer for herself to Mr. Farquharson, to whom she was immediately indebted, but as a mark—only a slight one, remember—of what I feel for the service you then rendered me, Lilly, I want you to accept this gift, and wear it for my sake;" and before the astonished Lilly knew what she was about she had quickly lifted off his hat and passed a gold chain, to which was attached a valuable watch of the same metal, over his head. This watch and chain, as I afterwards was informed, she had obtained from her father for this purpose. "Nay, Lilly," she said firmly, on his loudly protesting against the occasion for such a costly gift, "I desire you to accept this much from me."

"By the hokey, if you raily mane it, I will take it, and it is proud I will be to wear such a gift for the sake of a lady like you; but it's not for the value of it, but because you gave it to me that I'll prize it, as I would have prized it for the same raison if had it been but a silver one, and which same would have been more suitable for the likes of me anyhow—not that I wanted anything at all, but just because you want me to take it; and as for what I did for you, Miss, shure I would do the same for a black gin, let alone for Rachel Rolleston, the sweetest and purtiest lady in all the Murray district." In his unwonted emotion Lilly's feelings found vent in more of the latent Irish brogue than he was accustomed to use. For not only had Lilly been brought up in the neighbourhood of Dublin, where the dialect is purer than in other parts of Ireland, but, having been so long away from Ireland, the little brogue that had originally marked his speech had now almost disappeared.

In the meantime Mary and I were quietly conversing on the same interesting topic. "Mr. Farquharson," she said, "as I am not wealthy like Rachel, and consequently have no gold watch and chain to offer you, I am afraid you will think my bare thanks but a poor return for your bravery in risking your life for my sake, and I know that without what you did I should not be here now, but lying at the bottom of the river; but though I can now only thank you," she added feelingly, "I tell you that your kindness to me deserves more than I shall ever be able to repay."

"I must have something, however," I said, assuming a villainously mercenary look. "Mere thanks will hardly pay one for such an action as that. Consider the clothes that I spoilt in the wetting they got, and all that sort of thing. I am afraid, too, I shall have to get a new saddle."

The poor girl looked so piteously embarrassed under the infliction of this banter that she apparently seemed not to know how to construe it.

"I—I did not like to offer you anything, Mr. Farquharson, for fear of offending you."

"Offending me indeed, young lady, let me inform you that my delicacy is not so easily offended," said I brutally (for I was a brute to speak so, though but in jest, to the poor girl, whose face at this stage began to wear a shocked feeling; Miss Rolleston and Lilly at the same time looking at me with no small appearance of astonishment).

"What! do you think," said I, suddenly seizing her hand and holding it firmly, "that it would offend me to promise that you would always be one of my dearest friends? No, no; so just give me this promise forthwith."

"Your friend?" said the dear girl, her face, lately so distressed, now joyously brightening up; "is that all? I fervently hope that I shall always be a friend of yours, as know you will be of mine."

"What a shame, Mr. Farquharson, to torment poor Mary with that mercenary assumption of yours," cried Miss Rolleston. "I declare you looked as solemn as if you really meant what you said. It is you who are now in Mary's debt, to compensate for the momentary overthrow in her mind of the ideal of your character occasioned by your late words and looks."

"Then I understand that my character did have the honour of an ideal conception in your and Miss Campbell's mind," I quickly remarked, highly flattered at the thought. "Surely the knowledge of such a happy prepossession in your minds is in itself an ample return for the ducking I received on Miss Campbell's account."

"If you so much value such opinions you had better not again endanger their utter extinction, even in jest."

"I will certainly be most watchful over myself in view of such a penalty as that," I replied laughing, then continued:—

"But, ladies, and you in special, Miss Rolleston, while overwhelming both Lilly and myself with thanks for the services we were so happy in being able to render you, what about the black fellow, without whose aid, in the first place, I fear that Lilly, stout swimmer that he is, would have had the job of diving for you ere he could have reached you? As for Miss Campbell, it was most fortunate that she managed to seize the paddle, or I fear I should have had to leave Selim's back on the same errand after her."

"Oh, Bill," said Miss Rolleston, laughing merrily, "Oh no, we have not forgotten him either. When I found time to inquire after him he was discovered out in the Murray, where the canoe was swamped, diving for his tomahawk that had, of course, been lost when the canoe turned over in the water. Over this loss he was sadly lamenting, with vengeful abjurgation of 'no good lubra,' by the lubra meaning Mary, and myself, to whose perverseness he rightly ascribed the loss of his precious tool. So he was brought up to the house and arrayed at once in a suit of clothes with blankets out of Mr. Jamieson's store, and was promised the gift of a grand new tomahawk from town into the bargain, and provided, besides, with such a quantity of provisions that the whole camp have been holding a grand corroboree every night since. I am going to try and get him with his family to come and stay here from this time, and papa will see that the honest fellow shall not lack for anything."

The ladies and I shortly after this walked towards the house, and just then, at a summons from her mother, Mary left us. Left alone with Miss Rolleston, I remarked in an assumed manner of carelessness: "What would you think were I to confess that my genuine pleasure at being able to render the late service to your young friend is slightly dashed with the regret that Providence did not order that it should be her who fell to Billy and Lilly's share of the rescuing, and you to mine; and that the sense of pleasure I felt when enabled to lay hold of Miss Campbell, was little in comparison to what that pleasure would have been had it been Miss Rolleston's precious life that then depended on my care?"

"If I were to imagine anything at all, it would be the entire freedom from selfish considerations your action showed, for had you been guided by any suggestions of ambitious policy, doubtless, as a wealthy squatter's daughter, I was the greatest catch of the two, and you might thus have been moved to bend your efforts in reaching me, who was well cared for already, and so poor Mary would have been left to perish, but as it was, your conduct was as noble as it was disinterested."

This was spoken in a sudden gush of earnestness that seemed rare with Miss Rolleston, in whose playful humour this mood was seldom seen.

"You reproved me just now for my assumption of mercenariness with Miss Campbell, yet really I could be tempted to mercenariness in my desire for a reward from you, Miss Rolleston."

"Well then, I must find you some suitable gift to offer in acknowledgment of your gallantry likewise," she replied with a bright smile.

"Yes, but the reward I might covet might be of far greater value than you dream of."

"Indeed, and might I ask the price at which you seem inclined to appraise your services?"

"Did a greater equality of station sanction the probability of my demand being granted, it would not be represented by piles of gold or of rubies."

"What else?" she answered, looking intently at me as if to fathom my meaning.

"Your own fair hand perhaps," I said with assumed pleasantry, but the involuntary tremble in my voice too surely indicated the earnestness of soul within.

The archer had caught the woman off her guard, and the shaft struck home—that is, it struck home to her vanity ere she had time to hide its effect. She evidently had been wholly unprepared for such an avowal, else she might have gaily turned the dart aside. As it was, she coloured for a moment, then rallying from her confusion, gaily smiled and tossed her head.

Evidently however she was not displeased, and I breathed more freely, for she was now conscious of my soul's longings. As to her own feelings, I was soon to learn that they were of too mercurial a nature to be long retained by any powers of attraction I at least possessed.

"Tush," she playfully rejoined, "I thought you were too serious for such thoughts as these."

"Yes, Miss Rolleston, so serious am I, that but for the unfavourable circumstances of my present position, I should steadily urge you for a response to these thoughts. Ah! if I could but hope that improvement in my circumstances might yet admit of the prospect of my some day daring to openly press my suit upon you, what rapture would be mine!"

"Ah! come, Mr. Farquharson; you don't mean to hint that you would wish to enclose in a household cage a poor little wild bird like me, whose sole desire is its own sweet liberty for years to come? But see, here is Mary coming, and if she hears us talking such nonsense as this whatever will she think of us?" and with these words she tripped lightly away, leaving me almost confounded in a sudden sobering of thought, at my utter imprudence in so suddenly committing myself by such an avowal to her.

A few days passed, during which Miss Rolleston seemed totally unaffected in any way, by her sudden insight into my feelings towards her. We laughed and chatted as before, or if, when opportunities served, I ventured to advance my irrepressible suit, she, now on her guard, simply parried it with the address of a woman well accustomed to similar scenes—neither wholly repelling me nor yet yielding to my attacks. Evidently pleased with my devotion, she smilingly listened to all my ardent protestations of love, but when I essayed to get on to a confidential footing with her, she eluded my designs.

Alas! it was too patent that Miss Rolleston, an heiress and spoiled beauty, was also an accomplished coquette.

Matters were in this condition, when one evening a stranger rode up to the house. He was tall and seemingly very muscular, though but sparely built, with a yellow bushy beard and good features, though showing, in the lines that here and there furrowed his face, the wear and tear of time or of passion. He might have been some thirty-five years of age or so, but it was difficult in his case to come to a very definite conclusion on this head. His eyes were quick and peculiarly restless. One arm he carried suspended in a sling, as if he had just suffered by a fall from his horse. This surmise seemed substantiated by the traces of soil on the shoulder of the disabled arm. His horse, though apparently of superior breed, appeared to be much jaded and travel-worn; everything seemed to show that the stranger's hurt had been occasioned by the stumbling of his beast.

This at least was the conclusion that I, who happened to be standing at the door when he rode up to the gate, came to. Dismounting, he came forward and introduced himself by the name of Marsden, and described himself as being in search of back country for a run. "I have penetrated the scrub," he said, "betwixt here and Lake Hindmarch, that I left yesterday, but, happening to turn out of the proper track, I spent last night in the scrub, where I fortunately found some water, but of course no provisions for either man or beast. To-day my horse being jaded stumbled and fell with me and caused me to hurt my arm considerably, though I believe not seriously; there appear to be no bones broken."

After calling Dick, the groom, to take the wearied horse to the stable, I introduced Mr. Marsden, on the terms of his own account of himself, to Mr. Rolleston, who courteously welcomed him and expressed sympathy for his hurt, and then, after a few minutes' conversation, led him into a bedroom that he might there refresh himself with a wash, begging him to consider the room as his own till he felt himself sufficiently rested to continue his journey. To this room his valise, that seemed of considerable weight, was then brought. After refreshing himself with a rest, wash, and change of clothes, he reappeared in the parlour by supper time, though still nursing his arm, from which, judging by the tender way in which he seemed to try to relieve it with the other hand, he was suffering some pain.

CHAPTER XII.

THE stranger's manners were evidently those of a person accustomed to good society. This was seen by the easy, yet dignified manner in which he acknowledged his introduction to the ladies, his courteous demeanour to the latter, and his well-sustained part in the general conversation at the supper-table. In spite of this there was still something about him that seemed vague and unsatisfactory. Every now and then, and that often after one of his happiest remarks, an unquiet shadow, a look of suspicion, flitted across his face, so marked that on these occasions I could almost imagine him to be acting a part, as of a person who was other than what he represented himself to be.

Quietly watching him (for somehow a strange curiosity

strongly compelled my interest in the new comer), I noticed him, on the sudden slamming of a door, almost bound from his seat, and glance nervously around, whilst a sudden, fierce expression would contract his brows, as instantly all agitation would subside, and with quiet equanimity he would resume the conversation that such a trivial incident had almost interrupted.

Inquiries about his injured arm seemed distasteful to him, and Mrs. Campbell's kind offer of inspecting and bathing it he hastily and resolutely declined, remarking that it was a mere nothing. Then recovering his composure, in a more courteous manner he added, "Excuse me, madam; I have seen the world, and some rough service in it, and it has tutored me to make light of such a mere scratch as this. In India I have hunted both the tiger and wild boar, and on such occasions, when we did not always escape scatheless, to notice a trifle like this would have only provoked the laughter of my companions and my own."

Yet with all this show of courage, I noticed, that just as he spoke he involuntarily raised his hand to his disabled arm as if he had been visited by a sudden twinge of pain.

"You have been in India, then, Mr. Marsden?" remarked Mrs. Campbell, with some interest.

"Yes, madam, I was in the army there for over two years."

"Indeed, Mr. Marsden?" asked Mr. Rolleston. "May I ask how you liked soldiering? I should imagine the life to be as uncomfortable as it is hazardous at times. What regiment were you attached to?"

"Uncomfortable enough," replied Mr. Marsden, with the unquiet expression I have referred to suddenly clouding his face at the question with which Mr. Rolleston concluded his sentence, and which he parried by the hasty manner in which he seemed to attend to the first part of it, "but as for its hazard, why, young fellows such as I was then seemed to consider that one of the charms that compensated for the lack of comfort; but," he continued rapidly, as if desirous of diverting the conversation from a channel that was distasteful to himself, "for myself, on the whole I found the restraints of regimental duty and barrack life so irksome that I sold out, and for the last few years (here he laughed) I have been endeavouring, by an idle, wandering kind of life, to dispense with the remnants of a rather plentiful crop of wild oats that I seem somehow to have inherited with my temperament, and that even the occasional excesses of barrack life were insufficient to eradicate. Now, however, thinking it high time to take up with some definite mode of existence, the idea of a pastoral life, with its picturesque vicissitudes, in some undiscovered back country seems to fit in with my notions of freedom. My military experiences proved so distasteful to me in that respect, that I have conceived such a prejudice against it that I seldom care to speak about it."

The latter part of Mr. Marsden's sentence was emphasised in the manner that a man of determination uses when he resolves that his hearers shall not mistake his meaning, and clearly implied his desire to let us know that further reference to this part of his career would annoy him, so, as a change of subject, Mr. Rolleston remarked :—

"I own some back country that I have lately had thoughts of disposing of; it is excellent pasture land, and with but a moderate outlay of capital in the formation of dams, a permanent supply for sheep could be obtained. If you think that the land will be likely to suit you I shall be most happy to ride over with you one day, when you will have an opportunity of judging for yourself as to its qualities."

"I shall be just as happy on my part to take advantage of your offer to at least see the property. Indeed, I should prefer to settle in or about your neighbourhood, as the appearance of the country hereabouts pleases me," Marsden replied.

"Then I must contrive to amuse you here for a few days, till you have recovered sufficiently from the effects of your accident, and then we will ride out and have a look at the country."

"With the pleasing prospect of the society of these ladies, the task of amusing me will be but a slight tax on your resources, Mr. Rolleston," replied Mr. Marsden, laughing and bowing to Mrs. Campbell and the two girls.

"I am afraid your anticipations of our powers in that respect may be too sanguine," replied Miss Rolleston smiling, while Mary shyly cast down her eyes.

Whether out of consideration to the stranger's hurt or not I cannot say, but I observed that Miss Rolleston's manner to him seemed to be pervaded with a deeper tone of respect than was usual with her.

It might have been a lover's jealousy, but I had noticed from the time of Mr. Marsden's first entrance into the house that Miss Rolleston appeared as if fascinated by the influence of some secret spell in that gentleman's manner as she listened with rapt attention to his words, seldom venturing on a remark of her own accord, and merely speaking when addressed, a most unusual thing with her as far as my short experience of her manner went.

This feeling with Miss Rolleston appeared to deepen rather than to lessen as the days went by. Perhaps I was led to notice this more attentively because of the rarer opportunities that fell to me for *tête-à-tête* conversations with her now. Her attentions appeared to be wholly engrossed with this stranger, and as if the attraction was mutual, he seemed to chiefly devote his remarks to her.

I have frequently observed that there are some men who appear to be gifted with some mesmeric influence, by which they are able to exercise a powerful charm over nearly all of the opposite sex with whom they come in contact, and it was evident that Mr. Marsden possessed this gift in no small degree to judge by the influence he appeared to be so soon able to exert over Miss Rolleston. In less than a week, with this gay, skittish maiden, who on other occasions never seemed able to give a serious thought to anything, it was plainly evident he was on terms of even confidential friendship.

Yet, though more reserved towards me, her kindness to me had not lessened, and so much was this the case that she appeared to deprecate the suspicion of her partiality to the stranger, which she instinctively felt was present to my mind.

Two weeks rolled by and the time for our return to our station became imminent, duties there now imperatively demanding my immediate supervision. The interval had been actively employed among Mr. Rolleston's cattle, in daily mustering and draughting, for the selection of a herd, which with our late purchase were intended for the Darling Station.

By this time Miss Rolleston's and Mr. Marsden's prepossession for each other was so confirmed, that they were almost constantly in each other's society. Mary Campbell was of course generally present, but perfectly isolated on these occasions, so much were they engrossed with one another.

Mr. Marsden's arm being now almost recovered, four days preceding that fixed for the departure of Lilly and myself with the cattle, it was resolved that, in company with Mr. Rolleston and Mr. Marsden, we both should ride out to the land that Mr. Rolleston was desirous of parting with.

This land lay about forty miles from the Murray, away through the heart of the mallee scrub, through which thin strips of open country were occasionally interspersed.

We at length turned into Mr. Rolleston's back run, which we found to consist of a large stretch of open country, mostly flat, but occasionally undulating with sand hills, all covered with excellent salt bush pasture. Through the midst of it a pleasant little stream of water flowed just then, which, however, in the drought of summer was reduced to isolated waterholes at long intervals along its channel. As it was, Mr.

Rolleston merely occupied this country as a winter run, and for this purpose there were huts and yards that were then occupied by some half-dozen shepherds and their flocks.

We spent a day riding through the run, camping on both nights with one of the shepherds, having gone out provided with our own blankets for this purpose. On the third day we started on our return journey. Mr. Marsden appeared to be fully satisfied with the run; that is, if assenting words unaccompanied by the slightest look of interest in any of the qualities of the country, that Mr. Rolleston was sedulously pointing out to him as they rode along, can properly be said to express such satisfaction.

The morning on which we started for home was bright and inviting, but as the day advanced the sky suddenly lowered, and became quickly overcast with a dark covering of highly-charged electric clouds; and before mid-day we were overtaken by a fearful thunder-storm, during the course of which forked lightning played around us in startlingly vivid flashes, accompanied with such deafening claps of thunder, that the sky at times appeared to be fairly rent above our heads.

The sun became wholly lost to our sight, and although Mr. Rolleston carried a compass, through some inadvertence in marking our route correctly on starting, we, ere we had suspected it, had been inclining several points out of our proper way. It was Lilly indeed who informed us of this startling fact, by the sudden declaration that we were off our proper course.

Our situation was now anything but enviable, for the rain was coming down in torrents, whilst the roll of the thunder was something awful. But allowing Lilly to take the lead, we struck off at a sharp angle towards our right.

In this direction we had not proceeded very far, when the continued cawing of the crows attracted our attention. As in the bush such sounds are, in general, indications of a shepherd's hut, or of some dead carcase (for these are carrion birds), we immediately shaped our course in their direction, in the hope of finding some sheep tracks, if nothing more, that might guide us to some hut in the neighbourhood.

At length, where the mallee seemed to open slightly, we saw a number of crows, both on the scrub branches and on the ground, and looking more intently to see why they were collected there, we noticed an object that filled us with horror.

It was the carcase of a black fellow. But, though an aborigine, his clothes, and especially the cap lying beside him which was distinguished by a yellow band round it, plainly intimated that he was one of the more intelligent natives retained in the police force for their valuable services in tracking criminals. His body seemed partially decomposed, and his features mutilated by the ravenous birds round him, his eyes having been picked out and holes dug into his ears and face. Yet even with this disgusting mutilation, the manner of his death was still apparent by a hole in the middle of his forehead, evidently the track by which a bullet had entered his brain. His own revolver was seen lying close to his body. On examination, it was apparent that this weapon had been used, as one of its chambers was empty.

After attentively looking at him for some moments, Lilly remarked,

"I know him!"

"Indeed, Lilly," replied Mr. Rolleston, "who is he?"

"Black Billy the tracker."

"Black Billy, Black Billy," I mused, trying to recall a name that I seemed to know had been recently familiar in my ears. "By-the-bye, is that not the same, Mr. Rolleston, that you read of the other week in the paper at Euston, in the report of the bank robbery?"

"True," he replied, "and doubtless the brave fellow met his death by one of the bloodthirsty scoundrels whom he had overtaken here."

Some instinctive feeling caused me here to glance at Marsden. He was calm and silent; but, with folded arms and sternly-contracted brow, was gazing at the black, mutilated corpse.

And as I looked upon that strong man, whose features bore the impress of a ruthless will, I could not help thinking how kindred to the electrical tempest that was then reverberating overhead, seemed the passion that for some cause appeared then to darken that brow, and to gleam so luridly in those eyes.

But whatever might have been the cause of his seeming inward commotion, the words that here, in answer to Mr. Rolleston's remarks, he let fall, betrayed no conscious connection with that and the fate of him whose corpse he was then contemplating. "Doubtless it was so; the man who was fit to commit such a robbery as that was not one who would be inclined to hesitate much in suffering a black fellow's life to stand in the way of his escape."

The scornful manner in which this remark was uttered seemed to imply how little he, at least, would have hesitated in a similar situation.

After gazing for some minutes longer on this shocking spectacle, we resumed our journey, conversing as we went on the probable details of this tragedy.

Whatever speculative surmises had formerly tormented me regarding the mysteriousness of Marsden's character were now doubly intensified since witnessing his demeanour when viewing the black man's corpse, until wild suspicions began vaguely to fill my mind, to which I had not the courage to give form. One of our party, however, had less conscientious scruples on that score.

Reining his steed level with mine, and allowing the two others to draw slightly ahead, Lilly remarked, inclining his head towards Marsden :—

"Do you know who that cove is, Mr. Farquharson?"

"Well, only what he says of himself, that he is a gentleman on the look-out for a run."

"He wants no d——d run," muttered Lilly; then a little louder, "It's my belief, Mr. Farquharson, that that cove knows more about that dead black fellow than he cares to let on."

"What makes you think that, Lilly?" I replied, not a little startled at hearing him so suddenly give form to my own train of thoughts.

"Never mind what makes me think so, but I tell you it takes a smart man to throw dust in my eyes. I have noticed that cove ever since he came to the station, and there is something about him that I don't like, and it's my belief that he knows something more than he cares to tell about how yon corpse came to be lying there, and about that bank robbery too, I reckon."

I was the more inclined to pay attention to Lilly's suspicions, startling though they were, as I had had so many previous experiences in verifying the almost instinctive accuracy of his impressions in reading the characters of strangers. He was naturally shrewd, and from his own antecedents among the society of convicts and his knowledge of the traits peculiar to that fraternity, no disguise of manner could ever baffle his instant identification of any such characters that he came in contact with : that is, "old hands," as those who have once been convicts were styled in Australia. And often have I been indebted to this discriminative faculty of his in my selections of operatives for the station.

Lilly went on :—

"Besides, see how all this agrees with the accounts of the bank robbery. Black Billy was away on their track. Well, they were supposed to have gone towards the Northern Territory, but that would not prevent an old fox from doubling and coming in this direction. Well, now, didn't that customer there say when he came to the station that he had come through the scrub from Lake Hindmarsh? But as much of that yarn is true as he chooses to make you suppose. One thing is certain ; he came from the scrub. And another thing is certain, that Black Billy had tracked one of these bank robbers to that scrub, for there's the fellow's corpse to prove it, else what would it be doing there; and we know that the paper said that Black Billy was on the track of these fellows—and a plucky fellow, too, the same Billy was, although he was black: I knew him. Again, that fellow came to the station with his arm in a sling, and you know Billy had fired off one shot anyhow, as there was one of the chambers of his revolver empty, and what would you say if that shot wasn't the cause of that cove's arm being in a sling, and that it was that same gentleman's bullet that put yon hole in Billy's nut; did you notice the determined and savage way in which he looked at the corpse?"

Here was indeed a concatenation of circumstances that would have justified any magistrate in granting a warrant for Marsden's arrest for this murder at once.

But yet again, the counter evidence in the appearance of things against such a conclusion was so strong as to render the idea highly improbable. As, for instance, the idea of a common bushranger and murderer taking up his abode in a squatter's residence on the banks of the Murray, where he was apt to be visited any day by a passing trooper; the very thought seemed preposterous. Besides, the education and gentlemanly bearing of Marsden himself were so utterly out of harmony with a bushranger's character, that after a minute's startled reflection I said: "Nonsense, Lilly, you must be dreaming. There is something queer in these coincidences, I grant you, as well as in Mr. Marsden's mysteriousness, but the man is a perfect gentleman, so that it is impossible that your suspicions can be right."

"Very well, Mr. Farquharson, we shall see," was Lilly's quiet rejoinder. But, though affecting to think thus, I was still so far from feeling assured, that I resolved on the first favourable opportunity, ere I left for the Darling, to communicate Lilly's suspicions to Mr. Rolleston and so put him on his guard as to Marsden's possible character.

Shortly after this conversation, that was of course carried on in a tone inaudible to our companions, the rain ceased, and the thick vapours began to roll aside, revealing a blue sky in which the sun appeared now well on on his journey to the western horizon, and by his position we found that Lilly's surmise that we had drifted to the left was correct. It was not until sundown, that, wearied and soaked to the skin, we had the satisfaction of dismounting from our no less jaded horses at Mr. Rolleston's stable door.

Soon afterwards, a dry change, a blazing fire, and a cheerful supper, made us forget all our uncomfortable sensations, excepting the vision of that mutilated body, that haunted my memory at least all that night and for many a day afterwards; and the startling thoughts suggested by Lilly's words as to Marsden's possible connection with that deed of blood, acted not a little in increasing this disturbed state of mind.

While our supper was being discussed, the occurrences of our late journey were duly detailed, with the account of our discovery of the body of the murdered black fellow; a detail which shocked the ladies, especially Mrs. Campbell, much.

I noticed that Marsden, for whom I reserved the chief part of my attention during all that evening, not only sustained his own share in the conversation, but was even unusually communicative and chatty. As usual, his eyes and remarks seemed chiefly for Miss Rolleston, and her glances in return sparkled with singular animation as she listened to his vivacious description of the manner of his future home out in this back country. About that he said, "I am now resolved to come to terms with Mr. Rolleston".

"I assure you, Mr. Marsden," remarked Mr. Rolleston pleasantly, "that I, for my part, will throw no difficulties in the way of a speedy arrangement of this matter. By-the-bye, the police must be communicated with about this discovery of the black fellow's body: I expect it must be Black Billy."

"I believe it must be his body," here put in Mr. Campbell, "I read in the paper to-day that the police are beginning to suspect foul play, as he has been missing so long."

"As I am going to Melbourne to-morrow," remarked Marsden quietly, "I will report the matter to the police as I pass through Swan Hill."

"Surely you do not think of going off so soon," replied his host; "remain a few more days with us, you must be tired with your three days' ride."

"Nay," replied the other, "as I should like to bring our present negotiation to as speedy an issue as possible, and as that cannot be done until I have been to town, where my affairs are likely to detain me for a month or so, I should not like to put off starting for a single day longer than I can avoid; I trust however, at the very furthest, to be with you again ere Christmas." After this, he again so entirely monopolised Miss Rolleston's attention, that I felt myself thoroughly non-suited, and in quiet disgust, after passing a short time in conversation with Mary, I left the room, and walked about outside.

The storm was now over, and where its blackness and fury had lately reigned supreme the moon was shining calm and splendid. Meditating over the strange occurrences of the day and the patent hopelessness of my passion for Rachel Rolleston in the presence of such a dangerous rival, I paced for some considerable time to and fro before the house front.

Still, for the sake of the love that I cherished for her, I was animated by a chivalrous desire of being able to repay her coldness by a service that might prevent her from ruining her happiness by an alliance with a man whom I instinctively felt was unworthy of her. Somehow my mind recoiled against harbouring the suspicions of downright villany that Lilly did not scruple to charge him with ; but I did feel assured that he was not what he represented himself to be, but was sailing under false colours, and consequently was of suspicious principles, and much did I wish that it might be my fortunate lot to unravel the mystery that enveloped him, that so I might be enabled to give the thoughtless girl a timely warning against his wiles.

Musing thus, I walked past the kitchen (that, as I have said, was almost attached to Mr. Rolleston's house—that is, it was connected thereto by a covered passage), and the door being open I glanced in, and had an opportunity of appreciating how well Lilly was enjoying himself in his quarters there.

As I have said, with Tiny, the comely-looking serving girl, he was on the best of terms. At that moment the table was covered with heaps of linen clothing of every sort, that had already undergone the operation of washing, and were now being subjected to the equally necessary process of ironing and mangling.

On glancing in I was amused to note how Lilly was employing himself. I have already particularised his ingenuity in any matter he took in hand. I had no idea, however, that among his list of accomplishments was that of a launderer. But there stood the honest fellow, with a smoothing iron in his hand, busily passing it over the plaits of shirt fronts and other articles of linen finery in a manner that elicited Tiny's unqualified admiration and amused astonishment. I may state that this young maiden in general was rather shy and demure in her habits, and to see her thus on terms of the most chatty familiarity with Lilly was a proof of the discretion of his manner towards her.

It was now time, I thought, to return inside, and as I passed the sitting-room window on my way to the door, my feet were suddenly riveted to the ground by the sweet, clear ringing voice of Miss Rolleston warbling the melodious strains of one of Burns' sweetest lyrics, " Ye Banks and Braes of Bonnie Doon". This exquisite song she sang with a simple heartfelt pathos that fairly entranced me, and especially the last two lines—

"And my fause lover stole the rose,
But ah, he left the thorn wi' me " ;

which she appeared to repeat in such a strain of foreboding sadness that it rang in my ears for many a day after.

Poor Rachel Rolleston!

CHAPTER XIII.

MY first duty next morning, that broke warm and beautiful, was to take an opportunity of apprising Mr. Rolleston of my suspicions touching the genuineness of Marsden's profession, to which I added also Lilly's downright conviction as to the personality of his visitor. Mr. Rolleston listened to me with an air of extreme surprise. "Nonsense," he said at length, "you and Lilly are most ridiculously mistaken. Mr. Marsden's connections, some of whom I know by repute, are of the highest respectability. He has shown me a letter of introduction that he had to Mr. Goldsborough of Gelong that speaks of him in the most satisfactory terms."

After such conclusive evidence, I had no more to say, and there the matter dropped.

After breakfast, having first been to the stable to see our horses saddled, we returned to the house to bid the ladies goodbye. They were all standing on the verandah, and with them Mr. Marsden. Miss Rolleston, as usual, looked lively and enchanting. "Do you know, Mr. Farquharson," she said, as I joined the party, "I have been mentally cogitating how we could spend next Christmas in the pleasantest manner. I understand that Captain Caddell's steamer goes up as far as Menindie which papa says is within a very short distance of his station; tell me, are there many kangaroos up there? and if there are, could we not get up a grand hunt on Christmas Day? It would be famous fun. Besides Mary and Miss Campbell, there will be some other ladies and gentlemen whom I intend inviting from town to enjoy their holidays with us."

"You must be surely dreaming, Miss Rolleston," was my surprised reply to this strange proposal; "even if you all were to come up to the Darling station, there would scarcely be decent accommodation in our bachelor's hall for a bevy of young ladies."

"Pay no attention to what the madcap says," said her father, smiling. "She does not know what she wants."

"Now, papa, not only do I know what I want but I am determined that what I want I will have. Do you know, I have been thinking of this idea for a long time, but always put off mentioning it till now."

"The thing is absurd, girl," said her father. "How do you imagine you are ever going to get up there?"

"Now, papa, what is to prevent our all going up in Captain Caddell's steamer as far as Menindie? and then you say your station is only ten miles further on."

"And I suppose your gay party are all to walk over the hot sand hills all those ten miles! I think I just see you doing it," replied her father sarcastically, "for where," he added, "is Mr. Farquharson to find side-saddles for all your ladies to ride on, I should like to know?"

"Precisely in this way," replied his daughter triumphantly. "Surely what saddles we shall require we can take up with us, and if we have not sufficient of our own, I know we can borrow some from the neighbours here for the few days that we shall need them, and then Mr. Farquharson would only have to send a man to meet us with as many horses as were necessary. But even that plan does not quite please me; it is not picturesque enough. We can fetch the saddles to ride on when we get to the station, and whatever else we shall require for our comfort besides; but instead of riding to the station on horseback it would be much more fun to go in the bullock dray, besides being more convenient for getting the other things we should require taken on to the station. Now come, what is there so very absurd in that plan, pray?"

"And who is going to look after a madcap like you on an expedition like this? I am sure that I am not, for one, and besides, I am not sure but what I shall be away from home on other business at about that time."

"Now, papa, do not make any more objections," replied his daughter, speaking more seriously; "I am sure that I mean to conduct myself with the utmost propriety, for I should not dream of going without being under some responsible person's charge, I intend that Mrs. Campbell shall accompany us to keep us all in proper order."

"Me, indeed! you surely don't imagine that I am to let you drag me all that distance up country, to gratify your wild whims, you unreasonable creature!" replied Mrs. Campbell, smiling reprovingly.

"Now, Mrs. Campbell, pray do not say so, for I am resolved on having my own way here. We can return the day after Christmas Day, and I know Captain Caddell intends spending his Christmas at Menindie, for he said so to me, when I met him in Adelaide. And I know he will not be in a hurry to return till our party can get back to Menindie, so that we can have plenty of time to enjoy a kangaroo hunt at Christmas."

"Well, Miss Rolleston," I here put in, when I saw that the

strong-willed girl was likely to bear down all opposition to her pet scheme, " I am sorry to damp your ardour, but if it is merely for the pleasure of enjoying a kangaroo hunt that you are so determined to come up to the Darling Station, I fear you will have all your pains for nothing: in all the time that I have been there, I do not remember to have seen more than half-a-dozen or so of kangaroos altogether, and emus are almost as scarce."

Miss Rolleston looked greatly disconcerted at the threatened collapse of the scheme, on the furtherance of which she had apparently set her heart.

" Oh ! if Miss Rolleston wants to gallop after something," here remarked Lilly, jocularly, " we can manage to get up a cattle muster : I shall be branding some calves by that time."

" Oh, thank you, Lilly, that will do nicely," cried Miss Rolleston, " we can first ride out and have a pleasant picnic on the run, and after that bring home a mob of cattle; that will be fun !"

" Yes," replied Lilly, following up her joke. " And I will bake a damper for your picnic that you can have out at the lake, whither the black fellows can carry the things, and we can fetch the cattle home with us from there, for it is there that most of them are running."

" I declare I will write for Miss Campbell and Miss Brydone and a few other friends forthwith," Miss Rolleston replied in quite a rapture of excitement at the prospect that Lilly's idea had opened out to her.

" Am I to positively believe, Miss Rolleston, that you are in earnest in all this ?" I now seriously demanded.

" I positively mean it all, I assure you," she replied decisively.

I glanced towards Mrs. Campbell to hear that worthy lady's opinion of this wild plan.

" Upon my word, Mr. Farquharson," she smilingly replied to my look, " I really don't know what to think about it, but we had better leave Rachel alone at present with her hobby—in all probability her mind will be occupied about some fresh scheme long before Christmas comes round."

" Very well, Mrs. Campbell, we shall see if it will ; and, Mr. Farquharson, we will write you in time to apprise you of our coming, so that the bullock dray may be in readiness to receive us when we reach Menindie, and, Mr. Marsden, remember we shall expect you to make one of our party."

" Most willingly will I do so, Miss Rolleston, and I will make it my duty to arrange my business so as to be able to take advantage of your enchanting invitation—that is, if Mr.

Farquharson will not be too much inconvenienced in providing room for me among so many other guests."

"Miss or Mr. Rolleston's friends, sir, can always command what accommodation there is to spare, convenient or inconvenient, at Tappio Station," I replied somewhat stiffly.

"For that matter, for the pleasure of enjoying the society of Miss Rolleston, and, of course, Miss Rolleston's friends," bowing to the other ladies, "an old campaigner like me would be sufficiently recompensed with a rug spread under an umbrageous gum tree for his night's lodging!"

"Your chivalry, I trust, Mr. Marsden, will be submitted to no such trying test on our account, I hope," said Miss Rolleston laughing, then added, "and now, Mr. Farquharson, we will depend on your arrangements being proceeded with on receiving my letter apprising you of our coming."

"Very well," I replied, "I will promise to do my level best to receive you all with due honour when you arrive. It will certainly be an event to be talked of at Tappio."

We now went through a general hand-shaking performance preparatory to our departure, during which, on coming to Mary, I, giving her hand a tender squeeze, contrived to whisper in her ear, "Now, don't forget your promise to be my friend".

"No, Mr. Farquharson, indeed I will not," she replied, cordially squeezing my hand in return.

"Well, Lilly, what do you think of our prospective Christmas guests?" I asked of my companion as we rode along the road.

"Faith, I don't know what to think of it; the idea of such a party of ladies and swells at Tappio seems strange enough anyhow! Won't the other chaps stare when they see them! It will be something like a Christmas Day!"

Conversing thus we rode along till we arrived opposite the township of Wentworth—then merely represented by a small public-house and store, and a small collection of huts and tents —situated on the banks of the Darling, immediately above its confluence with the Murray. Here we overtook the station hands who were driving the cattle which had been made to swim the Murray at Mr. Rolleston's station. On Lilly's suggestion, that we must here "shout" for the men, we hailed the punt, and were straightway all, save one who was left with the cattle, ferried across the turbid waters of the Darling, then in a state of flood.

As the shearing season was now commencing, we were not surprised to see tied to the verandah posts of the "pub" several horses with "swags" strapped on their saddles, the occupation of their riders being also indicated by bundles of sheep shears fastened on the outside of these "swags". These riders, in various stages of intoxication, were seen crowding the bar and smoking in the verandah.

Some of them appeared to be fine, athletic fellows, with a daring, happy-go-lucky expression in their faces, while others again showed a more forbidding and ruffian-like aspect as they swung about the place blustering and blaspheming.

Entering the tap-room, we were waited on by the owner of the house, Mr. M'George, a civil and rather jovial-looking person, who at my request brought in drinks for all the party—we numbered four in all, counting the man left in charge of the cattle, to whom an ample share of the spirituous refreshment was sent across in a bottle. After all had emptied their glasses, Lilly, considering it absolutely incumbent on him to get rid of any loose cash that he might have about him, which, moreover, on the station at home he regarded as an almost superfluous encumbrance, not only repeated the "shout," but added to its quantity, inviting as many of the loiterers about the door as he happened to recognise to come and join us.

It was through this proclivity of reckless "shouting" that Lilly had contrived to get rid of the major part of his earnings year by year; yet Lilly himself was no drunkard, and when on duty no man knew better how to keep this lavish habit of his under restraint, as he never suffered any man when under his charge to get muddled enough to become incapacitated for his work.

While these drinks were being discussed I amused myself by watching the vagaries that some of the topers at the bar and in the verandah were displaying. Some were quarrelling and giving utterance to language that discovered the true complexion of their minds; others, again, seemed more bent on indulging their humour in an excess of mischievous sport.

It was court day, and consequently rather a gala day in the township. There were several troopers sauntering about the place; one of these, a sergeant, a stiff, consequential-looking man, with a very wiry and prodigiously curly moustache and finical goatee, in especial amused me by the evident self-importance he displayed in his manner.

This trait of his had also evidently attracted the notice of one of the athletic, dare-devil-looking shearers I have referred to, whose pugnacity was evidently rendered irrepressible by the sundry glasses of grog he had just been imbibing at the bar, and who apparently was bent on giving his humour full play by the exercise of a little "chaff" at the expense of the consequential-looking sergeant.

At this moment the equanimity of our own party was like to have been disturbed by the obtrusive insolence of a burly, pockmarked, ill-looking ruffian, whose quarrelsome disposition with his fellow topers I had already noted. This fellow, conceiving himself aggrieved in having been omitted from the invitation to join in Lilly's "shout," now began to give vent to his assumed contempt for all the party by various foul expletives, more particularly directed against one of the stockmen, a smart, spirited young fellow, named Billy Lorimer, to whom he offensively remarked, "What the —— are you jawing about?" Lorimer was at once preparing to retort when Lilly peremptorily interfered. Although the general effect of spirits upon Lilly was to render him genial and good humoured and inclined to practical jokes, though at other times naturally quick and impatient, yet he would at no time suffer anyone to take undue liberties with him, and much less would he submit to bluster.

On the present occasion he at once silenced the bully by rising up and sternly addressing him. "Look here, old man, your company ain't wanted here on no account, so you just clear out of here or I am the man that will take you by the scruff of the neck and chuck you out."

There was such an evident determination shown in Lilly's manner to make his threat good that the man at once cowered and took the particularly broad hint that Lilly had given him, to leave the room.

Meanwhile the consequential-looking trooper, not much relishing the personal remarks of the facetious shearer and his enquiries as to whether he "found the ramrod down his back rayther uncomfortable," suddenly stopped and stared haughtily at his interlocutor, fiercely twirling his prodigious and wiry moustache the while. Unabashed by this his tormentor asked him if he would part with his goatee for a consideration? This was altogether too much for the man of dignity. "Look here, my fine fellow," he said, "unless you are deucedly civil I will lock you up."

"Don't get your shirt out, old man! look here, what will you take for your white gloves? I want a pair for shearing," retorted the shearer.

The trooper deigning no reply, the mischievous fellow still further insulted his dignity by asking him if he was not going to "shout" for all hands, "for," he said, "it would be shabby for a gentleman like him not to do it".

This was really more than the guardian of the public could possibly submit to, and at once walking up with the intention of putting his threat into execution, he laid his hand upon the shearer's collar, an attention that the shearer resented by instantly tripping up the trooper, who was sent staggering out into the road, where he fell on his back. At once recovering himself, the outraged official applied a whistle to his mouth and again rushed forward to seize the enemy, when he was met by a well-planted blow in the face that sent him a second time staggering back into the road.

Several other troopers now made their appearance, and with their assistance the obstreperous shearer was held fast. His companions at this juncture rushed out from the bar, determined on liberating their friend from the clutches of the police, and a general free fight ensued, the police calling loudly on the bystanders in the Queen's name to assist them in quelling the disturbance.

"Here's a lark," cried Lilly, on hearing the commotion, and the cry of the Queen's name by the police. "In the Queen's name," he repeated, "come along, Billy," he cried to the young stockrider.

"In the Queen's name," repeated Lilly, rushing out accompanied by Billy into the midst of the mêlée, apparently a most determined supporter of law and order. Pushing in among the crowd and swinging himself about in a most extraordinary manner, Lilly's first feat was to push one of the combative shearers violently against another policeman, who at this moment was running towards the scene of the disturbance, and who, in consequence of the sudden impetus with which he was met, was sent reeling several yards in an opposite direction; following this feat up, Lilly next made a strange sweeping, circular kind of a blow, aimed at no one in particular, but which most inadvertently fell full on the unfortunate sergeant's ear, fairly knocking him down.

"Dear me, I'm so sorry, it wasn't you at all I intended to hit," exclaimed the penitent Lilly, picking up the luckless sergeant, and as if to give proof of this profession, he aimed next in thorough earnest a blow that fell so well planted on the nose of the blustering bully (who had also joined in the riot for the mere gratification of his brutal desire of being able to hurt some one), that the fellow reeled back several yards and fell.

Billy Lorimer, who all this time had simply kept close to Lilly, now received a whispered hint from the latter, in obedience to which he gave Lilly a sudden push that impelled him on to the policeman who still held the shearer, who had occasioned the whole disturbance, prisoner.

Lilly was a powerful raw-boned man, whom it would have been no easy matter for any ordinary person to have sent a yard out of his way against his own will. But in this instance the effect that the thrust of a slim-built man, like Billy Lorimer, had upon him, seemed magical. Not only did it impel Lilly against the policeman who was several yards off with his prisoner, but it occasioned him to come against that policeman with such a violent shock as to fairly overturn both him and the man he held ; nor was that all, for in the general overthrow Lilly was precipitated with such force, and fell so heavily on the policeman's arm, that he let go his hold of the prisoner.

"Run, you d——l, run," Lilly whispered in the shearer's ear; advice that he aided by the awkward scramble he contrived to make ere he was able to regain his footing, whilst the shearer, taking Lilly's hint, instantly sprang to his feet, rushed towards his horse, that was standing with others tied to the verandah posts, and immediately mounted and galloped off, followed by all his companions.

In the meanwhile, Lilly, having at length regained his footing and apparently greatly exasperated at what had happened to him, again attacked the bully, who, by this time, was standing quite quietly by himself, apparently suffering from the effects of Lilly's last blow and sick of further fighting. Rushing at him again, Lilly fetched him such a tremendous blow on the ear, that the luckless wretch once more measured his length on the ground; then, instantly seizing him and dragging him to his feet, Lilly gave him triumphantly in charge of the police as being one of the chief offenders in the late disturbance.

What the policemen thought of Lilly as an ally it would be impossible to say, but, by the dubious-looking countenances with which they received his condolence at the rough handling some of them, and especially the sergeant, had just received, it appeared to me as if they were not sure whether to regard him as a friend or a foe. However, they took possession of their only prisoner, for whom they were solely indebted to Lilly's prowess, and from their manner of securing him, it was evident that he was to be made the scapegoat for all the others.

Over this laughable scene, the shearer who had been the occasion of it often made merry with Lilly, for he shortly afterwards worked with Lilly in the shed, where he was looked upon as one of the smart men on the board. During all the shearing season, Lilly's zeal in the service of the Queen, that had contributed so much to the discomfiture of the Queen's servants, was the standing joke of the shed.

CHAPTER XIV.

THE busy stir incidental to the shearing had subsided, and been followed by the normal listless state of existence on a sheep station. The homestead lately swarming with men was now comparatively deserted.

The wool teams had made several trips, returning with the necessary station supplies for the ensuing year, and Christmas was near at hand when I received a letter from Miss Rolleston, warning me of the advent of her party by Captain Caddell's boat that was then making a trip, though not her first, in prospecting the navigable properties of the Darling, as far as Menindie.

From that station the river made a wide detour of about forty miles ere it reached Tappio, but instead of following the river, the road from thence struck straight through a sandy scrub to our station, shortening the distance by ten miles; therefore, in telling me of the coming of her party, Miss Rolleston in her letter desired me to be sure and have the dray awaiting their arrival at Menindie, from which point they would start on the ensuing day, so that the intervening distance between there and Tappio might thus be comfortably accomplished.

Besides herself and Mrs. Campbell, with her two daughters, and Tiny Sutherland, the serving maid, the party was to consist of a Miss Brydone and two young gentlemen of the names of Brown and Green. The first was a banker's clerk, and the second was attached to an insurance office. But I felt by no means sorry, from the absence of any reference to his name, to think that Marsden was not likely to be one of the party.

This letter I received about a week before Christmas, so that I had ample time, with the assistance and advice of Lilly, to fit up one of the bullock drays with an awning made out of tarpaulin, for the greater comfort of the ladies during the heat of their journey which would be very great, and especially trying, because of the sun's reflection from the sand, through which the road chiefly lay. Besides this, I had benches made for them, on which washed sheep-skins were tacked to act as cushions. The house was at the same time, at my desire, thoroughly overhauled by the cook; whilst he, with the assistance of Charles Knight, the men's cook, made a perfect store of sweets and cakes in honour of our Christmas guests.

On the day that I had despatched the dray to meet the expected party, my original intention of riding to Menindie and escorting them home had to be given up, as I had to assist Lilly with some calves that required branding.

Whilst thus employed with Lilly in his stockyard my attention was, to all appearance, concentrated on my work, but mentally my thoughts were busy with anticipations of the coming festivities, and, as the afternoon advanced, I could not resist the temptation of gazing in the direction whence the visitors would come, or, of keeping my ears on the alert to mark the echoes of the bullock-whip in the forest.

The prospect of enjoying the society of Miss Rolleston again, without the mortification of being compelled to witness her attention being monopolised by the more favoured Marsden, was peculiarly soothing to my still unrelinquished hopes of yet winning her hand; for although I ought to have seen enough already to have understood how utterly vain such hopes were in view of her evident preference for my dangerous rival, still it is wonderful how true love will hug itself and buoy itself up under the most adverse circumstances. The only shadow of hope I could now have, came from the thought of his absence, and the favourable opportunity I should thus have of being able to recover my lost ground.

Poor fool! when thus counting on the diversion in my favour of the thoughts of her whom I loved, when parted from him whom she loved, I should have remembered the line of the song,

"Absence makes the heart grow fonder".

I also thought with pleasure of the prospect of meeting Mary Campbell, for whom I entertained a genuine though, of course, only friendly regard. Then I found myself speculating on the probable appearance of her sister, and I remembered the caution I had received from Miss Rolleston about the danger of my heart when in company with her. "Not much fear of that, Miss Campbell," I sadly soliloquised, my thoughts shaping themselves involuntary into rhyme—

"However beautiful thou may'st appear,
Whilst peerless Rachel Rolleston is near".

At length, as the shadows of the afternoon began to lengthen considerably, the sudden clamorous barking of the dogs drew our attention to the figure of Billy Stack, the bullock driver, who had been despatched with the dray to bring our visitors from Menindie, and who, whip in hand, was now seen coming down the opposite bank of the river towards us, and, crossing over in the canoe, he soon reached the yard where we were at work. As we had surmised, he had got stuck in the muddy creek, an awkward, boggy crossing, about two miles from the station, so he had come on for reinforcements to help to drag the team out.

Lilly's team were at that time on my side of the river, so Billy had come either to get Lilly to take his bullocks himself, or to let him (Billy) take them to pull the dray out of the bog.

"Will you do it, Lilly?" he asked, "I am fairly bested, and the dray is fairly up to the axles in the middle of the creek."

Now Billy Stack was one of the very few men whom Lilly

allowed to be fit to drive bullocks, even paying him the extraordinary compliment of admitting that Billy was near as good a driver as himself, than which he could pay no bullock driver a higher. At any other time, therefore, Lilly would have been delighted to have taken the team himself, in order that he might be able to chuckle over such a proof of his superiority over Billy, in being able with his superior team (a contingency he would never have dreamt of doubting) to do what Billy had fairly confessed himself "bested at," but at that precise time Lilly was up to his eyes in work from which he could on no account absent himself, whilst yet the idea of trusting his beloved team to any other hands than his own, was simply revolting to him.

However, as matters were, there was no help for it, and having such confidence in Billy Stack as a judicious driver, who never ill-treated bullocks, punishing them only as a very last resource, he, although at first receiving the proposal with a resentful glance, at length reluctantly consented to let Billy take them; but he distinctly charged Billy not to yoke the two teams together as he reckoned his own were sufficient to shift the dray, however deeply bogged it might be. As the sun would now be soon setting, I directed Billy to return straight to the dray and have the bullocks unyoked by the time Lilly's bullocks arrived, and I at once sent away a black fellow to fetch and drive them to him.

It was easy to see that Lilly was dissatisfied that the bullocks should have to go at all. He growled to himself, but made no remark, while with renewed energy we both applied ourselves to the completion of our task. We had fortunately just finished as it was getting dark, though a bright moon was replacing with her beams the light of the departed sun, when the black fellow (Snowball, he was appropriately called) who had been sent away with the bullocks, came riding up with the astounding intelligence: "Lilly bullocks no good, no pullem wheel barrow" (*i.e.* dray). "What ———," I will not repeat what Lilly said, but he swore tremendously, and I believe, had he been alongside the black fellow, would have struck him, so greatly was he exasperated at this unlooked for slight on the honour of his team, the more so because I was there to hear, as he had frequently descanted to me in glowing terms on the staunch properties of all his bullocks, declaring that each one of them would drop in his tracks ere he would flinch at any pull.

As both of our horses were saddled, and feeding alongside the yard, we instantly caused them to swim across the stream, one at a time, behind the canoe; being accustomed to this, they did it without the least difficulty. Then, mounting, we rode straight away to the scene of action, where we found our party in a rather embarrassing plight.

Owing to late rains, the natural sponginess of the soil on the banks of this creek was increased, and that in the bottom, made much more so from the recent traffic of the wool teams; moreover, the water had been considerably widened by the rain. At the time at which we entered upon the scene, the bullocks had crossed so far, that some of them were on the opposite bank while the dray wheels were embedded in the mud near the water's edge. It was herein that the difficulty lay, for whilst the wheels were deeply embedded in soft gluey clay, the bank at the landing was now reduced, by the trampling of the bullocks in their past struggles, to a quagmire. A little lower down there was another landing, but so steep that it was deemed impracticable to a dray loaded as the present one was, that, besides its living load and their luggage, had also several cwts. of wire in coils, that had come up for the station in Caddell's steamer.

On arriving at the creek I at once rode forward and greeted the ladies, and from all of them received a very merry response, Mrs. Campbell, with her quiet smile, asking, "Is there any likelihood, Mr. Farquharson, of our getting out of this bog to-night?"

"Well, yes, Mrs. Campbell," I replied, "you may safely trust to Lilly's energy to relieve you from your present difficulty, but you must be fatigued with being so long confined in your cramped position in the dray. Lilly and I did our best to arrange it so as to make you more comfortable than otherwise you might have been, but I fear after all our efforts, it must still be very uncomfortable."

"Oh no! upon the whole I have found the dray very easy, although I shall be glad when the journey is over."

"Well, ladies," I said, glancing at the cluster of merry faces —merry still in spite of their present misadventure; "you seem to be happy at all events, whether you stick here all night or not."

"We always are," replied the ever-ready Miss Rolleston, "and have all the more reason for being so now; it looks so jolly stuck in a bullock dray in the middle of a creek on a moonlight night, and I am sure these two young gentlemen ought to consider themselves highly honoured by such close contact with so many young ladies, who have all contrived to do so much for their amusement through all this tiresome journey."

"Honoured, no doubt," I heard one of them rejoin; "still, for my part, I wish I was out of here. I am squeezed almost to death by so many of you. One can have too much even of a good thing, as the kitten said when it fell into the cream can."

"Oh, now, Mr. Green," exclaimed one of the ladies, Miss Campbell as I at once guessed, "it is too bad of you to say that, after the professions of loyalty you have been making, declaring you would go to the end of the world for me, and all that sort of thing."

"Good evening, ladies," here Lilly suddenly interrupted, "you are in a fine mess here, I must say. How would you like to have to camp here all night?"

"Oh, fie, now, Mr. Lilly," cried Miss Rolleston, "is that the tone with which you greet us, after we have come such a long way to see you? I expected more sympathy from you than that."

"Young ladies seem to take queer notions when they go riding in bullock drays, and it serves them right when they get stuck in bogs; why can't they drive in their own buggies or ride a horseback?" replied Lilly rather shortly, for his temper was up as things looked serious about our getting them out of their present awkward fix.

He now took the whip from Stack, who assured him that he had not pressed the bullocks much.

There was clearly no use in attempting to urge the animals to drag the dray through the hopeless quagmire in front, Lilly therefore turned his attention towards the lower landing place, with the object of getting the dray out of the creek by that means.

Now, although this attempt first involved a sharp turn to the dray, dangerous, from the imbedded position of the wheels, to the security of the pole and stays that might be wrenched off under such a strain, yet, as the mud was very soft, Lilly trusted by the steadiness of the strain to reduce that danger. He, therefore, addressed the bullocks, and accompanying his words with gentle lashes he wheeled the leaders back into the creek, not suddenly, but, as it were, in a semi-circular fashion, so as to keep a steady strain on the chains whilst the move was made. In this way the dray was gradually shifted round till the bullocks' heads faced the opposite bank, alongside of which the dray was then dragged until opposite to the lower landing. Then, again, at his steady orders, accompanied by ringing cracks of the whip, though now it never fell on any of the bullocks, the off-side leader, "Dauntless," with his horn instantly striking against his yokefellow, came round with his companions, until Lilly's "Gee, Fearless," suddenly arrested them, the latter bullock then instantly horning his companion off.

Thus alternately addressing each of these well trained

leaders, and giving accompanying orders to all the others, save the pole bullocks, the team in a straightened line began now to emerge from the water and up the bank before them. It was now that the tug of war commenced in good earnest.

As I have said already, this landing was a very steep one, and the ascent rendered all the more difficult from a deep grip in the bank just at the point of egress from the water; just there, however, the footing was much firmer for the bullocks, so that on this point they had the advantage of getting free play to any powers of draught that they could put forth.

By this time Lilly had every bullock well in hand, and firmly feeling the yoke, and now for the first time he spoke to his polers, whilst his orders to all the other bullocks rang out sharp and clear.

"'Dauntless,' 'Fearless,' 'Dainty,' 'Davy,' 'Roderick,' 'Bauldy,' 'Wallace,' 'Samson' up!" the wood meanwhile echoing at each word with the tremendous cracks of the whip, that now sharply admonished each bullock mentioned. "Samson," like his namesake of old at the pillars of Gaza, nobly seconded by his yokefellow, leant forward and bowed till his nose seemed to be almost touching the ground; the strain, evenly borne by all the bullocks in the team, seemed terrific. Something was bound to go, though for a moment the dray seemed immovable. At length both pole bullocks, as if in fury at the opposition they were meeting with, tossed their noses in the air and with a most prodigious effort seemed to lift the dray over the opposing grip.

This mighty feat of the polers being bravely seconded by the others, under the watchful eye of their driver, admonished by his whip or cheered by his voice, the dray began to slowly yield to the force laid upon it, and to steadily mount the steep ascent. Even "Wallace," Lilly's favourite bullock, did not now always escape the lash, so necessary was it that each bullock should keep up an unremitted strain.

When the last pinch of the ascent was fairly overcome Lilly's geniality of spirit was once more restored, and, stopping his panting team, he cried out in a tone of triumph to me:—

"Now, Mr. Farquharson, what is your opinion of that team of mine? I suppose you were suspecting I was blowing a bit when I used to boast to you of what they could do. Did you ever in your life see such a pull as that? I don't believe there is such another team as that in all the colony!"

"Nor do I. They are without exception the most magnificent animals that I ever saw in yoke, but I question if there is another man, Lilly, who could have made these same bullocks do such a feat as, under you, they have just accomplished."

Billy Stack, who had followed up, here remarked that he was blowed if ever he saw such a pull. "I reckon," he added, "that there isn't a man in the colony can drive a team of bullocks like you, Lilly."

The ladies seemed particularly elated at their extrication from their late plight, and both Miss Rolleston and Mrs. Campbell assured Lilly that they knew that they had to thank him for their rescue.

As for Lilly, he looked simply triumphant, and though certainly pleased at the proof he had now displayed of his skill as a driver, acknowledged by such competent judges as Billy Stack and myself, yet I believe the chief source of the good fellow's pleasure was in the manner that his team had staunchly acquitted themselves. He began to speak to each one of them as if he knew that they were perfectly conscious of the praise that he was giving them; and perhaps they were.

About an hour afterwards we reached the home-station, and glad enough, I believe, all the ladies were to get out of their stiffened and cramped position.

CHAPTER XV.

ON entering the house, where a smoking hot supper was quickly put on the table, I was introduced by Mrs. Campbell to the two ladies unknown to me, her daughter Miss Campbell, and Miss Brydone, and then to the two young men. Mary of course, I had long since shaken cordially by the hand, and was pleased to see her sweet face looking the very picture of health. Tiny also was there; as blooming and as shy as ever. She was just treated as one of the party by the other ladies, there being no means of separate accommodation available for her here. She, however, did excellent service in waiting on them, and in looking after the baggage that had been taken up for their use. It was of course for this that she had accompanied the party on their excursion.

In appearance, Miss Campbell was a bright, intelligent girl, rather below the average height of women, but well proportioned. She was rather pale in complexion, but with an animated expression of countenance that was enhanced by the sunny light that ever and anon shone in her expressive blue eyes.

Miss Brydone was also pale, but with a milder expression and, perhaps, more lady-like demeanour. Like her companion, she had also blue eyes and dark hair, but she appeared of a more delicate constitution than Miss Campbell, who, in spite of her paleness, seemed the embodiment of health and spirits.

Although lacking the personal beauty of Miss Rolleston, and consequently seen at a disadvantage by her side, yet but few minutes would elapse before any impartial observer would become convinced that, in the eyes of the opposite sex, Jessie Campbell, with her overflow of rich humour, and her contagious laughter coming straight from the heart, and shaking her little body in its passage upward, would hold her own, even against Rachel Rolleston.

The two young gentlemen I will dismiss with the remark that they were much like the average young men found in cities; they smoked cigars, wore short moustaches and neat neck-ties, and talked much about their fellows in town. They were both connections of Miss Rolleston. Mr. Green, in addition to the natural proclivities of the genus he represented, seemed to be a devoted admirer of Miss Campbell; Mr. Brown, on the other hand, I believe, chiefly admired himself.

And now my rude bachelor establishment seemed all at once to have undergone a singular transformation. Rooms that till then had been the scene of my own sober reflections only were now made bright with the presence of ladies, with happy, smiling faces. The wonted quiet of the house was broken into by a sort of musical Babel of voices and light, ringing laughter, and, to crown all, their light, many-coloured dresses added a picturesqueness to the scene as they whisked through the doors or sailed across the apartment.

I could scarcely realise that I was at home. In this very room where I had so often mused in solitude about her who absorbed so much of my thoughts, she was now seen in all her grace and loveliness—my visitor. And on these cushionless chairs, and on that plain wooden sofa were now seated graceful, dainty figures such as I had been accustomed to think of in connection only with well furnished mansions and refined society.

At supper, however, a brighter, happier, and at the same time hungrier circle of guests I verily believe never surrounded a table of tempting though plain viands. As to etiquette, it was simply lost in a confusion of laughter and the evident determination of all to put themselves at ease amongst their novel surroundings. Never have I seen anyone so determined to dismiss any nervous embarrassment I might be possibly feeling in my position as host to such fashionable guests, as Miss Rolleston and Miss Campbell seemed to be. They actually appeared to vie with each other in merry disregard of the usual forms of table ceremony, in which they were so well versed. Mary and Miss Brydone quietly smiled, and seemed to enjoy the scene. Indeed it took all Mrs. Campbell's authority to keep the two unruly ladies in order.

"Now really, girls," the good lady at length remonstrated, "Mr. Farquharson will be thinking you have taken leave of your senses, or that you have left all your manners behind you at the Murray if you go on so."

"Well, mother, you know that after having come so far to spend our Christmas holidays Rachel and I have made up our minds to enjoy ourselves, and if we don't do so, why," replied Miss Campbell very frankly, "Mr. Farquharson will be thinking that we are not appreciating his accommodation and the great pains that I know he has put himself to on our account."

"Quite right, quite right, Miss Campbell, I am quite delighted at your spirit. You could not please me better than by the way in which both you and Miss Rolleston convince me that you feel yourselves at home in my rude quarters. These other ladies, however, appear to have a quieter mode of showing their enjoyment of the fare of my rude bachelor establishment. As to what I have done with the means I had at hand to make that a little more presentable to such fashionable visitors, why, we must all fortify ourselves with the excellent philosophy of Burns' song—

'Contented wi' little and cantie wi' mair'."

Here the ladies all joined in a protest against such slighting references to the accommodation I had to offer, declaring that they were perfectly delighted with it, and that they had left home prepared to be contented with far less.

"Besides," Miss Rolleston remarked, "you must remember that we were too selfish to neglect ourselves, and the comforts that we expected that you would be unprepared to provide for us, we brought. Tiny there can tell you how many pillows, sheets, and other bedding materials we have packed away in that large box outside there, so, as Jessie said, we have come up here to thoroughly enjoy ourselves and cast starched dignity to the winds. Pray, has Lilly baked his damper yet? You know he promised to bake one for our picnic."

Lilly, who had stepped into the men's hut to have supper, was now summoned and interrogated on this important point. His good humour was now completely restored, and he answered with a smile:—

"Well, Miss Rolleston, no, not yet; but I will start about it now if you like, and if you want to take a lesson in baking dampers you can watch how I do it. When you get husbands all of ye it might come handy sometimes, you know, when you happen to run short of hops."

"Thank you, Lilly, for the hint," retorted Miss Rolleston. "As I am very anxious to get a husband soon, your idea suits me charmingly."

Miss Campbell also echoed the same sentiment most energetically, and, in view of the same contingency, seemed to be particularly eager to be initiated into the mysteries of damper baking.

Mary laughed, and Miss Brydone seemed greatly amused, whilst Mr. Green looked rather sheepish, as his companion shouted out "bravo" at Miss Campbell's pointed remarks.

So, supper being over, the whole party, with the exception of Mrs. Campbell, adjourned to the men's hut to watch Lilly begin operating with his damper. Such a merry group of visitors had never entered within the walls of that old hut before, nor in all probability ever did afterwards.

At once rolling up his shirt sleeves, Lilly began kneading the dough with such an air of concentrated determination as if the matter he had in hand was one of life and death, directing Charlie in the meanwhile to keep the fire well up in view of the plentiful supply of good embers he would require for the cooking of his damper. For fully half an hour did Lilly knead away at the dough, and that, too, with no small muscular exertion, ere it was wrought to his satisfaction; then, flattening it out into a circular shape, until it was about three inches thick with a circumference of fully four feet, he took the shovel and with it carefully opened out amongst the glowing embers and hot ashes a clear space that would admit of the great cake of dough being placed on the surface beneath. First carefully covering over all red embers on this bed that might blacken the damper's crust with a thin sprinkling of cold ashes, over this he next spread a sheet of newspaper that he had provided himself with for the occasion and then carefully taking up the dough with both hands, he deposited it gently on this paper. Over this, he then spread another sheet of newspaper, threw another sprinkling of cold ashes over that, arranged some round the damper's outer rim to guard it, and then finally turned all the red ashes and glowing embers over the whole, carefully arranging them, so as to distribute an equal degree of heat over all the immense cake beneath.

As Lilly laid the dough in the ashes and covered it thus, the astonishment, almost disgust, depicted on the faces of both of the young gentlemen and of Miss Brydone, who had but recently arrived in the colony, was most ludicrous in my eyes, as I quietly watched their features. Miss Brydone, however, made no remark. Not so Mr. Brown.

"Well, I'm blowed," said that outspoken young gentleman, if I eat that mess; it will be all over and stuck choke full of ashes and cinders."

"Well, young swell, we'll see about that," replied Lilly with quiet humour, "I reckon by the time you reach the lake you'll change your tune about that. Now then, ladies, you can all go away, and if you come back in rather less than an hour, you will see me take that damper out of these ashes, and if it ain't a good one call me a duffer if you like."

Accordingly, at about the stipulated time, we all returned to the hut: Miss Brydone and the two young gentlemen with curiosity evidently on the stretch as to what sort of appearance this damper would present when taken out of the ashes.

But when Lilly had fairly exhumed it, first tapping it all over on its surface, to assure himself by the hollowness of the sound that its soddenness had entirely evaporated, and had lifted it out, and with a dry cotton cloth smartly dusted it until not a particle of ashes could be seen on its smooth crust, he placed it carefully on its edge on the table, reclining at an angle against the wall, and then triumphantly tapping its echoing crust, he pointed to its appearance—completely baked, and without a single black or scorched mark on it—with the exclamation, "Well, young man, do you think you'll now be blowed before you eat it?" they all gave expression to their astonishment and unqualified admiration, and Miss Brydone frankly admitted that when she saw Mr. Lilly throw the dough among the ashes, she had felt convinced that he had wasted all that good flour.

By this time it was late in the night. The mirth and spirits of all seemed exuberant, yet at times I imagined I could detect symptoms of passing uneasiness in Miss Rolleston's countenance. I could have sworn on these occasions that she was listening as if in expectation of another arrival. She was flushed and beautiful, and yet, whether it was only fancy on my part or not, I do not know, but I thought that I could detect a shade of anxiety occasionally crossing her face, nay, that it frequently succeeded a merry laugh that had just rung from her lips. I also distinctly observed her on the final breaking up of the party for the night (a movement that had been by her means postponed so long) take a hurried step to the outer door, as if to assure herself of any possible chance there might yet be of this late arrival.

However, what beds we had received us all at last. I coiled up in a 'possum rug on the parlour floor, and the gentlemen beside me were provided with comfortable shake-downs, the beds having all been given up to the ladies.

I slept soundly till Tiny's tap at the door at six o'clock on the ensuing morning admonished me that she required the room to prepare it for breakfast.

I at once obeyed the summons by tossing the 'possum rug off me and getting upon my feet; but my two sleepy-headed young companions, who, in addition to their natural indolent habits induced by town life, were fatigued by their yesterday's journey in the bullock dray, were so drowsy that I had almost literally to pull them out of bed ere I could get any more sensible answer to my summons than a sleepy " all right ".

I then went out to arrange with Lilly about the horses we should be able to muster as mounts for all our party for their day's excursion.

CHAPTER XVI.

WHILE on my way down to the river I was suddenly warned by the clamorous barking of the dogs of the approach of a stranger, who, to my excessive surprise and chagrin, proved to be no other than Marsden. At once the cause of Miss Rolleston's apparent anxiety and expectant manner the whole night before flashed upon my mind. Evidently she had known by some secret source of communication that this gentleman would be here.

"Your servant, Mr. Farquharson," remarked Marsden, as he rode up, with a slight inclination of the head (while I fancied that I observed a sarcastic smile flit along his thin lips), "you did not expect to see me here this morning, I dare say; but having determined to make one of your Christmas party, on my return from Melbourne to the Murray finding that the ladies had already left for here I instantly followed them."

"You surely must have been early on your road to arrive at this hour of the morning," I replied in some surprise. With the same slightly sarcastic smile quietly playing on his lips he answered: "I am not in the habit of regulating my movements by the hours of the day, when on my travels, my dear sir". I could not help regarding him with some curiosity not wholly unmixed with admiration, as, with his powerfully built figure and daring look he sat his horse, a fine looking animal of a dapple grey colour, with an erect, military bearing.

Indicating to him where to put up his horse, he rode up to

the house, while I pursued my way across the river to Lilly, with whom I held some consultation with reference to the day's arrangements. A black fellow was at once despatched to fetch up the horses that were grazing about a couple of miles down the bank of the river, and then I returned to the house.

On entering the breakfast room I found all the ladies assembled and in high spirits, particularly Miss Rolleston, who was blooming, and perfectly radiant at the prospect of the fine day before her, the pleasure of which was doubtless now enhanced by the presence of him who was evidently in her eyes " the fairest among ten thousand".

As I watched her attentive bearing, her delighted manner, and her head continually turning towards him as he either addressed his remarks to her personally or to the company in general, the conviction that there was a secret understanding between them, and that his arrival that morning had been to her a matter of no surprise, but was the result of correspondence between them, and was, in fact, the event which she had been anticipating with such anxiety the preceding evening, was now firmly impressed upon my mind. What the effect of this on my mind was, the reader may judge for himself, and will feel no surprise when I say that I felt my sinking heart proclaiming the utter futility of all my hopes.

However, a truce to such fancies; here comes the black fellow, with all the mob of horses in front of him, tearing at full gallop towards the stockyard.

We found that Mavourneen, a pretty black mare, quiet and suitable for a lady, had, along with a few other horses, got separated from the mob. Now, as all the horses on the station that could be depended on for carrying ladies were no more in number than the number of the ladies themselves, exclusive of Mrs. Campbell and Tiny, who both preferred to remain at home, this mare's defection was unfortunate, and the unpleasant alternative threatened, that either one of the ladies must remain behind, or else the expedition be postponed till the missing mare was hunted up, and thereby considerable time lost.

A better expedient, however, suddenly suggested itself to me than either of these, in the shape of my own horse Selim, who was as quiet and docile as a lamb. I could let one of the ladies ride him, and take a young half-broken colt myself, that Lilly had lately been handling.

This colt was certainly far from quiet, but then I was not a very timid rider, and as Mary was not a very bold one, why, Selim would suit her exactly.

I proposed this, and so it was arranged. Already about a

dozen black fellows had been despatched to the Lake, some bearing provisions and others the crockery for the picnic. These were all carefully directed by Lilly to the place of our proposed encampment, about eight miles off. At about ten o'clock our party started, the ladies now all in high glee and satisfactorily mounted, their side-saddles and bridles having been brought with them from the Murray. The horses first being made to swim the river, the whole party with their saddles were ferried across in the canoe.

Miss Brydone being a very inexperienced rider, was provided with a very sedate old horse that did not seem disposed to hurry, so at first she had ample employment with a switch to make it keep up with the others.

Both Miss Rolleston and Miss Campbell on the other hand, being spirited riders, were mounted on active steeds, and right merry they seemed as they dashed along laughing like a pair of madcaps, as undoubtedly they were, in front of the rest. Mr. Marsden, mounted on a strong and fiery station horse (for his own was too tired after its late journey to be saddled again so soon), of course closely attended on Miss Rolleston, even amid the frequent spurts that she, in concert with Miss Campbell, took ahead of the party.

For myself, I kept close beside my sweet young friend Mary, with whom I chatted pleasantly as I rode along, and saw with pride the motions of my own steed, that, although chafing under his rearward position in the party, yet acknowledged the slightest pressure of his gentle rider's bridle hand.

Lilly was there on his own frisky dapple grey charger, Coleena. He said but little, though I noticed him keenly observing Marsden's motions, as I had likewise noticed him critically observing that gentleman's dapple charger that morning, as it stood in the paddock.

On seeing the lake with its broad expanse of blue water glistening brightly in the sunshine, with the luxuriant herbage around, though it was beginning to fade before the scorching effects of the summer's heat, the ladies fairly shouted with delight. The view was all the more striking because dotted over with mobs of cattle whose various colours added to the picturesqueness of the scene.

On all sides the plain was traversed with belts of timber, in whose deep shadows the recumbent cattle were at intervals seen, enjoying the relief they there found from the glaring rays of the sun.

By Lilly's strict orders, we avoided going near the cattle, as the unusual appearance of the ladies with skirts streaming behind them would have driven them in terror across the plain, but, as we approached the place where the blue smoke from fires kindled by the black fellows showed the spot chosen for our *fête champêtre*, all at once a loud cry from Miss Campbell in front directed all eyes to about half-a-dozen emus that were visible some distance off, with outstretched necks scudding across the plain. "A gallop! a gallop!" cried both Miss Rolleston and Miss Campbell excitedly, suiting the action to the word and whipping up their steeds after the birds. But against this proceeding Lilly loudly protested, crying out that they would only knock their horses up, and that the birds were too far away anyhow. After a race of a few hundred yards this was so apparent to the young madcaps themselves, as they saw the birds disappear into a belt of timber, that they reined up their panting horses, and, accompanied by the inseparable Marsden, returned flushed and laughing to the rest of the party.

On dismounting, we found, under the deep shade of some gum trees, a bucket for making tea suspended over a blazing fire, and, after committing the horses to the care of the black fellows, Lilly unpacked the provisions from their coverings and bags, and spreading out a large tablecloth on the ground, arranged a substantial cold collation, his own damper being in the place of honour in the middle; here was an apple-pie and there stood a plum-duff as large, and almost as dark, as one of the black fellows' heads. By their sides stood equally tempting viands of a more substantial character, such as cold roast mutton in various shapes, with a tempting looking round of beef and several brace of cold roast wild ducks. As Lilly emphatically remarked, "as jolly a feed as ever a white man need sit down to".

The tea, with milk in it for the ladies, was served out to them and the two town gentlemen in cups and saucers; but Lilly and I could find no way of drinking this pleasant beverage so suitable as from pannikins, where milk seemed strange and unaccustomed. "Where," demanded Lilly, "can you get tea equal to that made in a billy?"

"Well, young fellow," said he, addressing Mr. Brown, "what do you think about the damper now?" when he had handed round ample slices to all the party.

"By jove," said Mr. Brown, after tasting some of it with great relish, and hastily applying himself to the slice again, and speaking with his mouth full, "it is prime; really this is the best bread that ever I tasted—baker's bread is nothing to it." And Mr. Brown only spoke the truth, for Lilly's damper was simply excellent; and in spite of the closeness of its texture, it looked as white and as thoroughly cooked, and seemed so sweet and agreeable to the taste that really one scarcely knew when to have done with it. One slice but incited a craving for another, so that even the ladies, with their more delicate appetites, and wholly unaccustomed to such fare, declared it was actually delicious. Doubtless, too, the keen, healthy zest imparted to their appetites by the late exercise in the open, sunny air, contributed not a little to such thorough appreciation of the damper's genuine excellence.

Having all regaled ourselves to our entire satisfaction we rose, and withdrawing from our table on the ground and resuming the care of our horses, next amused ourselves by watching how our sable retainers attacked the provisions that were now entirely given up to them.

Short work did they make of them, as, each dressed in a white shirt and a pair of moleskin trousers, they squatted round the white cloth. The huge plum-pudding was first devoted to destruction, and then the other sweets. Next came the meat, some men sitting with an entire duck in their fists, others with large junks of mutton and beef and equally substantial pieces of bread (the remainder of Lilly's damper, however, was reserved as a special luxury for our own evening's tea) till finally they had stuffed themselves so that it was precious little that they reserved for further consumption at night in their camp.

At length we all were again mounted, but previous to this Lilly, with one of the blacks who was retained on the station as a stockrider, had ridden over to where the cattle were; and circling round an appropriate mob, was facing them in the direction of the station when we all rode up and joined him.

Marvellous was the effect of the unwonted appearance of these novel riders on the startled nerves of the cattle. On our near approach all the herd turned their heads round—some their bodies, too—to look at the ladies, then, sniffing fiercely, attempted to break past Lilly and his companion. It was only by the most desperate galloping and free use of the resounding stockwhips by Lilly and his man that the unruly animals could be kept in hand at all, and even then the close approach of any of the ladies (for neither Miss Rolleston nor Miss Campbell could be restrained by anything that Lilly could say from scampering at the top of their speed round the excited cattle), was the signal for a general stampede in the opposite direction.

The colt I was riding being comparatively unbroken, and therefore not sufficiently under control, was of so little service in enabling me to assist in managing the cattle that I could not help regretting my surrender of Selim to Mary, whilst, though evidently a good rider, Marsden was also of little service, being unprovided with a whip. He, however, had been unaccustomed to driving stock anyhow. Lilly, exasperated at the frantic efforts of the cattle to break away from him, seemed annoyed at the poor service he received from such a party of " greenhorns," as I heard him irreverently muttering. Mr. Green, however, burning to distinguish himself on this occasion in the eyes of Miss Campbell, made a gallant attempt to do something, and, spurring his horse, rode after a large red bullock, with wide spreading horns, that was threatening to break off on his side; whereupon the incensed animal suddenly wheeled short round on his pursuer, an event so unlooked for on the part of the latter that, halting abruptly in his stirrups, one of the leathers gave way from the strain of such a sudden pressure, and the gallant stockrider was precipitated over his horse's head and almost under the enraged bullock's nose, that, approaching and shaking his head menacingly, seemed as if it contemplated tossing the unfortunate youth into the air; but, made agile by terror, he sprang to his feet, and still clutching hold of the bridle that he had managed to retain in his fall, he sought protection on the other side of the horse, when the animal, as if in contempt of such a novice, took no further notice of him, and held on his original course. Lilly, roaring with laughter at Mr. Green's inglorious mishap, put a period to the bullock's farther career by a crack of his well wielded stockwhip, from the stinging effects of which the runaway was fain to seek shelter among the herd.

The appearance of Mr. Green darting for protection round his horse, was indeed so comic that all the party, including the ladies, notwithstanding the extreme danger in which he stood, joined in a loud peal of laughter, which so disconcerted the poor fellow that he attempted no more deeds of daring for the rest of the day.

Mr. Brown's efforts as a stockman were scarcely more auspicious. Being mounted on a good stock horse, but being an indifferent rider, in bravely galloping to turn in a wing of the cattle, a second laugh was raised when his horse suiting his action to the movements of the cattle, suddenly wheeled as if on a pivot, in one direction, whilst Mr. Brown, arms foremost, flew off in an opposite one, though, as it fortunately happened, without sustaining any ill effects from his fall. This occurred shortly after Mr. Green's catastrophe, and Mr. Brown retired disconcerted likewise.

But, in spite of these several mishaps, the cattle, gradually urged towards the yard, had now entered the deep belt of gum trees that from a distance, always, more or less distinctly, indicate the Darling's course; in spite of all Lilly's injunctions, however, nothing would induce Miss Rolleston to avoid pressing close on the herd. Even Miss Campbell, after witnessing Mr.

Green's rencounter with the red bullock, was inclined to keep at a respectful distance, especially as the same animal frequently wheeled round as if to threaten his pursuers. But Rachel Rolleston, with flushed cheeks, flashing eyes, and streaming ringlets, rode fearlessly on the heels of the cattle, menacing the refractory with her slender riding-whip, shouting and triumphant, but at all times closely attended by her cavalier, who never left her side for an instant.

Amid the mingled din and confusion of the lowing of cattle, cracking of whips, shouting, and a perfect cloud of dust, the cattle, now in a close body, were being driven towards the yard, whose friendly wings were seen stretching out on either side beyond us.

Until now, as my horse was but half broken, and unfitted for wearing stock, I had contented myself with keeping by Mary's side. I was moved thereto by fears for Selim, who, when stock were being driven about him, was apt to blaze up in a fever of excitement, and I wanted to be near at hand to seize his rein if he got uncontrollable under the weaker command of his inexperienced rider. But now, having got the cattle so far, when the tug of war in driving them into the yard was likely to commence in earnest, I requested the ladies who were near me, with Mary, to rein in their horses, whilst I used my best endeavours to assist in yarding the cattle. It was useless to ask Miss Rolleston, who was ahead and strenuously urging on the cattle, to remain behind, but an accident, that had almost proved fatal, was like to have made her pay dearly for her rashness.

Having such an inefficient mount for stock-driving, I had not provided myself with a stockwhip, as I deemed it would be but an incumbrance, the colt not yet being broken to the use of it. Therefore, on leaving home, I had contented myself, by way of carrying something, with the handle of a heavy hunting whip. The red bullock, by this time fairly infuriated by the clamour and the close quarters into which he was being driven, was now seen rushing through the herd towards the rear.

Lashed by Lilly, he turned aside, but still facing outwards, he was confronted by the black fellow, from whom he again received a stinging cut that only had the effect of making him incline a little further off and make a furious attempt at another point, where Marsden and his fair companion were riding and laughing. What followed only took a minute. Charging madly forwards, the red bullock made at Marsden, who, observing his purpose, had resolutely interposed his horse betwixt the threatened danger and Miss Rolleston, although, being himself unprovided with a whip, he was utterly defenceless against the bullock's charge. The animal rushed straight at him, and placing his long horns beneath the horse's belly, instantly overthrew both him and his rider; but singularly enough, without inflicting any further damage on either. Seeing Miss Rolleston's position of deadly peril, I galloped towards her, and knowing my horse was valueless for further service against the bullock, I had just time to fling myself off the saddle, and get between him and Miss Rolleston's horse, as the bullock recovering from his collision with Marsden's animal, again sprang forward with a fierce sniff. Grasping the lash end of my loaded whip with desperate energy and with both hands, and stepping quickly aside as the raging animal came blindly forwards, I dealt him a blow as he passed, with the hammer end of my whip, near the root of his horn, and sent him staggering on for several yards. It was lucky for me that the blow sickened him, as, had he renewed his attack, I was utterly defenceless. Like the Bruce at Bannockburn, I had broken off the head of my weapon by the violence of my blow.

But the bullock, seeing his way now open, made off towards the plains, passing on his way close by the place where the ladies who had stayed behind were assembled; but he only greeted them by another indignant sniff, without seeking their further acquaintance. But now another accident was like to have occurred, and that, too, on the very point that I had been so sedulously guarding against all day.

Selim, whose temperament with reference to stock driving I have already mentioned, had been for some time sorely chafing at the restraint imposed on him. He, seeing a single bullock darting past him, appeared to consider this slight upon his spirit as the last point of endurance that even his good nature could submit to, and, therefore, paying no regard to the desperate tugs on the rein made by his terrified rider, he, with a fierce snort, gave instant chase, and in a few strides had overtaken and headed the red bullock. The latter at this paused a moment, as if considering what he had to do with this new enemy, while Selim, suddenly wheeling round on the opposite tack, again confronted the bullock, which acknowledged his attention by instantly charging at him.

With a wild shriek, Jessie Campbell made us aware of the imminent peril in which she saw her sister placed.

The poor girl herself, almost beside herself with terror, dropped the reins and instinctively clutched hold of the horse's mane.

This proved her salvation.

It was not in vain that George Laycock had boasted to me of Selim's qualifications. Among many useful lessons that he had taught him was one for which Mary's involuntary action was the signal. As the bullock came rushing towards them with head lowered for the charge, Selim, on feeling Mary's hand touch his mane, wheeled like lightning, and with his two heels dealt blow upon blow full on the broad forehead of the astonished bullock, who was sent reeling half stunned on to his haunches from their effects; on recovering he seemed to consider that discretion was the better part of valour, and once more made off for the plains.

Not so, however, thought Lilly, who now arrived with his foaming steed, his long lash playing around his head till the forest echoes rang again as it fell on the bullock's flank, forcing him back again in spite of the frantic animal's desperate charges; and, ably assisted by the black fellow, Lilly eventually succeeded in securing him with the rest of the cattle in the yard.

"It is the last time I'll ever allow women to go stockriding with me," I heard him mutter as he was fastening up the rails of the stockyard with the wooden pegs that were attached by hide strings to the posts for the purpose.

CHAPTER XVII.

GREAT was our thankfulness in the providential escape of all parties in the late danger, and especially of Mary.

"Well, Miss Mary, you certainly owe Selim a debt of gratitude; this is the second time that he has been instrumental in saving your life, although in the present instance the old rascal was himself the cause of bringing you into the danger that he saved you from," was my remark to Mary after the cattle had all been yarded, as I helped her off her saddle preparatory to her getting into the canoe.

"Dear old Selim, I indeed owe him much," replied the girl, patting Selim's forehead, and he frankly acknowledged the attention by rubbing it against her hand, as if soliciting as much more of such patting as she would bestow upon him.

To Miss Rolleston I made no remark; but from her, nevertheless, I received a glance so full and even so compassionate that I felt myself amply rewarded for the service I had been so providentially enabled to render her. As for Marsden, who, as my successful rival, together with the unsatisfactory mysteriousness of his character, I could not avoid regarding with annoyance and suspicion, the manner in which he expressed his sense of my late action I could not but admire. As if conscious of my antipathy to him, he avoided all demonstrative congratulations on the subject. Neither did he show any jealousy at the signal service I had been enabled to render to the object of our mutual admiration, as, if small minded, he might have done towards a rival who had thus borne off the palm in an action in which he had received such a signal overthrow.

"Mr. Farquharson," he bluntly remarked, "you are a brave man, and I may yet have an opportunity of testifying to you how much I can estimate the spirit you have shown to-day."

"Thanks, Mr. Marsden," I replied shortly and almost sternly, "I have only done what any man would be expected to do for any lady, and especially for such a lady as Miss Rolleston."

"Right, Mr. Farquharson, right; for a lady like Miss Rolleston such a service needs no eulogy. Yet such service as you have just rendered her requires more nerve than ordinary men possess," and with this remark he abruptly turned his horse's head and regained Miss Rolleston's side, which he had just left to speak to me. This occurred as the whole party were riding down from the yard towards the river's bank.

We soon were ferried across the river, one at a time, in a canoe, a black fellow paddling. I had already, during the same operation in the morning, jokingly reminded the ladies of their former exploit, or rather misadventure, in such another craft on the Murray, and particularly cautioned Miss Rolleston against indulging a like dangerous curiosity, as on that occasion, about the motions of codfish. They laughed merrily at the recollection, Miss Rolleston promising to be more cautious in future, when voyaging in canoes. They were not nervous in venturing into a canoe again after their former narrow escape, nor was I on their account, as the present one was a particularly good canoe of the kind, being, from the natural curvature of the sheet of bark out of which it was fashioned, well turned up from the water both fore and aft, and otherwise perfectly trustworthy under the most ordinary precautions on the part of the passengers.

Mrs. Campbell turned pale when she heard the danger that had again so closely menaced her daughter and Miss Rolleston, and, although she made but few remarks on the subject, I could see by her manner that I again had risen wonderfully in her estimation for the fortunate conduct I had been enabled to display on the occasion.

"There must surely be some correspondence in the destinies of you three," she said after a while, "that you seem so mysteriously attracted to each other in the hour of peril and danger."

"Yes, mother," said Miss Campbell, laughing, "a marriage

destiny perhaps. I see it plainly: it is bridegroom, bride, and bridesmaid; so, whichever of these ladies is to be the bride, the other will be the bridesmaid." At this lively sally, Miss Rolleston coloured a little, but readily answered, "Then Mary is bound to be the bride, for Mr. Farquharson would never have me, for he knows that I am too giddy for him, and that I have set my heart on marrying some bold bushranger".

Mr. Marsden, who, while this lively dialogue was carried on, had been absorbed in thought, at the last word looked suddenly up, and, as I thought, with an almost startled expression, while he darted one of those swift uneasy glances at Miss Rolleston, that I had on my first meeting him so frequently noticed crossing and clouding his face, though, since his arrival here, they had been absent. The look was but momentary and scarcely observed, before it was replaced by that expression of careless daring, usually characteristic of him.

"A bushranger, Miss Rolleston!" he answered laughing, "surely that is a wild thought of yours. Ah! I see you have been reading about Robin Hood and Maid Marian; very picturesque, no doubt, but I fear such a life is hardly so *couleur de rose* in reality. Nay," he continued gaily, "a gentleman of fortune and rank would suit you better. Your true sphere in life—instead of that of a poor Maid Marian, catching rheumatism in some damp cave—is that of a titled lady, admired for her beauty and revered for her goodness."

"Thank you, Mr. Marsden, for the flattering prospect you open before me, but I fear it is as far beyond my deserts as are the silken restraints of high life distasteful to me; I think I shall have to come down to the bold bushranger or daring adventurer, forcing his own independent way and bursting the manacles of conventional forms that bar his path to glory."

"And mated to one like you, Miss Rolleston, to inspire him with your spirited sentiments, what would not such an adventurer dare?" replied Marsden, with a flash in his eyes and a look of pride dilating his nostrils.

"Oh come! this is mere romancing nonsense; you are up in the clouds, while we poor prosaic people are obliged to stick to the earth," interrupted Miss Campbell, rather petulantly. "We shall lose sight of you altogether if you go on at this rate much longer."

"We will come down again," replied Miss Rolleston smiling; "let us see, where were we? Oh! I remember; I was saying that Mr. Farquharson would never have me, so it is you, Mary, that the thread of destiny is to connect with Mr. Farquharson as bride, and I as bridesmaid."

"Well, I at least shall be well content with destiny for giving me such a wife as Miss Mary," was my quiet answer, with a smile at Mary. She blushed rather shyly, but her livelier sister came at once to her rescue. "Two dispositions of the same nature never get on well together; happiness can only be reached by blending two opposites—the grave with the gay. Now, Mary and Mr. Farquharson are both sober-sides. Why! they would never be able to keep one another awake: so the grave Mr. Farquharson and the gay Miss Rolleston make just the sort of blend that is wanted for married happiness."

I was beginning to feel rather awkward at these sallies of Miss Campbell's, the more so, that I suspected she had penetrated the secret of my passion for Miss Rolleston, and was now speaking by design, either from mischief or from sympathy.

To her last remark, however, Miss Rolleston retorted maliciously, "If gayness of spirits are essential for Mr. Farquharson, then, who so gay as Miss Campbell? And who more fitted to be Mr. Farquharson's help-meet?"

"Nay," replied the other, demurely, "to my sorrow, I am out of it altogether; your first acquaintance with Mr. Farquharson has given me no chance. It is plain that you are the first love."

"That's it," replied her quick-witted companion, turning the point of the other's fancied hit suddenly against herself; "but first love soon dies: it is always the second love that lasts. The love that I first awakened in Mr. Farquharson's heart has doubtless vanished now that he has had experience of my utterly impracticable disposition. At this most opportune moment, Miss Campbell, witty, affectionate, and above all, heart-free, enters upon the scene: Mr. Farquharson, disappointed in his former ideal, again feels his heart moved. This time the power that attracts him is permanent: thus I am cut out, and feel quite jealous of you. But see that you use him well, for I know that Mr. Farquharson deserves a good wife, and I believe you would make him a good one."

The laugh was now clearly against Jessie, whose rising colour was beginning to mark her confusion at the tables being thus so suddenly turned against her, when Mrs. Campbell interrupted any further badinage of this kind with

"You must certainly be losing your heads entirely, both of you. Whatever must these gentlemen think of you, when you talk so foolishly?"

"I crave the good gentlemen's indulgence," her eldest daughter replied with mock gravity, "and as for Rachel's disposal of me, I certainly ought to feel honoured, whatever Mr. Farquharson may feel on the subject."

"Nay, Miss Campbell, need you ask? but I fear there is a

young gentleman here," I said, glancing towards Mr. Green, "who may feel disposed to enter a protest against this mode of disposing of you."

"Indeed and that I will, and most strenuously," stoutly replied Mr. Green, amid a peal of merry laughter, in which even Mr. Marsden joined, though usually almost contemptuously reserved with either of the young Melbourne gentlemen.

"Oh, Mr. Green," said the laughing girl, "after the manner in which you let yourself be overthrown by the red bullock to-day, I could never think of allowing you to put in a claim for me again."

"Never mind her, Mr. Green, she is only jesting; she is very rude," said Mrs. Campbell.

"Indeed I am, mother," replied her daughter quickly, apparently fearful lest she might have wounded the young fellow's feelings too much, as he looked sadly abashed at the general laugh raised against him. "Mr. Green knows that I was only jesting, for I was very sorry to see him meet with his accident, and I know that it was through no fault of his that it occurred."

"How did you enjoy yourself, Mr. Marsden?" Mrs. Campbell now demanded of that gentleman.

"Oh, excellently, madam," he replied, rousing himself from one of those reveries into which he occasionally fell when his attention was not directly engaged by Miss Rolleston.

It seemed as if that young lady alone possessed the charm of rousing this man's genial mood. Therefore, when her attention was diverted from him, as if he considered the others unworthy of his deliberate attention, he immediately relapsed into silence, only rousing himself when, as on the present occasion, he was directly addressed by Mrs. Campbell. Then his conversation for a few minutes, though studiously polite, seemed to be as forced and unnatural as his previous silence had been oppressive.

"Excellently, madam, excellently; indeed, how could it have been otherwise with the charming companionship of so many young ladies? I could, indeed, envy the life of a stockman if this day is to be reckoned a sample of their general experience."

"Stockmen are not always so favoured I can assure you, Mr. Marsden," replied Miss Rolleston with a smile.

"Nay, I would not be so avaricious as to expect to be regaled with such a continual feast of enjoyment. A hut in the forest with any one of these fairies to brighten it with her smile could still make a stockrider's life preferable to that of the uneasy state of a king."

All this conversation—in which neither Mary nor Miss Brydone took any direct part, save in joining merrily in the girls' general laugh raised by the lively sallies of the other two—was being carried on during the discussion of an ample supper, that the day's stockriding and excitement had prepared most of us to enjoy.

Whilst the supper things were being cleared away, our attention was attracted to some wild sounds proceeding from the black fellows' camp. Upon enquiry we found that they proceeded from a grand "corroboree" that the natives were celebrating on the strength of the ample quantity of provisions they had been provided with at the store as a Christmas treat, together with such tit-bits as had been saved from the pic-nic.

At the suggestion of the ladies, we all adjourned to the scene of action, but as I feared as to whether the blacks would consider European ideas of propriety binding upon them within the limits of their own camp, I delayed the party whilst I stepped out to the kitchen to see Lilly about the matter, for ever since the advent of our Christmas visitors he seemed to have established himself there, merely going home to sleep. Doubtless the pleasure that he seemed to derive in conversing with Tiny, who was constantly passing to and fro between the house and kitchen, explained this sudden fancy of Lilly's to take his meals with old John. Communicating to Lilly the intention of the company inside and my suspicions of the unpreparedness of these primitive revellers in the matter of costume for the reception of such visitors, he at once strode away, and shortly afterwards a cooey from Lilly signalled to me that I could let the ladies come down to the camp without having their sensibilities shocked.

Upon this we all went down and witnessed one of those wild yet harmless scenes of savage revelry with which people located in the interior of Australia are so familiar, but with this exception. Instead of the wildness of the scene being enhanced by the sight of their nude bodies streaked and tatooed all over with pipeclay, as is customary on such occasions, they were now, by Lilly's peremptory orders, all covered with a shirt, only their faces and bare limbs being visibly hideous from the pipeclay as they leapt and grinned and quivered in the clear moonlight, men and women together, in the performance of some hobgoblin sort of dance. The music for this dance was an indescribable kind of chant, the variations of which seemed to be caused by the singer striking one hand on the pit of his stomach, while with the other, in the way of accompaniment, he kept beating a piece of folded 'possum rug with a club. Certainly a more rudimentary idea of music it would be impossible to imagine. Compared to this it must have been a civilised improvement when, as the song says,

> "Tubal in his oxter squeezes
> The blether of a sheep".

After surveying this wild scene for nearly an hour we all returned to the house, laughing and making merry at the ridiculous figures and extraordinary capers of these savage revellers. On their part they were so flattered and delighted at their performance being witnessed by such a distinguished audience that they appeared fairly to excel themselves, laughing and mouthing and gibbering like so many animated ghouls, and, indeed, no civilised merry-makers could ever have marked time more enthusiastically to the more refined strains of violin or castanet than did the untutored children of Nature to such primitive and truly barbarous instruments of music.

CHAPTER XVIII.

THUS closed one of the brightest Christmas Days I have known, and that still shines like a bright memory through a somewhat cloudy past.

But morning came, and I knew my fair guests must leave me to my solitary bachelor life again, and I knew how desolate I should feel when I returned after bidding them farewell and entered my house again, and should say with the poet,

> "How dark and silent are my halls!"

On the return journey to Menindie the bullock dray was dispensed with, the ladies thinking that they had had sufficient experience of such tedious means of locomotion on the day of their arrival, and preferring the more expeditious mode of riding back on horseback. But as there had been no more side-saddles taken up with them from the Murray than those used by the ladies on the previous day's picnic and stock-riding expedition, Mrs. Campbell and Tiny not having contemplated their necessity for themselves, or the idea of returning from the station to Menindie by any other mode of conveyance than that by which they had come, we, after some scheming, rigged up two saddles of sheep skin and bullock hide for the two young gentlemen, Miss Rolleston and Miss Campbell readily undertaking to mount the saddles the young gentlemen had used on the previous day. Meanwhile the black fellow had been despatched in search of the horses that had strayed from the main body, and, these being found, the black mare Mavourneen was saddled for Mrs. Campbell, while Tiny was provided with another old stager who would have scarcely been disturbed by the firing off of a cannon from his back, let alone the sight of a lady's riding habit fluttering by his side. Though, by the way, Tiny had no riding skirt, but for the sake of decorum a sheet was pinned round her dress and answered just as well, all laughing good humouredly at the figure she thus cut.

Mary, however, again rode Selim, to whom she seemed to be greatly attached, ever and anon patting his neck affectionately as they went.

On the road I observed Marsden on several occasions keenly inspecting Selim's points, then, leaving Miss Rolleston's side when he observed her engaged in an animated conversation with Miss Campbell about their mutual city friends, he rode to where I was, and, pointing towards Selim with his riding-switch, abruptly remarked, "A good horse, sir; what value would you be disposed to put upon him?"

"More, perhaps, than you would be inclined to give," was my somewhat stiff reply.

"Indeed, sir," he replied with an incipient sneer slightly curling his lip.

"What sum do you imagine it would take to balk me of any fancies that I choose to indulge in?"

"Of that of course I cannot judge, nor yet does it matter, Mr. Marsden; as it is not my intention to part with my horse upon any consideration, further parley on the subject is needless."

"Straightly answered and to the point," was his blunt, off-hand reply, and then more slowly he added, "and from witnessing your action yesterday I make no doubt that when you say and mean a thing you have sufficient determination to stick to it."

"Perhaps I have," I answered curtly, for I was always on my guard against this man; there was a mystery about him that seemed impossible for me to fathom, and until I had done so I could not trust him.

Lilly, who was riding alongside of me, and whose eye appeared to be constantly on this man when near him, here broke in with the remark, "You appear to be a good judge of a horse, Mr. Marsden; where did you pick up that one you are riding, if it is a fair question?"

"Perhaps not quite as fair as you may think, sir," replied Marsden with one of his sternly sarcastic smiles; "but in the present instance it may be answered fairly. My horse, ah! he is a good one! I bought him when in Melbourne, in the yard there."

"I didn't think that wealthy squatters were in the habit of parting with their favourite horses," replied Lilly drily.

"Indeed, you seem to be shrewd in your judgment, sir; but what has that got to do with my answer?" asked Marsden quickly.

"Well, I ought to know that horse of yours," replied Lilly in the same dry tone as before, "but I never thought that Mr. Tyson on the Murrumbidgee was so hard up as to be obliged to sell his best horse."

Quick and piercing was the frowning glance with which, for an instant, Marsden regarded Lilly, but, as instantly the strong-nerved man recovered himself. "Mr. Tyson, my dear sir, may, or may not have his scruples on these points—I have not heard of them, nor do I care about them; let it be sufficient for you to know that I purchased this horse at Tattersall's sale-yard in Melbourne, and a pretty considerable figure, too, it cost me," and with these words, uttered in a tone of stern decision, he rode forward, and joining Miss Rolleston, they both rode for some distance in advance of the party, and, still preserving this distance, for the rest of the way in close proximity, and with heads bent towards each other, they appeared to be engaged in deep and earnest conversation, until we had arrived almost within sight of our destination. Then, suddenly giving their horses the spur, they both disappeared round a turn of the road where it entered a belt of timber, and on the rest of the party doubling this point and riding on a little further, we found Miss Rolleston with a grave and almost sad countenance, waiting alone.

"Why, Rachel," exclaimed Miss Campbell, laughing, "here you are, are you, why, I thought that you and Mr. Marsden had run away with each other. Wherever has he got to now?"

"Mr. Marsden," replied the other, "has pushed on."

"Gone away?" cried several in a breath. "What a strange thing to do, without saying a single 'good-bye' to any-one!"

"It is his way," replied Miss Rolleston, "he says he hates having to go through the formula of leave-taking at any time; it does not suit with his blunt way of doing things."

"Well, I never," remarked Miss Campbell. "Where is he off to now? but as he intends purchasing that back country, I suppose we shall soon see him again at the Murray."

"Yes, he told me he was obliged now to ride to see a person who was expecting him at Wentworth; but there, please question me no further, for I don't feel disposed to give any more answers on the subject."

As we were now within about a mile of Menindie, there was

nothing more said about the matter. Miss Rolleston's grave look continued as she rode on in thoughtful silence, no one liking to inquire into the cause of it, though we naturally laid it down to Marsden's departure, for his influence over her had been apparent to us all.

Lilly, on hearing Miss Rolleston's account of Marsden's sudden leave-taking, made no audible remark, but I observed a bitter, meaning smile pass across the fellow's keen face. As he often remarked, he could see through a stone wall as far as most people, and with reference to the stone wall of Marsden's real character, I too soon had ample reason to verify Lilly's claim to this faculty.

On arriving at Menindie, there was but little time to spend in further conversation, as Captain Caddell, with his little steamer, was impatiently waiting for his passengers.

On shaking hands with Jessie Campbell, and looking into her open countenance, where no trace of a shadow of care could be seen lurking in the pellucid depths of her frank blue eyes, I could not help thinking of the badinage between her and her friend on the previous evening, and of the happy random shot, by which she had been so completely baffled by the latter. Viewing her thus, a strange thought for the first time crossed my mind as to the bare possibility of the fulfilment of the prophecy; and in speculating on the idea I discovered myself deliberately analysing my feelings, with reference to such a possible event. Further on, however, the reader will know what the result of the analysis even then seemed to show. Meanwhile, however, the commotion in my heart on Miss Rolleston's account was such as occasioned me to smile sadly at the idea. But, shaking hands cordially with her, I bade her farewell, with the hope of our meeting again before long.

With all the others I parted on terms of the highest esteem; but, in doing so with Miss Rolleston, who still looked sad and thoughtful, I could not repress a feeling of dismal foreboding.

Poor Rachel Rolleston, so young and so beautiful, and with so much of genuine worth, weakened it is true, by romance and enthusiasm, so free from vanity, yet with so many foibles, that had the effect of weakening her judgment! Poor Rachel Rolleston! Why was it your fate that your bright life of innocent enjoyment should so soon be eclipsed by the clouds of darkness and despair?

CHAPTER XIX.

ON the departure of the boat, Lilly undertook to drive the horses that had been used by our visitors back before him, whilst I turned Selim's head down the road for Wentworth, beyond which, on a station a few miles further down the banks of the Murray, there were some business matters that required my presence.

The fourth day after I had left Menindie was far advanced when I again entered Wentworth on my return journey, and I learned that the steamer, that seemed for some cause to have been delayed on its passage down the Darling, had only a few hours before passed by. From here it had again turned up the Murray, whither, for business purposes, it was bound before returning down stream to Adelaide.

That I regretted having missed this opportunity of again meeting my kind friends, by such a narrow shave, I need not say; but there was no help for it now, and I recollected that in the decrees of fate "a miss is as good as a mile". Instead, however, of turning my horse's head directly up the Darling road from Wentworth, crossing there on the punt, I rode on with a young man named Williams, with whom I had foregathered on the road on the further side of Wentworth, to a cattle station owned by him and some brothers, and situated a few miles further on up the Murray. My object in this was to inspect a young bull he had been telling me he wished to dispose of, and that from his description I thought might be suitable for Lilly.

By the time that we had reached Mr. Williams' station and duly inspected this bull, that seemed satisfactory enough, but whose purchase I deferred until I had had Lilly's own opinion on the matter, the setting sun had already begun to crimson the horizon of the western sky. Therefore, putting Selim into the stable to hearten him with a good corn feed, I followed my new friends, at their pressing invitation, to join in their substantial, but plain supper of bush fare, though their urgent request that I would pass the night with them I was obliged to decline, as I thought it better to ride on to Mr. Fletcher's station, situated about fourteen miles above the Darling from Wentworth. This course I thought the wisest, because it enabled me to better equalise the next three days' stage homeward, whilst, as the night was looking fine, and there would be a clear moon by-and-bye, this distance could be easily accomplished in two or three hours by Selim, who had only travelled a few miles altogether, that day.

Being a good horseman, and finding on inquiry that it was possible to ride straight through the scrub and bush to Fletcher's, I determined to go that way, thereby shortening my journey by four or five miles.

With this determination, I waited where I was till the moon began to rise; then, mounting Selim, and bidding all the brothers a hearty good-bye, without the least concern about road or guide, save that afforded by the moon, I set off on my night journey.

Though the country along my route was occasionally traversed by belts of mallee scrub, through which I had to force my way, and also occasionally by swamps, through some of which my horse had to wade, yet the directions I had received from my late hosts, who of course were conversant with every inch of the ground, had been so explicit that I was able to avoid all the worst places and to take advantage of what open forest and plain country there was along my route.

Thus I pursued my way with a light and confident heart— now through the open country stretching away at Selim's luxurious canter, at other times varying the pleasure by the slower, but equally enjoyable, sensation of his long, easy, walking stride.

The moon was shining so calm and clear, that my way was considerably lightened with its beams, and Selim, snorting in the pride of his power, and with pricked ears, was stepping out freely at a walking pace. The night was mild and balmy.

Reader, have you ever experienced the exhilaration of spirits that a ride on such an animal, and at such an hour, can give? If you have, and if you are naturally fond of a horse, you then will bear me out when I say, that such a feeling is the most bracing and manly sense of enjoyment that can be well conceived. This feeling was at that particular time greatly enhanced with me by the character of my surroundings: their natural wild and solitary appearances rendered more so by the transforming effect that moonlight always exercises on a landscape, while the weird howls of those denizens of the scrub, the Australian dingoes, could be heard all around in dismal harmony with the time, producing an effect that might well have affected the nerves of a traveller unused to such scenery. Not so, however, those of a bushman; in the ears of squatters, overseers, or shepherds the howls of these destructive pests create no other sensation than those of vengeance and the desire of being provided with strychnine-dosed bait, to scatter along his road for the especial benefit of the dingo with his keen scent.

Meanwhile I need hardly say that in the course of this ride my thoughts were chiefly occupied with those in whose society

I had spent lately a period of such felicity. Yes, and they were proud thoughts, too ; for healthy young blood was flowing in my veins, and, as its natural consequence, hope was buoyant in my heart ; for, with the proud consciousness of success crowning my efforts in my present sphere, prospects of a prosperous future seemed even then to loom large before me, in spite of my feelings of bitterness at my vain homage to my ideal love.

Such were my musings, when suddenly and rudely their course was interrupted.

I had just emerged from one of those scrub belts that occasionally traversed my path, into a patch of open country, when my ears were greeted with the deep, savage growl that proceeds only from the watch-dog, be it bull, mastiff, or retriever, but which I heard instantly silenced by a stern, quick " Hush ".

I had now struck the point of one of the swamps, or lagoons, that were thickly surrounded by lignums (*polygonum* is, I believe, their botanical name), a green succulent bush, in appearance at first sight not unlike Cape broom, though not of such a firm, woody texture as the latter. These bushes being rather thickly strewn over the open ground, I had not as yet seen the shepherd's hut or tent, that from the growl of the dog and the accompanying " hush " I doubted not was near. But I had already seen a further sign of the correctness of my surmise that some sort of party were encamped here—three horses that neighed on seeing mine—all securely tethered ; one of these, a rather fine-looking black animal, and the nearest to me, I was just in the act of glancing at, in passing, when my further progress was suddenly arrested by the stern command to " Bail up ".

Though, from my familiarity with Colonial annals, well acquainted with the terrible significance of this pithy order when applied to anyone in my lonely condition, in all the course of my wanderings, that had covered some of the most exciting years of Colonial history, I had happily never been stopped in this way before.

I now, in my surprise on hearing such a command, perceived what till then I had failed to notice, standing in the shadow cast by the reflection of the moon, by one of the largest lignum bushes, two men, one of whom was armed with a gun, that he now held levelled at my head, whilst he called out again in a brutal and savage tone of voice, " Up with your —— hands, or I'll send a bullet through your —— brains ".

I believe that at bottom and with time given me to collect my thoughts, I am master of as much courage as most men, but I will not attempt to deny that, thus taken by surprise, the thought of sudden death staring me straight in the face sent the blood tumultuously to my heart. I obeyed the order mechanically by holding up both hands. In my right I had been carrying a loaded hunting whip, that without dropping I still held, which singularly enough my "stickers" up did not appear to observe. This oversight evidently arose from a circumstance that from what followed I soon perceived, viz., that at the time the men were considerably muddled with drink.

Both men now approached my horse, the armed man still keeping his gun in its levelled position, now dangerously close to me.

"Get off your —— horse," was the next command, delivered in the same brutal voice as before.

Somehow this command, which showed me that they meant to take Selim from me, seemed to nerve me up to the determination of making a desperate effort to preserve my grand old horse in a way that the mere thought of danger to my own life or purse would have failed to arouse. Barring my life, Selim was indeed the most valuable property that they could have despoiled me of, as I believe I had only £20 in money about me in coin and bank-notes, whilst I should have considered the loss of five times that sum as cheap compared with that of my brave horse. The manner, too, in which the bushrangers comported themselves also favoured my design of resistance, for, instead of the armed man keeping his position, with his gun covering me until his companion had securely bound me, they both swaggered up to seize me on my dismounting. This action, and a movement that I immediately made of a rapid inclination of my body to one side, brought me to one side of the line of the gun. Ere the bushranger could rectify it I managed by a sudden spring forward to get inside of his weapon's reach, which I at the same time struck to one side with my whip handle. The oscillation thus given to it, by bringing an increased pressure to bear on the trigger from the fellow's finger, at once caused it to explode. At the same time as this occurred, with the butt-end of my whip handle I struck the unarmed man over the head, not with sufficient force to kill or even to stun him, as he was too close to me, and the blow too hurried, to admit of its being fetched with a sufficient force, but still with enough to cause him to be sent reeling backwards to the ground. I then repeated the blow violently at the second man, but he, with the barrel of his gun held in both hands, managed to avoid it. Fearing for the second barrel, I now let my whip drop, and seizing the gun with both hands endeavoured to wrench it from him. This, however, I saw at once I should be unable to do as my antagonist was a big, burly-framed man, so instantly closing with him and giving him a back-heel, I sent him over on his back, I falling on the top of him. Seizing him now by the throat, I attempted by strangling to compel him to let me gain possession of his gun. This I might eventually have accomplished but for the assistance of his companion, who, recovering from his fall and the half stunning effect of the blow I had dealt him, now rushed forward, and seizing me by the collar dragged me off my antagonist, although my fingers still held their grip of the miscreant's windpipe with bull-dog tenacity, he calling hoarsely to the other to get a stick and knock my brains out. That such would quickly have been my fate there is no possible doubt, but at that moment I became aware of the approach of the quick trampling sound occasioned by a galloping horse, whose rider's voice I heard almost at the same instant shouting out in tones of stern reproof:—

"What hell's game is this that you are up to now, you blundering blockheads? Are you mad, or drunk, or both? What do you mean by such work as this here?"

At the stranger's stern, commanding voice my assailants at once quitted their hold, when I sprang to my feet, and in the new-comer who had thus providentially arrived in the nick of time to save my life, and who was now on his feet and holding the bridle of his dapple grey horse, I at once recognised Marsden.

I might, under other circumstances, have been astounded at the sudden unveiling of the true character of this man of mystery, so wondrously in keeping with the accuracy of Lilly's suspicions, that had from the first pointed to such a probability; but at that time I was too excited for the sensation of any such emotion, and, perhaps my mind, by these very suspicions of Lilly's, together with my own vague uneasy thoughts about the uncertainty of this man's character, had so paved the way in my mind for the reception of the truth, that when it did flash full upon me, I received it as a matter of course.

Without, however, appearing to notice who I was, Marsden continued his stern interrogatories to my two assailants, who now with scowling countenances confronted him. Their features I had no chance of identifying, as their heads were closely cropped, and, with the exception of their moustaches, their faces were free from beard and whiskers and were, besides, smudged with charcoal, not very completely, to be sure, yet enough to make identification difficult.

"You two cut-throat devils," said Marsden, now addressing them fiercely, laying his hand at the same time on his revolver, "it is well for you both that you had not accomplished your purpose ere I came up, or I swear to both of you that you would each have got the contents of one of the chambers of this revolver" (he half drew it from his belt as he spoke). "What have you to say for yourselves? You, Morgan, you are always scheming about some devilment or other; why don't you speak?"

"I dunno, boss," said the burly villain who had been armed with the gun, "but I think, now we're on the road, it is right to pick up anything as falls in our way."

"To pick up anything that falls in your way! Do you call murder a thing that falls in your way to pick up? You know me, sir, and that what I intend doing I'll do clean; but when you want to start throat-cutting, you'll start on your own account, and not do it in my company. Anyhow, you knew that it did not suit my plans to do anything in this neighbourhood that would rouse the suspicions of the police as to any such characters as we are being about here. You knew these instructions, didn't you; but who can put any dependence on swill hogs like you?"

"Well, the cove came riding on to our camp and we had to bail him up, or he would have peached on us; and then he brought the mess he was like to have got into on hisself, for instead of bailing up quiet like, he fought, and was like to have killed Wilson with his whip, and he got me down and would have choked me had it not a been for Wilson getting up, and collaring him as he did."

"Ay! he may thank his pluck and spirit that he is not lying in that water-hole by this time. You would have treated him kindly, would you not, if he had given up quiet? I know you, Morgan, and have heard of what you have done before now; it would not be the first nor yet the second cold-blooded murder that I know you have been guilty of, under like circumstances. Now, hear me. I have told you already, that for your idle time here, I will pay you out of my own share of spoil; but, let me catch you at any slippery throat-cutting tricks again, and I can tell you that I will spare the hangman some trouble with both of you. If he is not able to lay his hand on you when he would like to, it won't take me long to do it. Now you had better go back to the camp, both of you."

At this command, with which he concluded, which was rendered all the more imperative by his sternness of tone, the two ruffians literally slunk, with the most hang-dog look, away back to their tent, evidently relieved to get away from their chief, and thoroughly cowed by his manner.

CHAPTER XX.

ON their withdrawal, Marsden first cast a keen glance in the direction whence the dog still continued to bark, and where I suppose, the camp, tent, or hut, was situated, though as yet screened from my view by the intervening bushes; then, mounting his horse, he briefly ordered me to do the same.

Feeling assured by his late conduct, that no foul play was intended on his part, I instantly complied with his order, first however, regaining my whip, and rode along by his side. For a few minutes he led the way, as if meaning to go towards Wentworth, then, suddenly turning round, and looking me sternly in the face, he enquired what errand had brought me there at that time of night. I told him as briefly as I could.

He looked keenly and suspiciously into my face as if doubting the truth of my account of myself.

"And your friends," said he significantly, "have you seen them to-day?"

I told him that I had not as the steamer had passed several hours ere I had reached Wentworth.

He then pulled up his horse and again searched my countenance with his swift scrutinising glance, and then, as if satisfied, I saw a peculiar smile pass across his face, whilst his stern, suspicious manner now wholly gave place to a more sarcastic expression, such as I had frequently observed on his features, and that so well became their strongly cast expression.

"Well, sir," said he in that tone, "you will now perhaps be able to realise the possible value that would be able to balk me of any fancy of mine that I lately hinted at, when expressing a desire to purchase that noble animal of yours."

"I realise it now only too plainly, Mr. Marsden," I replied; involuntarily addressing the bushranger by his former style of gentleman; "and also, that the horse is now within reach of your own terms," I added glancing at his revolver.

"Nay, sir, let your fears be at rest on that head, it is no intention of mine now to follow up my advantage over you; this I have less desire now to do on account of the opportunity it gives me of marking my sense of the brave action I saw you perform on the day of the cattle mustering, when you saved yon lovely girl's life from the charge of that ferocious bullock."

Then again resuming his tone of sarcasm he added—"And now, sir, might I enquire what your opinion is of the sudden transformation you find in me? That there was something about me mysterious and unsatisfactory in your mind your chilling reserve sufficiently indicated; as for your lynx-eyed assistant, whose eyes could not keep off me, I was such a source of interest to him, that he thought me some kind of black sheep of whose particular breed he might not be quite assured was plain enough, when under my assumption of the wealthy squatter, he had the audacity to attempt to scent out the common horse-stealer.

"And now, sir, having thus so unexpectedly unravelled the mystery that has evidently been exercising your brain on my account, and having now found but a common bushranger after all, may I ask if it is your intention (in return for the Quixotic idea of humanity I just now displayed in saving you from the clutches of yonder desperadoes) to have such a malefactor as I am delivered over to justice by giving timely notice to the police of my whereabouts?"

"Mr. Marsden," I replied firmly, "I am at present in your power, the power that a loaded revolver confers on a man able and willing to use it, over another armed with only a loaded whip handle, and so you can at your will prevent me from going where I can alarm any one, at least for the present. Yet, I assure you, were you now to release me to go where I wish, this much at least, in view of your evident influence over the affections of Miss Rolleston I should deem it my duty to do, and that is, to instantly write to her father my knowledge of your true character."

"Nay, sir, that would be but chivalrous on your part, and your action would, I am assured, be the more exalted from the entire absence of any such grosser motive as self-interest in prompting it, in your entire unconsciousness in connection with it of the grand opportunity such an action would afford you of getting a dangerous rival in your affections out of your way. Yet still, for thus permitting you to return by no rougher means than the common bushranging expedient that my friends yonder are so thoroughly familiar with—that is, tying you hand and foot to a tree and then leaving you on the slender chance of a passing traveller hearing your cooey, unless we choose to send back and release you—I say, for thus letting you go scot free, I should require only your word, as a gentleman, that you would allow twenty-four hours to elapse ere you communicate to any one what you have seen to-night.

"As for Mr. Rolleston," said he, again resuming his sarcastic and almost mocking tone, "well, I presume the path of your duty is also that of honour. So, as long as the condition I have stipulated for is not infringed, by all means write and unveil my lawless character to the poor old gentleman. Be sure you put it on strong so as the more effectually to put him on his guard against any more daring advances that I might still persist in making towards his lovely daughter.

"As it happens that the mail leaves Wentworth for Swan Hill, up the Murray, to-morrow, your letter will not be delayed any time on the road; and delays, my dear sir, are proverbially dangerous, you know.

"Now, sir, what think you of me by this time?" he added suddenly, at the same time scrutinising my face narrowly to gather from it some indications of my feelings towards him or the influence upon me of his tone of sarcastic banter.

"I know not what to think of you, sir," I answered gravely. "Outwardly you appear a man fashioned as I am myself, whilst judging by your conversation, you seem to be inwardly not unlike me, but moved with similar motions of right and wrong. You also know the certain consequences and ignominious punishment that must eventually overtake a lawless course such as you are now pursuing, and also the degradation of such a course in itself. Yet why, so unlike me, you can prefer leading such a life, and apparently glory in it, when with your education as a gentleman one would imagine that, whatever first occasioned your taking to it, you would at once make an effort to flee from it, I cannot understand. Most assuredly were such evil to befall me I should attempt to remedy it. Rather than go hand in hand with such villains as you now associate with, and rather than permit myself to be cheated into reconciliation with such a life by its false notions of glory, from the constant experience of hazard and escape associated with a life of crime, I should prefer the life of a hermit in the wilderness and to subsist on herbs. This much of myself I can say; but of you I can only say, Mr. Marsden, that now having discovered the true nature of your pursuits, you are a greater puzzle to me than ever."

My words, if plain, were also feeling; for whatever had been the nature of the original crime that had forced him into his present state of lawlessness, I felt that in his nature there were still some magnanimous points, that I fain would have appealed to. Therefore, by stripping his mind, as I sought to do, of the false gloss with which he might endeavour to justify the life he was leading, in his own eyes, and thereby placing it in its plain revolting nakedness before him, I tried to move his better nature to a feeling of shame at his deeds, and to incite in him a desire to return to the path of honour that he had abandoned.

He looked at me steadily while I was speaking. Whether my words touched a better chord in his nature or not, I do not know; yet he certainly did not appear to think any the worse of me for my plainness of speech.

"Mr. Farquharson," he replied, "you speak bluntly, and, as

I believe you mean what you say, I respect you all the more for it; but, as for the drift of your words, which I am not so blind as to miss, in one sense, it is useless for the present, in another, it is what I am bent on accomplishing—to wit—my restoration to my proper position in society. But you and I have been formed in such entirely different moulds, that we must naturally view a matter like this from an entirely different standpoint and come to different conclusions. You are calm and steady, and content to plod away in your obscure groove as manager of a sheep station; you are, in fact, content to serve; I am fiery by temperament and naturally ambitious, and would calmly submit to subordination to no man. Therefore, what, viewed from your standpoint of conventional sobriety, appears to be a life of degradation, to me, with a natural antipathy to all conventional notions, is simply a life of daring independence. You referred to going hand in hand in crime with yon brace of ruffians, but you must not think that such companionship as theirs is what I voluntarily choose; were it so, your terms of contempt would be but fittingly applied in my case; but, these men I but use as tools. Could I get tools of a better class, I would at once fling these from me; but none better being available, I simply have to make the best use I can of such as I can get, keeping a sharp look-out meanwhile, not to make the consequences of my work dangerous to myself, by the use of such bad instruments.

"And of myself, let me say only this now, that circumstances have combined to force Philip Marsden to lead the life of a hunted felon as you now see him, for such a life was none of his own deliberate choosing; but the legal meshes that would now confine him to a felon's life will yet, by the force of his own will, be burst asunder. This career I will yet forsake, though not," said he scornfully, "for the hermit's life and herbal fare. Such a prospect might well agree with the milk and water spirit of some spiritless cur—yet it is strange," said he checking himself, "you are no spiritless cur, yet you can calmly brook the idea of a man of daring adapting himself to a lot in life that would better befit the craven soul of some canting methodist."

"Why should you consider courage to be a quality antagonistic to a life of sobriety and social order?"

"Don't I see the proofs of it everywhere, in the constant checks imposed by society and local polity? where are the road and trespass boards, and game laws, by which the free motions of decorous citizens are schooled and held bound like those of so many children? Your decorous, law-abiding man sees the board that warns off trespassers, and meekly goes two or three miles out of his way; whilst the man of spirit leaps the wall, and goes straight to his goal by the straightest route that leads there, in scorn of the legal menace—the decorous man goes to church on Sunday in superstitious fear of future damnation, the man of daring chooses to think for himself, and uses Sunday —like any other day—for his own supreme pleasure, and scoffs at superstitious consequences. What does all this argue, if not an inferior spirit and manliness in your law-abiding, canting hypocrite, to that of him who can boldly dare to place himself beyond the pale of all law, whether human or divine, in the regulation of his conduct?"

"Sir, having been a soldier you ought to be better up in proofs on this subject than you seem to be; for you must know, that the soldier who is most amenable to discipline in the camp, is the most reliable in action on the field."

"A statement I utterly dispute; the reckless, dare-devil, happy-go-lucky man, who concerns himself with nothing about this world, and the imaginary one to come—except indeed to make the most of the one he is in—is far less likely to shirk the bayonet of an enemy, than your psalm-singing, strait-laced, crawling hypocrite, who is continually praying to the Lord to bring him out of the battle with a whole skin."

"The facts of history completely belie your assertion; as, witness the actions of Gustavus Adolphus, and of Cromwell's ironsides; what soldiers ever transcended the deeds that they accomplished?"

"A set of fanatics, they are no rule to go by; as their superstition was wrought up into a frenzy by the hypocritical cunning of their leaders—just as that of the Saracens was by Mahomet, who accomplished even greater feats than did Cromwell's bigoted, canting psalm-singers."

"Mr. Marsden, by reading history a little more impartially, you may find that Cromwell was much less of a fanatic, and a great deal more of a genuine man than he has generally had the credit for being; but leaving him alone, take such a case as Havelock in our own time: what say you to the example of fortitude and gallantry that that Christian warrior, and his Christian troops, so lately gave to the world?"

"Havelock I knew personally," he replied thoughtfully, "and his courage as a soldier, and ability as an officer, were indeed undoubted; but, however, this discussion is getting too metaphysical and abstract for my taste. I will now leave you. You will find Mr. Fletcher's about a mile further on (we had been moving on during the chief part of this dialogue). Your honour and love of fair play will both engage you to observe the terms that I have stipulated for, to enable me to get a fair start from here, ere my friends the police are on my track; but I only ask for that brief term to place myself beyond all fears of their interruption. As I strike camp immediately on my return, parting now, my best desire is that you and I may never cross each other's paths again. Your sympathy is not likely to be with me in my present course, nor do I desire it, being quite willing to take all responsibility for my actions upon my own shoulders. I have your word on those terms? that is well. And now, good-bye, sir."

Uttering these words, and without observing the conventional, but British usage, of emphasising his parting words with the offer of his hand, he abruptly turned his horse's head and galloped back towards his camp.

The inmates of Mr. Fletcher's house were on the point of retiring for the night when I arrived; for it was then considerably past eleven o'clock.

Mr. Fletcher, an unmarried man, received me hospitably; but after some light refreshment, I felt so little inclined—owing to my late adventure—for general conversation, that, pleading fatigue, I hastily retired to the bed assigned to me. This was not indeed for the sake of sleep, which I knew I should seek in vain, but for the opportunity I desired for quiet meditation on the night's events, and the nature of the engagement I had entered into with Marsden, together with the possible suspicion that such an engagement might expose me to in the eyes of the police. However, on this score, I troubled myself but little, as under the circumstances I could not in honour have done anything else.

But a greater weight oppressed me, and that was as to the advisability of at once riding to Mr. Rolleston's station, to put them on their guard against Marsden, lest he, reaching there before the arrival of my letter, might contrive to induce Rachel, by stratagem or force, to accompany him.

However, I recollected that the mail would reach Mr. Rolleston's station within a few hours of my arrival, as Mr. Rolleston's place could not be less than sixty miles distant from where I then was, and the idea of such a daring abduction occurring in the short interval between now and when my letter would reach there on the ensuing day was too improbable to be entertained. Yet there was something in the sarcastic and knowing expression in Marsden's face when he assented to my determination to write to Mr. Rolleston that disquieted me not a little. Apparently the best I could do in my peculiar position was to write two letters, one to Mrs. Campbell and the other to Mr. Rolleston. Thus, in the event of the latter being from home at post time, Mrs. Campbell would be in a position to put Miss Rolleston on her guard against Marsden. I disliked, however, the idea of writing on such a subject to Miss Rolleston herself.

I dozed off towards morning, but owing to my anxiety, I was awake again and up by six o'clock.

The next morning I asked and obtained permission of Mr. Fletcher for one of his men to ride with my letters to Wentworth to be in time for the mail. On this head, therefore, I was made easy.

On my return to our station, my first action was to make Lilly acquainted with all the incidents of my adventure. Though naturally surprised at the sudden discovery, he was indeed but little astonished at this verification of his own suspicions with regard to Marsden's true character, whilst he listened with strong interest, not unmixed with admiration, at Marsden's resolute manner with my would-be murderers and subsequent conversation with myself; for Lilly was an admirer of force of character even in a bushranger.

"It was lucky," he remarked, "that you happened to drop upon him as you did, before he had time to get the young lady away with him; and mark my words, it was only for that purpose that the daring villain came up here; and he was skulking about down there watching for a favourable opportunity, either to get Miss Rolleston to go with him quietly, or else to carry her off with him by force. But now she will be safe anyhow, for since you have discovered him, he will see his game is up in that direction, and so clear out of that neighbourhood, while your letter will put the strong-headed girl on her guard against him for the future. There was a wildness about that fellow's look that made me certain he was not *jonick*, and I just put the black fellows up to keep their eyes on the horses, and if he had tried any capers with them, I should have nabbed him at once. I knew that was Tyson's horse he was riding, and I guessed as he had just helped himself to him, for Tyson refused a lot of money for that horse, I heard when I was over that way last year, looking out for a young bull he wanted to sell."

CHAPTER XXI.

ALTHOUGH I had exacted a promise from Mary to write to me occasionally, I was certainly surprised when, in a fortnight's time after reaching home, on sending for the fortnightly mail to Menindie, I received a letter from her. This Menindie, by the way, was a small township whose only pretention to that distinction at that time, was a modern hotel of sun-dried bricks, and another smaller building—a wooden one, if I remember rightly—in which stores were retailed.

I at first thought it might be in reference to my letter to Mrs. Campbell, and this was partly true, but I quickly found it charged with tidings of a far more startling nature than even my worst fears would have dared to anticipate, for the villain Marsden had actually succeeded in abducting Rachel Rolleston; but I will let Mary give the information in her own words.

"Dear Mr. Farquharson,—

"When I promised to write to you I had no idea that such a sad necessity would oblige me to take up my pen in fulfilment of that promise so soon, as that which impels me to make you acquainted with the sad event that has just befallen poor dear Rachel. Both mother and Mr. Rolleston received your kind warning letters revealing that vile Marsden's true character, though he certainly was very kind in saving your life, as you mentioned, and for which both mother and Jessie and I are very thankful. But, dear friend, your warning came too late. Poor Rachel had gone away with him two days before we got your letter, and, indeed, we are surprised that you did not see her that night at Marsden's camp; but perhaps he was keeping her in some other place then. But let me tell you all that I know of this strange event.

"You remember how he and Rachel rode on together just before we reached Menindie the day that we left, and how, when we came on afterwards, we found her alone and looking sad? I believe that it must have been then that that wicked man had succeeded in persuading the poor, foolish girl to consent to go away with him. All the way down—and we were delayed for more than a day through the steamer getting fixed on a snag—Rachel looked very dull, and she just seemed to get angry with us if we tried to find out what was the matter with her.

"Well, when we reached Wentworth who should be there but that Marsden again, and he was leading a beautiful black horse with a side-saddle, and said he had come to take Miss Rolleston home, as it would be much nicer for her to ride than to be kept cramped up so long in that little steamer. Whether Rachel really knew at that time what that man's true character was or not, of course I do not know, but I hope she did not, for although the poor dear girl had always some romantic notions of having a bold bushranger for a lover, yet I don't think she could have been so deceitful as to have kept from our knowledge such a deliberate purpose of going away from all her friends in that strange manner, with us all about her all the long time we were going down the river. But mother thinks that this man had perhaps only partially revealed his character to her as well as his desire some day to get her to go along with him, and it was this that made her look so sad, and that he had really deceived her as to his intentions when he persuaded her to mount and ride away home with him, and that afterwards, he then having her in his power, when they got out of sight of the people and houses he compelled her to go with him. I, too, think this was very likely the way in which he accomplished his purpose.

"We were all in a state of great consternation on arriving at the station when we could hear no tidings of her, and so we continued till your letter put the certainty of the step she had taken beyond a doubt. Mr. Rolleston was like to go out of his mind, and he immediately sent word for the police, and he has been away with them ever since. As for the rest of us, we have been all distracted with sorrow over our dear girl's dreadful misfortune; however, we can do no good now, but pray that the Lord may help her wherever she is.

"Now, dear friend, whilst I know well that you too will be greatly afflicted at the news about Rachel, I have some other strange news to tell you. Mother has received a letter from some anonymous friend enclosing a bank draft for £2000, and in the letter the writer recommends father to come over to New Zealand, where, near some lake, whose name I now forget, but which is in a province called Southland, there is some good unoccupied land, well adapted for sheep, that could at present be easily obtained. And father is so pleased at the idea of having a run of his own that he says he will go at once. Mother is greatly excited about the contents of this letter, for although the writing is in a strange hand, she is sure that it must have come from her dear, long lost brother Malcolm, whom you may remember we once told you had disappeared so mysteriously many years ago, before I was born.

"And so, dear Mr. Farquharson, we are all going away next month, and I feel so sorry, and so do mother and Jessie, that we may perhaps never be able to see you again. And I shall never have the chance of patting dear Selim again, tho' he has saved my life twice. But I hope to write to you and tell you what sort of a country New Zealand is. I am afraid I shall not like it, for I am told it is very cold there. We are taking Tiny with us, at her own desire, as her parents are settled somewhere in Southland, and she wants to go over and be near them.

"And now, dear Mr. Farquharson, we all join in wishing you good-bye, but with the hope that sometime hence we may meet you over in New Zealand. We also particularly desire you to give all our kind loving remembrances to Lilly.

"And ever believe me to be,

"Your affectionate friend,

"MARY CAMPBELL."

I believe that the contents of the latter part of this startling letter, though I read it all mechanically at the time, I only understood some hours afterwards, after a careful reperusal, so much was I confounded by what I had read at the beginning. But on re-reading where it mentioned about Marsden taking Miss Rolleston away on a *black horse*, there flashed through my disturbed mind three visions, viz., the black handsome animal I had noticed before being ordered to "bail up"; Marsden's keen glance towards the direction of his camp as he ordered me to mount my horse and come away from where we had been standing; and the peculiar, sarcastic, expression of his countenance, when recommending me to write at once to Mr. Rolleston, as delays were proverbially dangerous.

"Good heavens!" I ejaculated, "can it be possible that I was then within a hundred yards of that unfortunate girl, for whose sake I would have gladly gone a hundred miles and laid down my life to save her from dishonour?" I was on the point of starting to my feet and hurrying across the river with my dismal tidings to Lilly, so as to obtain some relief by giving vent in words to my thoughts that almost maddened me, when Lilly himself chanced to enter the house. Seeing my perturbed countenance and the open letter still in my hand, he inquired,

"You seem put out with that letter, Mr. Farquharson. Nothing gone wrong, I hope?"

I looked at him for a moment ere I was able to rally my scattered senses, and then replied,

"That villain Marsden has accomplished his object after all!"

"What's that?" cried Lilly excitedly, "what's that? What do you mean?"

"I mean simply that in this letter Mary Campbell says that he has persuaded Rachel Rolleston to go away with him."

Lilly responded with a tremendous oath, "The murdering villain! Oh! but that I could have guessed that he would have tried that game on, either he or I should never have left that scrub alive, that last day by ——! How did he get her away?"

I told him, as also the discovery I had just made, that she must have been in Marsden's camp and within a hundred yards of where I had been standing, when I was attacked. Nay, very likely might have wondered what the noise of the struggle—that she must have heard—was about.

"And to think, Lilly, that I was so near her, and not to know it, whilst talking to that polished villain."

"Yes, it is bad enough to think of; yet, after all, what could you have done against him with his revolver, if you had known?"

"What could I have done, Lilly? I could have hurled my whip handle in his face."

"And he would have shot you like a crow, as he did the black fellow in the scrub. Yon man is no blunderer with a revolver, you may be sure. So, after all, although the thought is bad enough, that you have been so near her and not known it, when you might have tried to save her, yet it is better as it is; you would only have been shot, and that wouldn't have made matters any better for her, but only worse, if she were to know that you had lost your life in trying to save her. But what could have possessed that girl to go away with a man like that, and chuck away all thoughts of home and respectability at the same time, when she could have made—with her good looks and money—one of the best matches in the colony?"

"I am afraid with all her fine qualities, Lilly, that Miss Rolleston was flighty, and I believe that even her knowledge of the fact of Marsden being a bushranger would only have been an added incentive to the self-willed girl, to cast in her lot with him, if fully persuaded in her own mind that the man really loved her."

"Flighty or not," replied Lilly warmly, "a sweeter or a better-hearted girl than Rachel Rolleston there is not in all the colony, and a lady too at that. But of pride she had none, perhaps not enough for her own good for that matter; many's the poor, hard-up man she has given five shillings, and ten shillings, and a pound, to help him along his weary way; and she was never such a proud lady as to ride past a poor swagger without giving him a kind smile and a tender word; and many's the poor traveller who has blessed her in his heart for the same bright smile and kind welcome that was always there for him, if she happened to be at the door when he came up; it was always the same word with her, no matter what the time of day, 'You must be tired, stop where you are and rest for the day'. And if ever I come across that unprincipled, two-faced villain, who managed to get round the trustful nature of that sweet girl, and seduced her into leaving her pleasant home to go along with him, and if he has injured her honour by so much as the breath of reproach, it will be my life or his. I swear it by ——."

Lilly uttered these words with a passionate earnestness that betokened how truly he meant what he said.

CHAPTER XXII.

MY experience of life since I arrived at manhood had been on the whole a dreary one. Shut in by the rude surroundings of bush life without the refining effects of female society, whose gentle associations form by far the happiest experiences in the life of man, the deeper feelings of my nature had been left to harden and contract from the want of sympathy to exercise and keep them expanded; and my late experience of such happy fellowship with a high type of womanhood, such as I had lately enjoyed, was so unlike the general monotony of my previous life that, in contemplating it now, it appeared like a bright star shining through the bank of dark clouds, that elsewhere obscured the sky.

"Truly," was my sad meditation, with reference to that bright view, "life's clouds had then a silver lining for me."

And now the brightness of the firmament was again overcast, in the sudden and final eclipse of my hopes. And she, whose beams had shone the brightest of all that fair group that I had so lately contemplated, had now by the clouds of fate been finally withdrawn from my gaze.

A black-edged letter that I about this time received from Scotland announcing the death of my only remaining parent, my noble-hearted mother—for whose sake, with the hope of one day making her in her declining years a partaker of my comforts, I had conjured up many a bright vision—added considerably to my depression and sense of deep loneliness.

In this unhealthy state of mind I continued to go through the dreary round of station duties for some time, until I was at length enabled to procure a sovereign remedy for my mental distemper in more active employment.

Although the lambing season would not come on for months, I roused myself to set about getting all my preparations afoot for developing the plan I had in view for the better management of this operation when the time arrived. I, therefore, at once despatched the station hands, that is, the two bushmen, Billy Stack, the bullock driver, and two other extra hired men, with Charlie Knight, the cook, and with them the bullock team loaded with provisions, tent, and implements to the back country where the ewe flocks, four in number, were to be all lambed.

Macalister, one of these bushmen, a short, stoutly built man, with red bushy whiskers, had served his full term of apprenticeship at a carpenter's bench, and was even accounted an adept at his trade, though preferring the rough-and-ready work of a bush carpenter, at which work, moreover in those days, there was fully as much money to be made in this part of the country as at the more regular work in one of the town workshops. His companion was a tall, muscularly built man named Crawford, or fighting Tom Crawford, as he was more familiarly called from his prowess in pugilistic encounters. He had curly light brown hair, with rather well-formed features, and a complexion that, though at one time probably clear, was now bronzed from long exposure to the Australian sun, and a rather intelligent expression in his face. The pleasing effect of his features was, however, rather neutralised by a recklessness of manner that, almost amounting to rowdyism, made him usually assert a sort of leadership over whatever group of men he happened to be associated with, unless, indeed, with Lilly, whom even Crawford slightly deferred to. He was indeed a splendid specimen of a colonial frontiersman with his reckless daring; a man who regarded with supreme contempt the feather-bed and law-protected amenities of civilised life, and was ever to be found in the foreground of newly occupied territories, securing to himself thereby a higher wage from his life of hazard in such a post, exposed as it was to the truculent attacks of untamed savages; and thus living a life more congenial to his own natural love of adventure and impatience of restraint. He was at bottom a generous-hearted man; such a one as would readily part with all that he possessed to relieve the sufferings of another; yet with his impetuosity of temper, on the other hand, easily engaged in a quarrel; the kind of man, in fact, who, under an overbearing explorer, would be swift to head a mutiny. He was, moreover, a handy, all-round workman, good at any sort of duty required on a station, the chief of these being sheep-shearing, bullock driving, and bush work, as rough carpentry is termed on a station.

Amongst other needful articles I sent out two double-barrelled guns with some powder and shot, more with the object of shooting wild ducks than from any idea of protection against any treacherous proceedings on the part of the natives, from whom, although till then perfect strangers, we indeed apprehended no sort of danger.

This absence of fear on their account was chiefly owing, I believe, to the appearance of the natives themselves. Not only did they appear to be insignificant as to numbers, but those that were there were also insignificant specimens as regards physique, as is indeed usually the case with the Australian aborigine who inhabits regions where game, and consequently food, is sparsely distributed and difficult of access.

Along with the team there also went a station black fellow to guide them to the place where I designed that they should pitch their tent. This place he knew from having been out there already with me some time before, when I marked the most favourable position for a sheep yard and camp by the side of a deep water-hole, of which there were several others, all being parts of a chain of several deeper bottoms along the line of a shallow watercourse that, with these exceptions, was unable to survive the evaporation caused by the summer sun.

On the third morning after the departure of the dray and men Lilly and I started after them. We had waited until then so as to give them time to reach the camp as soon as we did, for, as the distance was only about forty miles, we concluded we could easily do so in one day.

On the way there a rather amusing episode occurred. We had taken another station black fellow along with us on horseback, with the idea of making him assist in building the brush-yards and other useful jobs that he could turn his hand to. Well, we were approaching the hut where Bellamy lived, at about the middle of the run, rather more than twenty miles from the home station, where we intended stopping to get some luncheon from Mrs. Bellamy, when Lilly, the corners of his eyes deeply wrinkling, as their manner was when any humorous idea crossed his mind, proposed to me to halt in the timber whence a view of anything going on in front of Bellamy's door could be easily obtained, whilst we should remain screened ourselves by the intervening trees, and that we should then send the black fellow forward whilst we watched the scene that Lilly's acquaintance with Mrs. Bellamy's habits made him confidently anticipate would occur between that good woman and black Billy.

Nor was his confidence in this expectation misplaced, for, acquainted with the inordinate bump of curiosity and love of gossip that Mrs. Bellamy was possessed of, the amusement he proposed to me and to himself was to listen to the torrent of questions that she was sure to put to the black fellow, whose command of English for the carrying on of such a conversation was of the most limited range. Hence the baffled curiosity of the woman, and the spectacle of ludicrous imbecility in the man, constituted the very cream of the entertainment that we now prepared ourselves to enjoy. Nor were we disappointed. Shut out, as Mrs. Bellamy was, from intercourse with other women, of whom there were two in other parts of the run—shepherds' wives—the sight at her door of a station black fellow on horseback, whom she could closely interrogate as to her neighbours' concerns without the necessity of veiling her eagerness by an assumption of indifference, was an occasion for indulging her *penchant* in that respect that she there and then gratefully availed herself of.

Without even waiting to see if there were any one in Billy's company, or even waiting for the black fellow to dismount from his horse, her exclamation on opening the door as the sound of Billy's approaching horse fell on her ear, was, " Good day, Billy, you come from the station ? "

" Yes, me come from station, missus."

" You want 'em some tea ? me give 'em you directly ; you see white *lubra*(woman) Mrs. Campbell ? "

" Yes," said Billy smilingly, " me see 'em."

" And you see white *lubra* Mrs. Crow ? "

" Yes, me see 'em," said Billy, now grinning very broadly.

" Me hear 'em Mrs. Campbell no like 'em Mrs. Crow. You hear 'em that ? " pursued Mrs. B. eagerly.

" Yes," said Billy again, smiling approvingly at Mrs. Bellamy.

" What for Mrs. Campbell no like 'em ? Mrs. Crow been making some remarks about her, as usual, I suppose," answered Mrs. Bellamy again, in her keen desire for full particulars on this important subject gradually advancing into a current of English that was completely beyond the depth of Billy's comprehension ; but he, feigning the clearest perception of all her ideas on the subject, nodded his head in gracious assent, and with as broad a grin as before.

" Mrs. Crow been saying Mrs. Campbell no good ? " Mrs. Bellamy ejaculated, interpreting this nod in the direction of her own thoughts, and putting an eagerly suggestive question in the hopes of information.

" Mrs. Camly no good," repeated Billy, with evidently very little conception of what the conversation was all about.

" What Mrs. Campbell say ? " queried Mrs Bellamy, now moved to an intense pitch of friendly interest in the matters at issue between these two worthy females.

" Mrs. Camly say," replied Billy, with a rather disappointingly vacant look.

" What ? " inquired Mrs. Bellamy in keen expectation.

" Yes ; " hereupon Billy again nodded in a highly intelligible manner.

" Mrs. Campbell said something about Mrs. Crow ? " interrogated Mrs. Bellamy here, rather sharply.

" Crow," again repeated Billy, with the evident idea that

Mrs. Bellamy was desiring him to pronounce that word as a slight exercise in English.

"You know Mrs. Crow white lubra—ca, ca," cried Mrs. Bellamy in a tone of remonstrance, and endeavouring by the utterance of these sounds to direct a ray of light into the darkened mind of Billy, whereby he might apprehend that she was speaking of the woman whose name had a strong connection with the bird.

"Ca, ca, cr-r-a," answered Billy, giving a decidedly more accurate imitation of the voice of the bird that these sounds represented, and smilingly pointed with his finger to a number of these sable plumaged fowls that were roosting and cawing lustily in some of the neighbouring tree branches, thereby showing how clearly he apprehended what Mrs. Bellamy was talking to him about.

For a moment Mrs. Bellamy's face, as she looked at Billy, presented a good picture of hopeless defeat, then quickly rallying, and assuming a look of great determination, she seized hold of the astonished black fellow by his bare shoulder, for he had by this time dismounted from his horse, and actually shook him, as if with the frantic idea of literally shaking all the information she desired out of him, whilst she shouted to him almost at the top of her voice :—

"What for not you tell me news? What for not you tell me all about white lubra? What for not you tell 'em me what Mrs. Crow say about Mrs. Campbell, and what Mrs. Campbell say about Mrs. Crow? You hear? What for you big one stupid? You tell 'em me what Mrs. Crow ——"

Here a loud guffaw from Lilly, unable any longer to control himself at the ludicrousness of this scene, with the spectacle of Mrs. Bellamy thus shaking and shouting to the bewildered and half-frightened black fellow, made her for the first time aware of our proximity, at which she was so much disconcerted, that, quitting her hold of Billy, she hurriedly retreated into the hut. But not for long did she remain in that mood, for, like most talkative persons, she was not easily extinguished; so she soon rallied her courage, and received us with sufficient urbanity as we both entered the hut, having first slipped off the bridles and saddles to let our horses have the benefit of a roll and a feed in a small paddock alongside, whilst we were regaling our own inner men with Mrs. Bellamy's tastefully cooked if but simple fare.

"Well, Mrs. Bellamy," Lilly remarked on entering the hut, "you seem to have had a hard job in getting any news out of Billy?"

"Dearie me, yes," she answered, though still a little red in the face from her late confusion. " I feel so lonely here all by myself and not another white woman within miles of me. So I just wanted to know how Mrs. Crow and Mrs. Campbell were getting on, but I might just as well have talked to a post as to that black fellow, there was no getting him to understand one word that I said to him, till I felt quite provoked with him at last, and got hold of him, and was shaking him, as you and Mr. Farquharson saw me do when you rode up. But I really could not help myself. My friend Mrs. Williams used to say that I was one of the meekest of God's creatures until my spirit was roused, and then, says she, it is ' stand clear,' for Mrs. Bellamy's spirit will not be kept in a bottle always, meek as she is generally. ' And then as for silly gossiping, I do detest it. I never saw a woman like you in my life, Mrs. Bellamy,' Mrs. Williams used to say ; ' you never seem to concern yourself about any strangers' business at all, as long as you know that your friends are all right in their healths and their private affairs ; you might as well live the life of a hermit in the midst of a wilderness, for all the interest you seem to take in the people round you.' And that's just the way with me, sir, and I am now living in a wilderness sure enough, little as Mrs. Williams thought as I ever should, when she made that remark. As long as I know that my friends and neighbours are well, and not falling out with one another, I never trouble myself about them. Although when I do hear of anything being amiss with them I like to know all about it, so that I may be able to put in a neighbourly word like, or speak my mind out when I know any one is acting contrary to her duty and not keeping herself to herself, and having untidy houses and dirty children. These are things I don't like, and I say so too to the women's faces, as well as behind their backs. Mrs. Williams always used to tell me, ' I can never go to any house that is so tidy and well kept and the children so clean as in your house, Mrs. Bellamy ' ; and of course I always try to keep both my house, and children, and husband tidy, for I think this is a woman's duty, and not gossiping, and idling about, and neglecting what they ought to be looking after at home."

"You are right there, Mrs.," replied Lilly, when Mrs. Bellamy's voluble exposition of her various intrinsic qualities, as verified by the testimony of her friend, Mrs. Williams, had at last been exhausted; "you are right, Mrs.," he continued, though in a tone of light sarcasm, that the other however did not observe, " a man might well excuse any amount of talking in a woman, who can show him such a clean wife to look at in herself, and such tidiness in her house as this;" and he glanced at the same time from the spotless figure of the comely but loquacious matron, to the apartment in which we were then sitting at table; and the room was well worth looking at; for every article in it was clean and tidy, and the tinware and even the iron hoops on the wash-tub, and barrel-churn were scoured and shining like polished steel. In every direction, indeed, our eyes were gratified, though their pleasure was paid for by the trial to the ears. Mrs. Bellamy had three children, quiet, chubby, curly-headed little urchins, whose sameness of dress and heavy, stolid phlegmatic features prevented there being any visible difference between them, though I believe one of them was a girl. Bellamy himself was out in the run. After thanking this good woman for her savoury meal, Lilly and I, accompanied by Billy, who had not been neglected by Mrs. Bellamy, in spite of her late sense of exasperation against him, mounted our horses and resumed our journey.

Instead of holding a straight course from Bellamy's hut towards our destination, we inclined our horses a little to one side; as there was a shepherd's hut about four miles from Bellamy's place that I wanted to visit, as I was desirous of having some conversation with the hutkeeper.

This hutkeeper was none other than Lilly's simple-hearted friend, William Lampiere, to whom the reader has already been introduced; and for whom I had, shortly after his coming to the station, procured this occupation. The occupation of hut-keeping, that is in reality the laziest and commands the least pay on a sheep station, had been chosen by Lampiere in pre-ference to that of a shepherd (the choice having been offered him) and that too by Lilly's express advice, who recommended that course to him for a short time, so as to have a proper op-portunity of getting his hand in at cooking, and thereby gaining some sort of practical knowledge of a subject, that makes an important item in the education of a good all-round bushman, as Lilly expressed it.

My idea was now to promote Lampiere in his wages and office at the same time. He had now been over six months at this employment, quite sufficient, with Lilly and Cabbagetree Jack's previous instructions, to enable him to have put into practice their rules for all the requirements of good bush cooking. I knew, therefore, that he himself was by this time anxious for an opportunity of gaining some other experience of the duties that help to constitute a handy man on a station.

As the shepherd for whom he was hutkeeping was tending a flock of ewes that were soon to be shifted out to the back country to which we were then journeying to prepare for their arrival, I had resolved to ask Lampiere to take part in the coming lambing operations, and as the work I designed for him there—viz., tending the ewes with lambs—was one more suited to qualities that Lampiere seemed by nature endowed with, that is, patience and gentleness, rather than to require any amount of previous experience as shepherd, I thought he would be very well adapted for this post. His weakness of eyesight was indeed a serious drawback, but this unfortunate defect of nature I felt assured would be amply compensated for by his conscientious attention to his duty in other respects.

Moreover, Lilly, with whom I had first spoken, ere deciding upon this step (for the reader must know that for the coming operations I was taking with me none but carefully selected hands), warmly approved of my suggestion, and ventured to guarantee that after Lampiere had received some hints from himself for his guidance in the management of his new charge that I should have reason for feeling as fully, if not more, satisfied with my choice of Lampiere than of any other man I should have with me at the work. At this point, I may state, without troubling the reader with details, that for my success in my lambing operations, my chief dependence was on the security of all my yards, which with great pains I had made thoroughly dog proof. I also conceived the original idea of having all ewes with twin lambs tended separately, from which in the event of a death, I readily supplied the loss with a twin lamb, covered with the skin of a dead one. Also the four ewe flocks were lambed at two separate stations, that is, two at each— the flocks being timed to lamb in rotation to suit this arrangement. These lambing stations were put each under the charge of an efficient manager. Of these two managers, one was a new arrival, named Mulray, who had been highly recommended to me for the post by Lilly, who was well acquainted with him; the other was one of the married shepherds on the station, named Campbell. With Campbell was the other married man, Crow, in attendance with one of the ewe flocks, and another station shepherd with the other, while Burrell and another shepherd were with Mulray— with whom also Lampiere went to tend the stronger lambs—while a new man named Harvey did the like for Campbell. Besides these, there were blacks engaged to look after the younger lambs at both stations.

A half-hour's ride brought us to Lampiere's hut, on nearing which we observed him sitting outside on a log, and closely engrossed in the perusal of a novel. So interested was he in it that he did not observe us until we were close up to where he was sitting. On his attention being at last aroused by the sound of the trampling of our horses, when we were so near him that, short-sighted as he was, he was able to at once recognise who we were, he first gave a hurried glance over to his water bucket, which, proving to be empty, he instantly seized, and sprang down towards the water-hole with such nervous haste, that his foot caught in the root of a tree, and he was sent sprawling headlong down a rather steep bank to the water's edge. "Evidently Lampiere has not forgotten your injunction to endanger his very neck rather than be slow to attend to a traveller's wants," I remarked with a smile to Lilly, on witnessing this extraordinary feat. "The soft greenhorn," replied Lilly, the crow's-feet again deeply indenting and radiating from the corners of his eyes, in his intense amusement at this proof of Lampiere's faithful attentions to his injunctions on the matter of hospitality to all callers at his hut, "he swallows everything like gospel truth that a cove chooses to stuff him with, but he's a good-hearted, honest chap for all that." "Hilloa, young fellow, what do you mean by going head-foremost down the bank like that? do you want to break your silly neck? Take your time, man, we're all right. I am glad to see that you were paying such attention to the directions I gave you, about slinging your billy the moment you saw any callers coming to your hut; but I did not mean you to break your neck altogether, though I told you to do it. How are you getting on, lad?" Whilst shaking him heartily by the hand, Lampiere's face bore testimony to the unfeigned pleasure which Lilly's presence gave him. Lilly continued, on entering the hut: "Oh, you seem to have things pretty tidy and clean: that's right, you will make a good all-round hand on a station yet, but you needn't bother making tea, Mr. Farquharson and I have just had dinner about half-an-hour ago, at Bellamy's hut. Bellamy was away with rations to one of the huts; we didn't see him. Let me see what sort of bread you are making."

On this Lampiere handed him a cut loaf of hop manufacture, that Lilly critically examined and tasted, and then said approvingly, "Yes, that will do; not bad at all; no man need complain of such bread as that; you are doing very well, lad".

Lampiere's hut, like the generality of huts of this kind, was simple and rude enough in its description. It was in size 10 × 12 feet, built of logs laid horizontally upon one another, and plastered in the crevices with mud. The ample fireplace, almost like another smaller apartment joined on to the end of the hut, and its chimney, when seen from a distance, together with the hut's retreating hip-bark roof, really appeared to have no connection with the body of the building. The floor of the hut was simply the clayey ground, worn into large hollows by the friction of feet and other causes, but kept cleanly swept with a brush broom. The furniture, that consisted of a table and two forms, was of the rudest description, being made of slabs smoothed by the adze, and further polished by long usage. I spoke to Lampiere about the purpose for which I had come to see him, and asked him if he was agreeable to this change of duty. For this work the pay that I proposed giving the men was £1 5s. a week; besides, by attention to their work, they had the opportunity of making, in the liberal percentage I was allowing them, 2s. for every lamb over ninety to every hundred ewes. My offer Lampiere very thankfully accepted.

Although I always avoided any appearance of haughtiness in my manner with the men, always making it a rule to speak to them quietly and straight to the point in connection with their duties, yet I always preserved a due distance between us such as became my superior office, though the line that marked that distance was never of so frigid a character as to prevent a kindly greeting or the exchange of an occasional jest between us. But I observed of Lampiere, with some inward amusement, that he never seemed to be able to overcome a sort of respectful awe in my presence, and never ventured to address me or reply to my remarks save in a tone of the most deferential respect; yet I shrewdly surmised that this deference was occasioned as much by gratitude for the boon I had been able to confer upon him in providing him with work when he was in such indigence, as from any feeling of inferiority or meanness of spirit. Of course, mentally and by education and family he was quite on an equal footing with me, and, indeed, in the latter respect, rather above me. But he seemed to be naturally of a diffident and sensitive disposition. This, however, only the more recommended him to my inward regard, and made me resolve to quietly encourage and favour him whenever I saw an opportunity of so doing. With Lilly, on the other hand, to whose kindness he in reality owed the most, as it was he who had brought him first under my notice, and whose recommendation had prevented my regarding him as only one of the many hard-up swaggers to whose earnest solicitings for a job I was obliged to turn a deaf ear, with him, I say, Lampiere experienced no such feelings of diffidence. But Lilly, having no official position to maintain (he would have ignored the necessity for it if he had), assumed airs of superiority over no one. A plain man by nature as well as in his surroundings, the highest flights of fortune would only have left Lilly as he was, as he himself expressed it when observing any indications in a new man to show him any deference in virtue of his confidential position on the station. "I'm a working man like yourself, mate, and I want to be nothing else but a working man." And, indeed, a per-sistent disposition in any man to "Mister" or "Sir" him only provoked Lilly's contempt, as he regarded it as but a sign of toadyism or "crawling," to which he had a peculiar aversion. Out of the sheer force of his character he could make the most refractory looking man, when under him, do his bidding without a word.

With Lilly, therefore, Lampiere was always quite at home, whilst with me he was always on the very tip-top of respectful attention.

"What is this, Lampiere?" I demanded, on concluding my business conversation with him. This sudden query referred to some sheets of writing I happened to observe at the head of his bunk, and on which, as being softer than the form, I had taken the liberty of sitting. By the regular appearance of the lines of the writing I at once concluded it to be some form of verse. "Has the muse been visiting you lately?"

"Oh, sir," replied Lampiere shyly, "I don't know that it is worthy of such a pretentious character; it is just some verses I have been writing lately to while away the time."

"May I look at them?" I asked, at the same time taking up one of the sheets in my hand.

"Oh, certainly, sir."

"Yes, read it out, Mr. Farquharson," Lilly eagerly requested.

"Well, just allow me to glance over the lines myself first, that I may be able to read them better on a second trial."

This I did, and found it no very difficult task either, as Lampiere wrote a very legible hand. The matter, too, I thought by no means indifferent. However, the reader may judge for himself.

As I read, Lilly signified his satisfaction at the contents by sundry slaps with his hand on the table and such comments as "Well, that's jolly good!" "Capital!" "Why, Burns couldn't beat that!" etc.

The subject was indeed one that Lilly could fully appreciate, the poem being entitled

A HUTKEEPER'S ADDRESS TO A TRAVELLER.

The sun is sinking in the west,
  Then, mate, lay down your swag;
Here you will find both food and rest,
  No thanks for that I beg.

My ration bag just now contains
  Both plenty and to spare;
But while one single pound remains,
  That pound I'll freely share.

Then if your things with dirt look done,
    In going be not rash :
For I have soap, if you have none ;
    Then stay with me and wash.

Tobacco, too, I have to spare,
    With which you may make free ;
Then come, sit down, draw in your chair,
    And fill your pint with tea.

Affairs look worse now every day,
    Wages are falling fast ;
Australia's sun has set for aye,
    Her golden days are past.

The rich have now the upper hand,
    A power that they will keep ;
That they will use throughout this land
    To make the poor man weep.

The grinding, avaricious crew,
    Their fellow-creatures' bane,
Who still their mammon god pursue ;
    All hope from them is vain.

Hear ye not how the heathen rage,
    And mark their bitter frown—
Their endless cry, " The poor man's wage,
    Still, still let us bring down ! "

But rest you, mate, awhile with me,
    And shake off your fatigue,
For you will travel yet, I fear,
    Full many a weary league.

" Well, I be blowed," remarked Lilly on my concluding, " if that isn't as fine a poem as ever I heard. I have only one fault to find with it, and that is where the hutkeeper asked the traveller to draw in his chair. Now there are no chairs in shepherds' huts, and consequently that remark is not natural, and looks like an idea that has no right to be there. But all the rest is so natural, that I could fancy I could hear the hutkeeper speaking, and saying things that I should most likely have said myself—only a heap nicer put—to a poor, hard-up traveller. Now, Mr. Farquharson, you who are a good scholar, and know when a poem is properly made, don't you think that there is true genius in that poem ? It seems to be so well metred, too."

I replied that I certainly thought this performance very creditable, and, for the occasion, very natural.

" Did you ever send any of your verses to the newspapers for insertion, Lampiere ? " I asked.

"I, sir?" replied Lampiere with an astonished look, as if to such a pinnacle of ambitious distinction he had never dared, even in his fondest thoughts, to soar. "No, I've never dreamt of such a thing. You surely don't think, sir, that an editor would put such verses as those into the poets' corner of his paper?"

"Well, if you have no objections to trust them to me, I shall be better able to answer your question a few weeks hence; meanwhile, I think them good enough for the poets' corner of the *Sydney Morning Herald*, to which I will send them when I return from the back country. I think I have seen worse verses than these in the same journal."

"I shall feel extremely obliged to you, sir, if you will be so kind."

"Then you had better take a copy of them in case they should be rejected and consigned to the waste-paper basket, if you do not wish to lose them altogether after your trouble in composing them."

"Oh! no danger of that, sir, I have them securely stowed away in the safe store-house of memory."

"Why, Bill," demanded Lilly in surprise at this statement, "do you mean to say that you can keep all your poetry in your head after you have written it down?"

"Well, yes, Lilly, I can mostly do so."

"You must have a wonderful mind then, sure."

"Oh no, I don't see how that should follow. Don't you see I never commit my verses to paper, till I have them thoroughly fixed in my mind; so that all the time I am composing them I am actually committing them to memory, just as I should do any other poetry, by repeating it over, and over again, till my mind is made thoroughly familiar with them."

Lilly still seemed to think it a wonderful proof of the power of his mind in spite of that, and he said so.

"What does your companion Burrel think of your taste for poetry? I suppose you show him your verses sometimes?" I now asked Lampiere.

"Oh, yes, sir, Burrel and I get on very well together on this subject; he is a poet himself."

"Burrel a poet!" I repeated with some surprise, as I recalled the grave, but shrewd looking countenance of the person in question; "why nothing would be further from my mind than to imagine him guilty of such a weakness, with his dry, caustic, way of speaking; I should be rather inclined to think he was made out of the stuff that cynic philosophers are fashioned from."

"Oh, that is only his way, sir," replied Lampiere smiling,

"but below that unpromising surface, there is a soft enough vein of poetry for all that. I was very much surprised on my first discovery of his poetic taste, that he kept very quiet, until he noticed me scribbling away at some verses one night, when on looking over what I had been writing, he all at once came out of his shell, and we have got on capitally together, with our mutual weakness for poetry, ever since then. But he affects more of the Tennysonian style, and writes a good deal about old ruins, and other such like dreamy subjects."

"Stuff and rubbish!" ejaculated Lilly contemptuously, "give me something that has some grit in it, and that I can understand, like your 'Hutkeeper's Address'; that's the sort of poetry I like, such as Burn's 'Man was made to mourn,' or better still, his 'Tam o' Shanter,' or his 'Twa Dogs,' before all the moonshine that these dreamy poets rave and sicken about. It may be all very well for young boarding-school misses, to read such sentimental stuff as that; but there's not enough grit in such subjects for the brains of common working men like me to lay hold of."

The effect of this interesting discussion, of which the part here recorded is a fair sample, was, that on our attention being at length called to the business we had on hand, we found that we had insensibly allowed the afternoon to wear away to a considerable extent—as, on going to the door, our lengthened shadows (the bushman's clock) testified—so that the idea of continuing our journey that day, where there was no road, with a moonless night coming on, was voted impracticable.

Lampiere, to ensure us ample comfort, if we would consent to pass the night where we were, instantly volunteered to surrender his own bed to me, and ventured to promise as much for his friend, on Lilly's behalf, as he said they could make up a shakedown on the floor for themselves, with the blankets we had on our saddles.

This course was eventually agreed upon, both Burrel and Lampiere over-ruling all our objections to the idea of their vacating their comfortable beds for our sakes.

Having decided upon staying, we at once ungirthed our saddles, and hobbling the three horses, let them feed about the hut, giving Billy a charge to keep his eyes on their movements the last thing at night, so as no unnecessary time might be lost in the morning in seeking them.

The sun was dipping beneath the glowing horizon as Burrel put his sheep in the yard and shortly afterwards entered the hut, and he was not a little surprised at finding guests for the night. He, as I have already remarked, was a quiet, but extremely shrewd looking man, of a slight build, with rather a warm complexion. He might have been about five and twenty years of age. Looking at him now with more interest, in the light of the knowledge of his latent tastes, I observed that his forehead, though not uncommon in size, was well proportioned, with the organs of ideality well defined.

Well, we passed the night there very agreeably. Burrel was one of my favourite shepherds, whom I had always noted for his regularity, and the uniformly thriving appearance which his sheep always presented. During the evening he was induced to recite some of his poetical compositions. These appeared to be of a very thoughtful cast, and the lines particularly smoothly measured, a feature not so characteristic of Lampiere's productions, though these latter to my mind seemed to lay claim to a greater degree of rugged strength than those of his friend.

Even Lilly, despite his repugnance to this class of poetry, inclined at last to think favourably of Burrel's talents, though still far and away preferring Lampiere's "Hutkeeper's Address" to the best of them. But then of course Lilly judged by mere impressions. Of artistic rules of criticism he was wholly ignorant, "and willing to remain so," he emphatically remarked. Yet, I doubt if he really spoke what he thought when he ventured to compare the last verses of Lampiere's "Hutkeeper's Address" to Burns' wondrous dirge, "Man was made to mourn".

We in time retired to our beds—Lilly and I to Lampiere's and Burrel's comfortable bunks, and the latter in our blankets, and stretched on some sheepskins on the floor, while at the further end of the hut, next the fire, Billy, to his own entire satisfaction, lay coiled up on two more sheepskins. Sheepskins, indeed, at that period might have been termed the staple material for bush mattresses, though bushmen since then have awakened to the knowledge of their unhealthy properties, and have begun to object to their use.

CHAPTER XXIII.

SHORTLY after sunrise the next morning we were again in the saddle, and leaving the belt of the forest in which Lampiere and Burrel's hut was situated, we held a bee line for the end of a distant blue mountain, round which our route wound. By mid-day we rode into the camp, just as the men were preparing to sit down to their dinner, which was spread on the ground, in front of a large tarpaulin, improvised into a tent, that had just been erected and was to serve as a sleeping apartment for all hands, until some of the huts could be finished.

Charles Knight, with a white cap on his head, looking every inch a cook, was there, attending to the wants of the diners, with an air of profound importance, as though presiding over the arrangements of a civic feast, or agricultural dinner. His erect carriage was supported by the proud consciousness of Self's resources for provisioning the party by which he was enabled to spread before them such a choice repast of meat pastry and fruit pastry prepared at home before starting. All this now was partaken of with great relish, together with some fragrant tea, of which an ample supply was steaming in a large American bucket on the ground beside them—all this excellent feed having been improvised ere any conveniences for his cooking arrangements had been made.

He now kept replying with the most imperturbable demeanour to the constant, vociferous summonses to "Self" for attendance on the varied wants of the party, which wants appeared to be almost voracious in their character, with the calm replies of "Yes, sir," or "Self is here at your service, sir". Tom Crawford, as usual leading the conversation, was entertaining the party by the relation of some of his wild stories of adventures with the blacks, or experiences with noted pugilistic champions among the whites.

All seemed hearty and happy; nor did the sudden presence of their overseer, or Cove, as the bush term for such a personage is, put the least damper on their spirits. Among these roving, independent bushmen, Jack is always reckoned just as good as his master—and indeed I would not have envied the overseer's lot who would have attempted to assume an air of superiority over fractious and facetious Tom Crawford. Not that either he or any of the others were disrespectful, quite the contrary. Looking to me for orders for their work they addressed me with simple respect when they spoke, but this respect on their part was based on the assumption that what work I ordered them to do, I could judge of when done, and that I gave no other orders than what I meant should be carried into effect. For of all the evils that such men as these find it hardest to submit to, it is that of being under the orders of an incompetent director who interferes from a love of meddling, and frets and fumes with dissatisfaction when they are conscious of having honestly executed their work.

"Well, Tom," remarked Lilly, as we joined the party at dinner, "fighting some of your old battles over again? What a terrible warrior you have been in your time, man!"

"Well, yes, Lilly, I reckon I have been in a few scrimmages

in my time; I think you have seen me in one or two yourself, old man!"

"I should like to see the man who has known Tom Crawford long and not seen him in a scrimmage," here remarked Mulroy, speaking in that tone of solid emphasis that, without the ear being offended by the inflection or any peculiarity of accent, at once indicates the speaker to be a Scotchman.

"Well, I believe that is true too," replied Tom, joining good humouredly in the laugh raised by Mulroy's dry remark, "but you know for all that, Mul, that I would quarrel with no man without a cause."

"Somehow or other, you seem to have a cause pretty often for quarrelling, for all that," replied Mulroy in the same dry tone as before, thereby eliciting another peal of laughter from the party.

"Well," replied Crawford apologetically, at the implied charge of his readiness to take offence, "I am a man that will take nonsense from no one, and it makes my blood boil when a man speaks to me in a domineering tone, and especially, too, when I see a bully trying to bounce a weaker man than himself."

"Just so, Tom," replied Mulroy, in the same tone of dry though good humoured banter, "you're like a carpenter that is pretty handy with his tools, and so likes to be always using them whenever he sees a crooked job that he reckons he can put straight, and so when you see something going crooked between two men, you are bound to try and straighten the matter with your tools."

"Well done, Mulroy," cried Lilly, laughing heartily at the illustration, "but people may be too handy with their tools, as Tom was when he tried to straighten the job that had got crooked between the swagger and the six 'Tips' who were going to mob him, and who all turned on Tom himself, and the man whose part Tom had taken left him to fight it out by himself whilst he showed them a clean pair of heels."

"Aye, Lilly, old man, I had some cause for thanking you for helping me out of that scrape that time."

"Yes," said Mulroy, "I heard something about that, but I never heard the particulars; it was down at Menindie wasn't it?"

"Yes," replied Crawford. "As Lilly says I went to help a cove who was drinking at Tom Payne's, and who had been shouting for these 'Tips,' and having said something to them that they did not like, as soon as he went outside, they all followed him. It was just about dusk, but I just saw them as they went round the corner, and I rushed out among them, for I wasn't going to stand by and see one man mobbed by six; and then as Lilly says, the sods all turned on me, whilst the cove I had gone to help bolted and left me to do the best I could with them! Well, I had knocked one of the 'Tips' sprawling, when the other five rushed at me, and just then Lilly came at them with his stockwhip. My word, he soon cleared the course of them!"

"I just happened to come out of the hotel," said Lilly, "and hearing a row round the corner I went forward and saw at a glance what was up. Well, I was just going to rush in among them to help Tom with my fists, when I happened to think of a better plan; so I just ran back and unhooking Coleena from the verandah post, I jumped into the saddle, and unfastening my stockwhip I just went at them with the lash! Talk of rounding up a mob of cattle, it was nothing to the way that I made my stockwhip play about those dogs."

"Yes! and didn't they soon clear away, and bolt into the house and lock themselves up in one of the rooms! I would have smashed the door open with the axe, if Tom Payne hadn't begged me to go away and leave them alone. But you saved my bacon that time, Lilly!"

"Well, I don't know but what you would have cleared off the whole bang lot of them yourself, Tom. When I came up, you were going at them with your feet and hands; talk of a cat being raised, I never saw a man looking as wild as you were in my life! But I should have cut the eyes out of the heads of the sods if they hadn't bolted when they did. I am an Irishman myself, but curse foul play! I can never stand the sight of that in Irishmen or Englishmen."

"Aye, Lilly, it would be well if all Irishmen were like you: there wouldn't be so much said against them if they were," replied Crawford, who himself was a Scotchman, and known for entertaining a violent prejudice against Irishmen in general, from the frequent collision he had with such unfavourable specimens of the race as the "Tips" in the squabble he had just alluded to.

"I shall always consider myself in your debt for a broken head, Lilly, since your service to me that time."

"Never mind the debt, Tom; unless indeed you find me being mobbed by a half-dozen wild 'Tips,' when you may then pay a little of the debt back in the way of interest if you like."

"I pity the half dozen or dozen Tips that interferes with you, Lilly, when you are sitting on Coleena, and have your stockwhip in your hand!" This sally produced another loud laugh from all the party.

"Did any of you chaps," inquired Crawford, "ever hear how Lilly announced dinner to the governor, when he had been stopping at Tappio, on his way down the river, when Emerson was there, the year before Mr. Farquharson arrived?"

Most of us replied that we had not heard of the circumstance, whilst Lilly laughed, and one or two of the men who had been acquainted with the affair, smiled in amusement at the remembrance of it.

"Well," continued Crawford, "it was one day as I was splitting posts on the river bank, just at the station, and Lilly was beside me and lending me a hand, as he said he had nothing to do that day, when Captain Caddell's steamer, on his way down the river, stopped and was moored to a gum tree just beside where I was working; and then out of the steamer came several swells, one of them being no less a person than Sir Richard M'Donnell, governor of South Australia.

"Now Mr. Emerson was naturally a fussy stuck-up sort of a man, and down he comes when he saw the steamer stop, and starts bowing and scraping to this viceregal swell, as if he were in the presence of royalty itself, instead of merely its shadow.

"I stood looking straight at them and should not have cared if it had been Prince Albert, with the Queen to boot. I wasn't going to touch my hat to them, nor, I reckon, did Lilly, although I could see by the angry manner with which Emerson looked at us, that he considered we should have both done so. Well, Mr. Emerson then begged that the governor would do him the great honour of dining with him, although he said he had but poor entertainment to offer him. But the governor, who appeared to be a rather jolly old cock, told Emerson not to put himself about: he could do very well with what he could offer. Then he and Captain Caddell, and two or three other swells, one of them the governor's secretary I believe, and the other perhaps his snuff-bearer for all I know, went all the way up to the house.

"In a very short time afterwards Emerson came down and begged Lilly to go up to the kitchen and cook the dinner, for he knew that Lilly was not to be beaten as a cook, and he wanted to give the Governor and his friends as grand a feed as he possibly could manage. At first Lilly swore and declared that he would see Sir Richard M'Donnell in pandemonium, before he would cook for him; however, he at last took pity on this poor fussy crawler of an overseer, as he could see his heart was set on doing the grand to the governor in the little tin pot way that he could manage, so he went up to the kitchen.

"As soon as Emerson had got him fairly started, he got old John into the house, and had him fairly rigged up in one of his best suits of clothes, to wait at table, whilst he fossicked up some white tablecloths, and had one cut up into sizes for dinner napkins, and lent a hand himself in laying out the dinner to make things look as grand and genteel as possible, all to do honour to this Governor. Governor indeed," ejaculated Crawford contemptuously, "why, I was told afterwards that on his way down the river, at a station where he called, he saw a parrot in a cage that was hanging outside of one of the huts, and he actually had the cheek to walk away with the parrot! Yes, the Governor of South Australia actually took a parrot. I reckon he should have been put in quod for it, for my part. Well, among other things that Emerson thought of doing to have his dinner to this Governor served up in as grand a style as possible, was to send in a bullock bell that Lilly was told to ring when the dinner was ready. I was in the kitchen where I had gone to have a yarn, when the last order came in. 'Ring be d——d,' said Lilly to John who had brought in the message, 'I'm going to ring no bell!' Well, dinner shortly after this was ready to serve up, and Lilly soon had it on the table. Emerson was out in that bit of garden behind the house airing his importance before his Excellency.

"He could be heard coming across this title at almost every other sentence that he uttered. But I guess his dignity got a fine down-come, when, instead of the tintinnabulation of the bullock bell announcing dinner, he heard Lilly's rough voice shouting from the kitchen door, 'Now then, swells, the grub's on the table; go in and get a feed,' and with these words off Lilly marches away down to the river to his own hut. Laugh! I thought I should have split my sides with laughing when I heard Lilly shouting out like that, and I laughed all the more when I saw how Emerson looked when he heard it!"

We all laughed very heartily at Crawford's narrative, that was told with a spirit very hard to convey on paper, as we realised the inglorious collapse, that Lilly's unceremonious announcement must have occasioned to my predecessor's air of importance in the eyes of the Governor of South Australia.

Lilly returned home next day, but I remained where I was for several days longer, to personally superintend the initial operations of my arrangements for the lambing. The weather being good, the work in the bush, in which I took a personal part, was very pleasant and invigorating, with its continual chop, chop, of axes, and see saw of the cross cuts, and the sound of falling timber; though as this (save a little for the huts) was not required of any great size, it did not fall with anything like the shock or noise occasioned by that of the more important trees.

At night, seated on logs around a blazing wood fire, in the open, or by the ample fireplace of one of the huts, the first of which was completed in about three days, there was no lack of entertainment to keep our spirits from flagging in our solitary encampment.

What with songs from the one, and performances on the accordion, on which instrument there was more than one ready player—and even step dances, the time between the completion of our evening meal, and that for seeking our beds some hours later, was as cheerfully and pleasantly occupied as if the idea of our lonely encampment in this Australian solitude, so far removed from any other civilised settlement, was a source of disquiet to no one.

Then, when there came a lull amongst the singers and players, there was always a standing source of entertainment in the reminiscences of someone of the party, though in such tales it was usually Tom Crawford who took the chief part. Yet, though free enough in talking about his own prowess, there was but little of what he told us that could be set down to the score of mere gasconading, and many a tale of wild adventure and of fighting had he to tell, all of which, from the daring and decisive manner of the narrator, I doubted not had actually occurred. But of all his narrations, that which excited me most at the time, and which in the light of subsequent events I had most reason to remember afterwards, was his relation of a tragedy that had happened some weeks before (of which I had already heard some rumour), at a creek called the Warego, situated farther up the country towards the Queensland Border, in which two white men had been murdered by the blacks.

This incident, Crawford, who had a naturally dramatic and impassioned way of telling a story, was able to describe more vividly, from his having viewed the scene of action a few days after the tragedy had taken place.

Crawford, by the way, had only been a few weeks with me at this time, for although he was frequently engaged in one capacity or another at Tappio, yet, being a thorough bird of passage, it was seldom that his stay at any place was prolonged beyond two or three months at a time. Being a professional shearer, it was seldom that he failed to quit whatever employment he chanced to be engaged upon, on the first dawn, so to speak, of the shearing season, so as to catch the early sheds in the very earliest districts. Thus by following these sheds up till they were finally closed for the work in the latest shearing districts, he reaped the full benefit of what money was to be made at this work, by which he contrived each season to turn over no inconsiderable sums, even as working men's wages were at that time rated in the colony.

I use the term advisedly when I say that he was able to turn all this money over. In truth, Crawford did little more than turn it over, when he had received it—to wit, into a publican's hands at the inn or shanty nearest to the shed at which he had just earned a good cheque. This money, notwithstanding the toil he had undergone in securing it, he there, in the midst of a group of kindred roysterers and fellow-shearers, contrived to wholly dissipate in a surprisingly short space of time, by the simple method of continual "shouts" of "nobblers" for all hands at a crowded bar.

By this mode of fast living on the part of the shearers it was soon evident, even to themselves, that the fastest and surest shearer of them all was the shanty-keeper himself.

As the result of all this, at the end of the most prosperous shearing season Crawford generally turned up in search of employment at Tappio absolutely penniless; flyblown, in fact, as a very unsavoury but expressive bush simile expresses the condition of things. On the present occasion, however, Crawford had agreed to stay and shear at Tappio shed, being provided with ample employment in the shape of a contract for a dam that was to be constructed after his present work at the yards was at an end, and which would keep him and Macalister going until the shearing at Tappio was ready.

As to the narrative in question, however, we had been sitting round the fire in a newly-finished hut one night after supper, and Crawford had just finished a description of a desperate encounter with the blacks in some newly-occupied country in Queensland, where the natives were particularly daring and turbulent, when Billy Stack inquired:—

"Did you hear, Tom, of those two white men who were murdered a few weeks ago up in the Warego? I heard something about it, but not the particulars of the case. Do you happen to know anything about them?"

"Do I know anything about them? I should think I do, Billy. I came past the place a fortnight after the murder was committed."

"Then you might tell us all about it. I heard that one of the men had been a captain in the army."

"And so he was a captain," replied Crawford, "a Captain Bruce; a man of good family; but I suppose that won't make his body taste any sweeter to the worms now they have got it. And his aristocratic friends will have to come a long way if they want to see the grave in which he now lies in a bend of the lonely Warego. Mrs Hemans's lines,

'The Indian knows his place of rest,
Far in the cedar shade,'

only wants the alteration of two or three words to make the comparison perfect, as thus—

> 'The Australian knows his place of rest,
> Far in the boxtree's shade'.

And, indeed, very few besides the Australians do know it. I suppose the man must have given way to gambling and drinking, like the rest of us, or he would not have come so far down in the world as to be carrying his swag when death, by means of a savage's tomahawk, overtook him in that out-of-the-way corner of God's earth. You want to hear the particulars. Well, you shall have them as I got them from some of the men at Bloxham's Station, within six miles of the place where the cold-blooded murder was committed.

"This Captain Bruce and a German named Raynor had come the night before to Bloxham's Station, and staid there all night. They were travelling up the creek in the hopes of getting on at some bush work at a new station that was being formed a good way further up the creek. I knew the German myself. He and I were at the same station last winter. I was making some sheep yards where he was shepherding some ewes and lambs. He was a stout butt of a man, a native of Frankfort, and could speak very good English. He also understood the sword exercise, as he said most of his countrymen, or at least fellow-townsmen, did, for I saw him one evening showing a cove the different passes and guards, and I remember he surprised me by saying that most of the guards are now performed by a simple turn of the wrist, though I can hardly see well how that can be so. But, however, about this German. I always, before I met with that man, used to look upon Germans as only soft, slobberly sort of men. Big enough frames, and strong enough, for that matter, but altogether wanting the hardiness and pluck of us Britishers. However, I admit that this man, at least, was an exception to that rule. I partly thought so when I was in his company, but I am now convinced of it by the way in which he behaved himself in the scene that I am about to describe to you.

"The next morning as this German and this Captain Bruce were rolling up their swags outside, there were several of the station blacks looking on and watching them. As these blacks were all reckoned quite harmless, no one dreamt of the villainy that was at that very time filling their minds. These two white men had a good stock of tobacco with them, and evidently these black devils had noticed the fact. Now, in my opinion, a black fellow, no matter how quiet he looks, is never to be trusted, for in their eyes a fig of tobacco constitutes as great a temptation to knock a man on the head as £100 would be to a white murderer. Besides their tobacco, both men had very good swags of clothes and blankets with them.

"Curiously enough, it would seem as if a sudden presentiment of foul play had flashed across the German's mind; for he looked up and remarked with a smile to one of the station hands who was beside him, 'I hope that the blacks won't take a fancy to my red blankets!' Little did the poor fellow really imagine what these devils were planning in their minds at that time, or he would not have left that station so happily as he did that morning, if indeed he would have left it at all that day.

"Well, these two men putting their swags on their shoulders, went along the road quite easy in their minds I suppose, and without the slightest idea that close on the tracks of each, death's dark shadow was silently stalking, in the persons of two of these blacks, who followed them along the way they went, without the white men having the least suspicion as to why they were doing so. If they thought of them at all, they would probably have only fancied that they were accidentally going along the same way as they themselves. Another thing would have confirmed them in this idea, that is, if they had happened to give a thought to the matter at all: on the other side of the creek was another black fellow keeping up with them, in company with a gin.

"I have always understood that when blacks mean mischief they leave their women at their camps, and doubtless these travellers had heard the same thing, so that the sight of this gin in company with the black fellow would have at once quieted any suspicions if they had happened to arise in their minds; a thing that is very doubtful, for no one suspected the least danger from these station blacks. Of the two customers that were following them, one was armed with a tomahawk, and the other had a *nulla-nulla*. This weapon they either fight with, you know, in close quarters, or throw at anything they want to hit, from a distance of from ten to twenty yards. I forget what arms the black fellow on the other side of the creek was provided with, but I know he had one of those long sticks, that the gins use for digging up roots or yams with.

"However, the two doomed men went carelessly along till they thought they would like a spell, when, throwing down their swag, and undoing their pannikins, they went down to the creek and had a drink, then, sitting down on their swags, filled their pipes and had a smoke. It is to be hoped that they enjoyed it, for it was to be the last smoke that either of them was ever to have on this earth. Their smoke being over, they had both turned round and were in the act of fastening their pannikins on to their swags again, when the two bloodthirsty black villains who had followed them so far, now suddenly stole behind them and both struck their victims a deadly blow at the same time. As it happened, it was the black fellow with the tomahawk who had gone behind the Captain, who by the blow was killed instantly; but the German, while bending over his swag, just happened to catch a glimpse of the tomahawk in the act of descending on his companion's skull, and threw himself up immediately. This caused the blow from the *nulla-nulla*, intended for his skull, to light full on the nape of his neck, which it did with such crushing force as to fatally injure the sinew of the spine, as he afterwards, but not then, died from its effects. There was still sufficient energy left to enable the brave fellow to turn and grapple for dear life, with the two murderers. Two, did I say? yes, three of them, for the third black fellow now crossed the creek and rushed to the assistance of his two mates; and in the deadly clutches of these three black devils, the gallant German now rolled over and over among the scrub.

"I saw the marks of that terrible struggle while they were still fresh, a fortnight after it occurred, and I tell you, lads, that there were saplings there as thick as my leg snapt off in the course of that death struggle, as if a bullock had been struggling for his life, instead of only one brave man in the clutches of three devils.

"At last the German wrenched himself free from the three blacks. Would to God that I had been beside him to help him at the time, for my heart warms for a brave man whatever country he belongs to. Drawing his knife he challenged his murderers to come on. That man must have had the heart of a lion, German or no German: there was no white feather or craven blood in him.

"Instead of venturing to take up his challenge and come at him again, one of the cowardly blacks, evidently the one that had come across the creek, threw his yam stick at the German's right hand, which it struck with such force, that the knife dropt out of his grasp, but instantly picking it up again with his left hand, he dared them once more to come near him. On their shirking this challenge, he then for the first time seemed to think of flight, and turning round, he fled back with all his speed, in the direction whence he had come. He ran until he fainted; when, on recovering his consciousness, and memory of what had occurred, he again ran with the energy of despair, until he made the station that, a few hours before, he and his now murdered companion had left in such high spirits.

"When he arrived, he was able to tell to the horrified people there all that had occurred. As he spoke, it was observed that from the injury done to some of the sinews of his neck by the force of the black fellow's club he was unable to hold his head erect: it always had a tendency to fall back. He was able to tell all his story, but shortly afterwards his consciousness left him and never again returned. This event occurred on a Saturday, and on the Saturday following the German died, and on the Saturday following his death one of his black murderers was shot. The other two had also been secured and handcuffed together, but as a trooper was bringing them along through the scrub, they suddenly made a bound into it and disappeared, and though closely followed, no trace could be found of them afterwards. It is supposed that they must have managed to slip their hands out of the handcuffs, and that they have come down in this direction.

"These three blacks were all powerfully built and particularly ugly looking, and it seems strange that each of them had a name corresponding to his forbidding appearance. He that was shot was called 'Tim the Butcher,' and the other two are called respectively 'Billy the Bull' and 'Tommy the Turk'."

The relation of this tragic event, particularly affecting in itself, impressed me vividly enough at the time, though I did not know then, which nevertheless proved true, that I was shortly destined to reap the fruits of this deed of blood.

I had indeed soon reason enough for believing the correctness of Crawford's suspicions that the escaped black murderers had come down in the direction of our present locality. Could I have dreamt at that time of the possibility of such a contingency, I should hardly have owned a guiltless conscience in leaving a party of men with but one double-barrelled gun among them as a protection against blacks, among whom two such desperadoes were known to be at large.

But of course I was innocent of any such knowledge, the distance from the scene of the outrage being too great for me to attach any weight to Crawford's remark about the route that the escaped murderers were supposed to have taken. But the sequel proved that there was no small measure of truth in Crawford's other remark, that "blacks were never to be trusted".

CHAPTER XXIV.

THE completion of my preparations for the lambing was just in time for the commencement of operations. In the meanwhile I had passed several times to and fro between the back country and the home station.

On one of these occasions, on entering my house, my eye fell on the table, where was laid my mail—several letters and a whole bundle of *Sydney Morning Heralds*. My first feeling on seeing so many numbers of this journal was only surprise, until a pleasing thought suggested to me the cause, as I recollected my letter with stamps to the Editor of that paper, requesting him to forward their value in papers, in the event of the accompanying verses being thought worthy of a place in his poet's corner. What these verses were, the reader may recollect. Therefore in eager anticipation, I hastily opened one of the papers, and turning to the poet's corner my eyes were gratified with the sight of Lampiere's "Hutkeeper's Address". Reading them carefully over, in this new form, I really thought that they became their position in the paper very well.

Feeling impatient to see the effect upon Lilly, of this proof of the merit of his friend's production, I went to the door and desired John to call to one of the black fellows, to cross over the river and tell Lilly that I wished to speak to him.

In a short time Lilly came stalking into the house. He was always keen to see the papers at any time, for although not much of a politician, he still had a desire to know how the "Coves who were at the helm of state were shaping," as he rather quaintly expressed it. Besides this, he naturally wished to know all the news of the colony at large, and the main items of European news as well. In preparation for his coming I had the newspaper with the sheet in which the verses were, laid conspicuously open upon the table. "Oh, you have got the paper, Mr. Farquharson," he said, his eye falling upon this as he entered the room. "Yes, Lilly, you can just take a look at it till I am ready to attend to you," I replied, pretending to be engaged with my note book. He did so. In a surprisingly short time afterwards, he exclaimed with great eagerness—although usually Lilly was anything but an impulsive man—"Jeehosaphat, Mr. Farquharson, why Bill Lampiere's 'Hutkeeper's Address to a Traveller' is in this paper!"

"Yes, Lilly," I answered, laughing, "I guessed the sight would surprise you."

"Now, didn't I tell you there was real genius in that poem, and doesn't it look splendid now in print? I knew there was something in that young fellow's 'nut,' although he does look so soft and simple; he'll be having a book of his own yet! I'll get this poem and frame it, and hang it in my hut, when you are done with the paper."

"Oh, you can have the paper away with you, see—I ordered a dozen when I sent the verses away; I thought perhaps that

Lampiere would like to send copies of his composition home to his friends."

Lilly's satisfaction was indeed unbounded, and I felt persuaded as I saw him marching away with the prized newspaper, that a present to himself of £20 would not have made him more happy than he then felt at this proof of his friend's talent. Lilly was indeed a warm-hearted, unselfish man.

My next source of pleasure in connection with these verses, was on the occasion of my presenting the paper in which they appeared to the author himself. It was positively a treat to watch the expression of almost childish delight and pride, that flushed Lampiere's cheek and kindled in his eye on thus seeing the production of his own brain actually staring him in the face, in the form of print. He read and re-read his own verses, as if he were desirous of familiarising his mind with words that, as he formerly assured me, " were already safely stowed away in the store-house of memory".

When I handed him the extra number for transmission to his friends, he thanked me as much as if I had conferred some extraordinary favour upon him. Doubtless he regarded it as such; but I was amused at the importance that he attached to an action with which, with such little inconvenience to myself, I had been enabled to confer so much pleasure.

But to return to my lambing operations in the back country.

I still kept my whole staff of hands employed there: the two odd hands and a couple of black fellows, as assistants, at the lambing. One of these odd hands was Mulroy, to whom, as I have already stated, I purposed giving the charge of one of the lambing stations; the other, a man named Harvey, I sent to look after the strong lambs, at the lambing station under Campbell's charge. The two black fellows had the duty assigned to them of each looking after the young or green lambs, close to the hut, at both stations, their principal duty being to watch against the attacks of hawks and crows, especially the latter, upon the young lambs; for crows with their cruel beaks make short work of any weakling they find strayed, or lying asleep at a distance from its mother, by quickly digging out its eyes, and killing it. For the two bushmen, with Billy Stack, who attended on them with his team, and Charlie as cook, I had still plenty of employment in building huts and other yards, for the ewe flocks, in other parts of the run, when the lambing season should be over, as their present stations were to be kept free of stock, until required for the same work in the following season.

So determined was I that my plans should receive the benefit of a fair trial, that after the lambing had fairly set in I spent my whole time out in the back country; sleeping at the bushmen's hut, but during the day constantly in the saddle riding from one station to the other, going into the yards in the morning, and viewing how the lambs were attended to on the run, and seeing with my own eyes that none of the men were negligent of their duty.

As Lilly had predicted, I found in Lampiere a most conscientious shepherd for the lamb flock, when once he had been put into the way of managing these wilful young creatures, and of being prepared against their peculiar habit of making a break for their camp, whenever their shepherd left their rear.

When this occurs, any of the young creatures who chance to miss their mothers instantly and instinctively make a bee line back to their camp. If this stampede once begins it is extremely difficult to check it, for an ordinary dog is then utterly powerless to arrest them, as the senseless things pay no heed to his appearance at their front, and either double past, or bound over him.

Never shall I forget my intense amusement at the ridiculous spectacle poor Lampiere presented, on his first experience of this difficulty.

It had been explained to him, that he was to keep close upon the rear of the lambs, beating diligently among the bushes as he went along, to rouse up those who were asleep, for these are in such cases apt to be left behind, when they soon become a prey to crows and hawks and prowling dingoes.

So far, so good; but unfortunately Lampiere had not been also coached up in the further lesson, of first beating up the rear, with the quiet dog he had been provided with, when he found it time to go round the flock and head them for home. To trust this latter duty to the dog, was to run the risk of some of the lambs being cut off in the course of the career round the flock and darting in terror into the bush, and so being lost. Not knowing anything of this, Lampiere, when he at length deemed it time to head the flock back into the yard, had for this purpose left his hitherto careful station at the rear, when his attention was suddenly roused by a clamour behind him. Looking round he then saw with dismay a long string of lambs and ewes — lambs seeking their mothers, and mothers seeking their lambs, and more lambs following up these again — all making off as fast as they could for their camp, the lambs dancing and flinging capers in the air as they went along. Off ran Lampiere at the very top of his speed to head the runaways, but it was one thing to get at their heads and quite another to stop their progress when there. He ran, and yelled, and sent his dog in front of them, ordering the animal to speak up at the same time; but all to no purpose: the lambs only bounded higher in the air and ran all the faster. Just then I happened to ride up to the hut, and hearing a loud, splitting laugh, on looking for the occasion of it, as I came out of the scrub that came close up to where the hut stood, I beheld Mulroy literally shaking with laughter, and then observed poor Lampiere, hanging with one hand for support to a tree, and holding his hat in the other gasping for breath, whilst the mischievous lambs that with the ewes still kept streaming along on each side of him were, as if in sheer derision, actually kicking their heels at him as they passed.

I really could not help laughing myself on beholding the spectacle of helpless despair that poor Lampiere then presented, but we soon explained to him how to prevent a recurrence of the same accident, after which he got on famously, entering with heart and soul into his work, and putting himself to great pains to save any distressed-looking motherless lamb that he noticed, whose plaintive bleating, he declared, went to his very heart. As the final result of these laborious pains on my part, I found, to my infinite satisfaction, when the time for marking the lambs came, that while at Campbell's station the tally of lambs' tails represented a percentage of 99, that at Mulroy's station amounted to 103, the best percentage of lambs, as Mulroy remarked, that had ever been recorded in all the Darling district. The higher percentage at Mulroy's station, although Campbell's was also excellent, I readily ascribed to the greater care and attention paid by the former, combined with Lampiere's painstaking and unremitting attention to the lambs under his charge, by which hardly any, if indeed any at all, had been lost in the scrub or neglected in any way. With the man Harvey, on the other hand, though on the whole attentive, the same pains-taking care had not been so conspicuous; while in his flock I had frequently observed lambs whose pinched-looking noses and tucked-up appearance betrayed unequivocal signs of having been lost or deserted by their dams. However, the men had now no reason to regret my vigilance, because of the increase in their wages that this large percentage now secured to them. And thus this great business of lambing was at length satisfactorily dispatched, and we may turn to more eventful records.

The nature of my arrangements with the men whom I had, after the closing of the lambing, left at the back country, was as follows:—

Lampiere I left in sole charge of the flock that he had tended so well. Campbell and Crow I took in with me to the river station, to give them flocks in their former quarters, dispensing with Harvey; his flock was consigned to Burrel's charge, who, now that Lampiere was placed in charge of a flock of his own, preferred doing without a hutkeeper. This arrangement is common among shepherds, who, in consideration of the additional wage thus earned by doing their own cooking, etc., voluntarily lead the life of hermits among the back country solitudes. Such monopoly, however, is frowned upon by many bushmen, as tending to reduce the number of station offices that would otherwise be open for the employment of hard-up swaggers.

Charlie Knight was left to cook in the hut where Lampiere was now stationed until I should require his services, at no distant date, at the men's hut during the shearing season. At the hut with Lampiere the two bushmen, Crawford and Macalister, were still staying to complete some fences with Stack and his team; but the three latter, shortly afterwards, on the completion of their task, were to remove to the scene of operations for the dam, in constructing which they were to employ the interval between that time and the shearing.

Mulroy had now left me, to my great regret, I having no employment for him that would induce him to remain longer with me.

After Lilly, I never saw a man whom I could have liked better to have with me on a station. He was a man of about forty years of age, rather slight, but active, and very energetic, with a clean, smart appearance, and frank, open face. Although his education was limited to a mere knowledge of reading, for it was with the utmost difficulty that he could scrawl his own name, yet he had a keen, intelligent mind, and was acquainted with the writings and histories of some of the most famous British authors. In that particular, indeed, he was greatly superior in intelligence to Lilly. Like the latter, he could be depended upon like steel in matters of trust, and, like him also, could turn his hand to almost any sort of station work, everything he tried succeeding; although in point of artistic taste and downright mechanical skill, taken all round, in these particulars he was decidedly inferior to Lilly. Though small of stature, he was as firm in manner as he appeared to be courageous in spirit, and I had often admired the way in which in discussions with the fiery and more powerfully built Crawford, who, when checkmated, was at times inclined to bluster, the stern decision with which he could maintain his own ground until the other, who, as the saying goes, was "big enough to eat him," thought it wise to retreat from the high ground he was on such occasions inclined to take. If anything had occurred by which I should have been deprived of Lilly's services, I should at once have offered Mulroy the vacant post, for I knew of none who seemed so fitted in every way to fill up the void that Lilly's loss would have occasioned in Tappio.

As things were, however, I could offer no sufficient inducement to prevent his going away to other places where his good qualities were known and prized.

Like Lilly, although no drunkard, he was seldom encumbered by a superfluity of cash, owing to a like happy-go-lucky way of parting with his money. He left me, but for the short time he was with me I was strongly prepossessed in his favour, and lament that I never saw him again; although I have often wondered since as to what had become of him. Such are the wandering and unsettled habits of Australian bushmen. I only knew that after leaving me Mulroy went towards Queensland.

CHAPTER XXV.

"Few and short were the prayers we said,
And we spoke not a word of sorrow."
— *Burial of Sir John Moore.*

IT might have been about a fortnight after my return from the back country, when, a little after midnight, I was suddenly startled out of my sleep by a loud knocking at my bedroom door (in those days no one thought of barring their outer doors on the Darling), and on asking who was there, I was startled to hear a voice I at once recognised as Charles Knight's calling out. "Get up, please, Mr. Farquharson, the blacks have murdered Bill Lampiere, and I have had a close shave for my life."

At these awful tidings, I was on the floor with a bound, and striking a match, I sprang out with my trousers in my hand to the sitting-room, where Knight then was, asking hurriedly, "Bless my soul, Knight, what's this that you say? the blacks have murdered poor Lampiere! what has gone wrong with them? how did it happen? But dear me, my poor fellow, what a dreadful plight you are in yourself! have you travelled in barefooted all the way from the back country? how your feet are bleeding!" Knight appeared indeed in a most pitiable plight: his face was very pale and his lips bloodless, and his eyes had still a scared look in them as if from recent excitement. Yet in singular contrast to all this, his jaunty white cook's cap was still on his head. He was dressed in his trousers and flannel shirt, both torn, especially the trousers, almost to tatters, while his feet were bare and covered with blood as if they had come in contact with many a prickly shrub and ground thorn, in a desperate cross country flight. Yes! a cross country flight of forty miles, which, in that condition (such is the latent energy that despair can call into action), he had traversed since about four o'clock on the preceding after noon.

"Yes, sir," replied Knight to my question, "I have come in this way all the way from the back country." He then proceeded to give a detailed account of the disaster, which he did, however, in the same even measured tone, that no excitement seemed sufficient to alter; nay, even now his narrative was given as gravely as if his present trouble were not of any more importance than his ordinary cooking dilemmas, one of which he had once related to me in precisely the same tone of voice.

"It was just after sunrise," he said, "and Bill had just let his flock out of the yard, and in fact he was still standing at the gate, when a mob of about twenty blacks came up. I had just gone out to cut a sheep down from the gallows, with this knife that I have inside my belt" (pointing to a naked butcher's knife there), "and that I stuck to when the thing I am going to tell you of, happened. Well, of course you know that the sheep gallows is on this side of the hut, and that the sheep yard is on the other side of the hut from here. I at first didn't notice the blacks coming, on account of the dust that the flock had raised on going out of the yard, but the wind happening to blow the dust on one side, I noticed them when they were within a few yards of Bill, for the dog started barking then. Well, I just saw that they were all naked and had spears and *nulla-nullas*; but I hardly had time to think about what they might be after, when I heard a shout, and then I saw them all make a rush at poor Bill, and I could see one man strike him over the head with a club, whilst Bill seemed to cover his head with his arm. I heard the blow and saw him fall, and all the blacks round about him, and striking at him, when I turned and darted for dear life into the scrub that comes close up to where the sheep gallows is.

"Although the scrub is pretty thick there, yet I came up to a big box tree growing in the middle of it, with long, low-set boughs, spreading out from every side.

"Now, I knew very well that the blacks would follow my tracks in that soft sand as well as I could follow the tracks of a bullock, aye and better too for that matter. But as soon as I saw that tree, I says, 'Self, old man, you may be able to throw these black sods out yet if you are slippy'—at least I hadn't exactly time to say all that, but such thoughts flashed through my mind—so I just caught hold of one of the branches next me, and swung myself up into the tree, and when there, the first thing I does is to take my knife and rip open the laces of my bluchers, and off with them, and then ramming my socks inside of them, I took and flung them as far away into the scrub as I could. Then running out as far as I could on a branch that was leaning that way, I drops gently from it, on to a log, that by good luck happened to be lying there, and running along that log as far as it went, I strikes out again through the scrub. Now, I knew that the blacks would keep hunting for my boot tracks about that tree, and if even they did see my bare feet marks, at first they might think it might only be one of their own tracks.

"I was always reckoned a pretty smart runner as a boy, and now with my bare feet (for I was too excited to think about their getting hurt) I went through that scrub, and across the plain in the other side, just like a kangaroo. Now this plain, sir, is about a mile and a half across, and I was hoping I should get into the timber on the other side, before the blacks would be able to see me on the plain, but just as I was about entering the scrub on the edge of that timber, I heard a great yelling behind me, and looking round I saw that all the black devils had sighted me from the other side, and were now coming across the plain after me like so many warragul dingoes. Well, I just ran on into the timber, to the edge of the water-hole, and then stopped, for I began to think that the game was all up with me. So I says to myself, 'Self, old man, it's all up a tree with you now. It isn't no use your running into the water, they will just see where you go into it, and follow your tracks out of it again; it's no use, Self, you will have to die here, and as well die like a man, won't you, Self?' 'Of course, Self, I knowed you would; and you'll leave your marks behind on their ugly hides with this knife, won't you, Self?' 'Of course you will. You'll die game, Self, won't you?'

"'But stop, Self, look round, old man, and make sure that there is not a way out of this trap after all. Never say die if there's a chance to live, Self, you know!' Well, with that I just cast a swift glance, first round me on my own side of the creek and then across the water-hole, when I at once saw within about half a dozen yards of the bank a big tree growing in such a slanting way, that with my bare feet I could have easily climbed up its trunk. Its boughs hung a long way over a lot of scrub that was growing thick all round that side of the water-hole. 'Bravo, Self,' I said. 'You see your way out of the trap now, of course you do! These cannibals should have got up earlier if they had intended to cook your goose to-day, Self!' and I dashed into the water-hole, and swam across to the other side with my knife in my mouth. Then I walked straight up to this tree, and makes a gash in the bark as if I had been making my knife help to draw me up; but instead of trying to do that, I goes carefully back on my own track until I got into the water again, and again putting my knife between my teeth, I struck out as fast as I could to get round a bend towards the other end of the lagoon, before the blacks could get up and see me. Well as soon as I turned round this bend, I saw at the end of the water-hole, where a lot of scrub came down to the edge of it, a big log lying half in and half out of the water; so I strikes for this log to get out on the top of it into the scrub, so that I should leave no tracks to show where I had left the water. Well, as soon as I got my hands on to this log, I found it was only a half shell of a tree, which, on lifting up, I saw had space enough in its hollow to cover me over nicely, while I was lying along in the water, yet with my head on the ground at the edge of it. I thought this would be the best hiding place I could get, for they would never dream of lifting it up: and getting under it from the water, there would be no tracks that they could see on the bank that would betray my hiding place to them; whilst even if I got into the scrub, as I first had planned, even if they did not see tracks at the time, yet by making a circuit at a little distance from the water, as they would be sure to do, they would be bound to drop on my steps at last; and so run me down to death, as the hounds do a fox. So I just laid down in the track of the log, in the water, with my face resting against the bank and letting the log shell carefully drop into its place again, I could scarcely see anything, for the shell had lain there so long that there was a lot of grass and rubbish that had choked the space up in front of it. This, of course, I was only thankful for, as it would prevent the blacks from getting a sight of me if they passed by the end of the log, and so keep them from fancying that I might possibly be stowed away below there. Well, by and bye, sir, and not long after, neither, I could hear them yelling and jabbering at the farther end of the water-hole, and splashing through it. They seemed to stop there for a long time, as if they were hunting for my lost track. I knew at the time, that as the whole crowd would dash on after my tracks to that tree, they would just make them in such a mess with their own, that they would not have a chance of noticing my back tracks between them, as they might have done otherwise, for they have eyes like hawks. By and bye I could hear them hunting about and coming my way, as if they were looking for tracks leading from the water-hole among the scrub; whilst two or three came coasting round the water's edge, and walked across the very log under which I was lying. However, as the log lay solid, they never had a suspicion of my being under it. After, I should think, about three hours hunting about for me, I at last heard the sound of their voices dying away as if they were going towards the hut again. But I lay for a long while where I was after that, for fear of some of them having stopped behind to watch for my re-appearance. Then slipping out from under my friendly log, I crawled away into the scrub, for there was not much fear of their noticing my tracks now even if they did come back to look, among so many of their own. Then walking quietly and listening for any signs of the presence of my enemies on the watch for me, and keeping always where the scrub was thickest, I at last made a straight run for the river, where I knew the station lay. For a long part of the way I have come at a run, and by good luck I found some water in the salt-pans as I came along; but for the last few miles my feet have felt so dreadfully tender, and I began to feel so weak, that I could almost have lain down. I suppose the reason of this was that the sense of danger from the blacks had now left me; so that, with the absence of this fear, there came a consciousness of pain, that I never noticed while thinking that my life might be in jeopardy at any moment, for the first twenty miles or so. I was determined to continue as long as I was able to keep moving, so as to tell you what had happened, that you might get out at once and be in time to save the other men's lives who are there, that is if it is not too late now. God grant that it is not! Now, sir, I can hold out no more." And with these words the brave little fellow, whose sense of duty, on behalf of the men exposed to the peril he had just escaped from, had still nerved him to persevere in his efforts to reach the station, even after the spur of immediate danger to his own life had ceased to urge him; here sunk in almost a fainting condition into one of the chairs.

Although always abstemious as regards alcoholic drinks, I was not then, as I have since become, a total abstainer, and a plentiful supply of spirits was always kept in a keg in the store, for the station requirements. In the present emergency, even the staunchest blue ribbonite would admit this to have been a fortunate occurrence. Of the contents of this keg I usually had a bottle in the cupboard, to which, on seeing Charlie's sudden weakness, I went hastily and bringing it from its recess and filling a glass to the brim applied it to Charlie's lips, and he after one or two sips of the stimulant, that past habits of rather free indulgence had made only too grateful to his taste, took the glass from my hand, and drained off the remainder at a single draught. I next helped him off with his tattered trousers and, placing a basin on the floor, and filling it with water, began to wash the poor fellow's lacerated feet. The smart of their numerous wounds caused him to flinch a little, but after rubbing them gently, I took them out of the water and rolling them up in my towel left them, whilst I went again to the cupboard, and took out some cold provisions, in the shape of roast mutton and the remains of an apple tart, and bade him at once partake of them and afterwards roll himself into my bed. I then replaced the bottle in the cupboard, and turned the key, telling Charlie that I did so lest he might be induced to help himself too freely to its contents, which, in his present weak and excited state, I thought might do him harm instead of good, and he thankfully acknowledged my care for him.

These hasty offices discharged, I at once went out and roused up old John, and telling him what had occurred, bade him at once get breakfast ready for Lilly and me, and as soon as his tea was prepared, to carry in some for Charlie, and also to keep his eye on the latter during the remainder of the night, lest he might be suffering from any feverish symptoms consequent on his late excitement and terror.

From the kitchen I passed rapidly down to the river's bank, and getting into the canoe, as rapidly paddled myself across, and springing out and hastily mooring it, ran up the bank and on to Lilly's hut, whom my loud shout as I entered instantly roused up in astonishment, that at once gave place to an expression of horror when he heard the cause of my presence there at that hour of the night. But as there was no time now for useless sorrow in face of the instant action that was required of us, Lilly's horrified look quickly gave way to one of determined energy and grim purpose, for, for this young man, Lilly's first feelings of regard had now begun to deepen into those of positive affection. But beyond the utterance of a few fierce and excited expressions, he made but little comment on the matter, although it was apparent that the tidings, over and above their natural horror, went to his heart with the poignancy of a personal bereavement. But our only thoughts now were as to what was the best and quickest thing we could do under the circumstances. Selim, without whom I could never imagine myself properly equipped for any emergency, was, unfortunately for me, out on furlough, running at large in the bush with the "mob," as all the horses when together are termed on a station, and in his place in the paddock was a horse I had been lately using. This horse, though of good enough average mettle, was a little flighty, and occasionally required the control of a firm hand when suddenly confronted by anything unusual, for he was easily alarmed and then inclined to be restive. Lilly and I, however, at once determined to ride out to the back country, leaving word for Cabbage-tree Jack and the black fellow Snowball to come after us as soon as it was light enough for them to secure a horse apiece, and that each was to bring a gun and ammunition with him. We, on our parts, armed ourselves with a revolver apiece, besides taking a double-barrelled gun with plenty of ammunition with us for the use of the men to whose assistance we were going. Then, leaving Lilly to secure his mare, that was always kept in a paddock close to his hut, I recrossed the river and went in the paddock for my horse, that, after a little manœuvring, I managed to lay hold of, and led up to the house just as Lilly rode up on Coleena.

Partaking hastily of the breakfast now ready for us, we mounted, and pricking on at a rapid rate, were on our way to the back country within half-an-hour from the time that I had first been roused by Charlie Knight, who, when we left, was in that profound slumber that only utter exhaustion can produce.

We were very silent during our long but rapid journey to the back country, which was not so rapid, however, as to leave us with blown horses on our arrival. We carefully guarded against that result, and throughout our long journey, the major part of which was performed in the darkness of night, made still more profound by the many belts of timber and patches of dense scrub that traversed our path, whilst going at a hard gallop, or swinging canter, in places where the country along our route was open and level, in other places where it was uneven, and we thought it necessary, we carefully breathed our horses with long walks.

It might have been about eight o'clock on the ensuing morning, when we reached about the centre part of the back country run, that was as yet undistinguished by name—a few miles still intervening between us and the scene of the previous day's tragedy. Here, at a point where the dray track divided into two, one proceeding on to Lampiere's hut, and the other to that where Burrel was staying, Lilly suddenly pulled up and remarked: "What about this man? they may have knocked him on the head too when they had their hand in ". Singularly enough the thought of Burrel's danger had never up to that moment crossed my mind, as in his solitary situation it might well have done, but now, upon Lilly's remark there flashed through my mind the dreadful probability of such a fate having overtaken Burrel also in his lonely and defenceless position, without any firearms to defend himself with. Unhappily, in the absence of any idea of danger threatening us from the blacks, I had taken home one of the double-barrelled guns, on the conclusion of the lambing season, leaving the other with Crawford and Macalister at their own request, for the sake of shooting game. I therefore, at Lilly's word, cried in despair, "Dear Lord, yes! they may have done so indeed".

After a few moments of hurried consultation on this new aspect of affairs that the thought of Burrel's danger brought to our view, we resolved that Lilly should continue straight on to Lampiere's hut as he had originally intended—to learn the certainty of Lampiere's fate, and then to either wait there or to push on to where Crawford and Macalister were now encamped (at the place where the dam was being constructed, and to which they had removed a few days previously) according as circumstances might direct, whilst I meanwhile would ride on to see what tidings I could find of Burrel's fate. With these resolutions we went along our several paths, and soon were lost to one another's view.

I had left Lilly about a quarter of an hour, and was hurriedly posting across a scrubby sand hill, when on my entrance into a clear space in the middle of it, I was suddenly confronted by the spectacle of about twenty black fellows springing to their feet, all entirely nude, but armed as if for a foray, whilst with scowling looks they all appeared to resent my sudden intrusion upon them. At the sight of them suddenly bounding to their feet, my nervous horse snorted and half reared up.

Had there been really less danger than there evidently was, I believe I should have been startled enough at this unlooked for encounter myself; but I believe I am speaking within reason, when I assert that though on occasions of slight danger, nervous fears are as incidental to bold, as to timid spirits, yet with those whose nerves are really sufficient for the strain of the severest peril, the consciousness of the necessity for immediate action enables them to at once recover from the thrill of fear that the first sight of danger naturally engenders; the greatness of their danger leaves them no time for the indulgence of fear. Thus was it with me.

The imminence of my danger from an immediate attack was so plainly evident from their scowling looks and menacing gestures, that, with a powerful effort, I at once controlled my nerves, and with my bridle hand firmly restraining my agitated horse, and holding my revolver in readiness with the other, I sternly demanded of the foremost of the black fellows—a burly, but brutal looking savage—in the strange idiom, by which these people are familiarised with the English language: "What name you want?" In his reply he showed that he was at least conversant with some of the lowest forms of our language, as uttering, " Go to —— you white ——," he hurled his *nulla-nulla* straight at my head. Only by a swift inclination of my body to one side did I save my skull from being crushed by this weapon in its unerring aim ; but while in the very act of this inclination, my horse, already shaky with fear, was altogether upset by the whirring sound over his head and bounding suddenly to one side he plunged so violently that I, already off my balance, was thrown clean out of my saddle to the ground ; with the tenacity of despair, I clung to my horse's bridle, and was at once on my feet, and, as the savage, now triumphant at my fall, that he doubtless fancied had left me completely at his mercy, rushed towards me with the intention of impaling me with his hideously jagged spear, I levelled the revolver that I had fortunately still retained in my hand, and shot him through the centre of his breast. He fell on his face and expired without a groan. At the sight of their companion's sudden fate, all his cowardly companions instantly turned tail and darting into the scrub, vanished out of sight.

Giving a glance at the prostrate black fellow, and seeing that he did not move, I drew my horse towards me, although still palpitating violently (the horse, not me—I was now as cool as a cucumber), and throwing the rein over his head, I again mounted, and rode rapidly forward on my way, now doubly vigilant from my late encounter.

This vigilance of mine was soon rewarded.

At the edge of a belt of scrub a little to the left of me, I suddenly observed, as I cantered along, the figure of a black fellow squatting on his hams. Instantly pulling up my horse I regarded this new object fixedly for a moment, then, struck with the fact that, although he was eyeing me, yet he made no motion or demonstration of alarm, I rode up to where he was, my revolver held ready with my finger at the trigger in preparation for the first motion of treachery from either him or any concealed companions in the background.

On riding close up and regarding him attentively, what was my surprise to observe that he was severely wounded, mutilated, in fact. The imperfect covering of earth with which he had been attempting to staunch the flow of blood from a hideous gash in the thigh, did not prevent me from seeing that the flesh had been literally torn away from the bone. Recovery from such a wound was plainly impossible ; even with the most skilful surgical treatment, mortification could hardly for long be kept at bay where there was displacement of so much living flesh ; how much less so by this miserable savage, absolutely destitute of liniment or bandages for the binding up of such a wound save such virgin properties as might be found in a salve of mother earth. In reply to my fixed look of inquiry he said nothing, though sullenly meeting my gaze.

He had the appearance of having been a burly savage, with looks malignant and brutal enough to have made him pass as the brother of him whom I had just shot.

"What you name?" I demanded of him briefly. Let the reader judge how my heart thrilled at the reply—I say, judge it —in the light of the tale so dramatically told some weeks before by Tom Crawford, of the cold-blooded murder on the Warego.

"Billy the Bull."

"Billy the Bull, ha! you rascal! What for you kill white fellow? Who shoot 'em you there?" I sternly asked, pointing at the same time to his mutilated thigh.

"White fellow."

"You kill em 'nother one white fellow over there?" I asked, pointing over in the direction of Lampiere's hut. He made no reply, but merely regarded me with that dour, sullen look, that such savages assume, when conscious that their deeds have cut them off from all hope of mercy.

For a moment I grasped my revolver with the idea of at once despatching him : but for the certainty of the impossibility of his recovery from his wound I most assuredly would have done so ; but to me there was something so peculiarly horrifying in deliberately taking human life, save under the most dire necessity of self-preservation, as in my recent encounter. To the reflective mind, this must naturally be always the case, as he remembers that from the eyes of the most brutal savage there still gazes at him the soul of a fellow man. But with this man's complicity in the cold-blooded murder of two unoffending men some months before, and his hand also probably reeking with Lampiere's blood (what other blood I knew not, for his mutilated thigh gave evidence of a scuffle with the bushmen or Burrell, the results of which I had yet to learn), what I would have otherwise shrunk from doing, I would now (but for the above cause) most assuredly have done. With a clear conscience, too, would I have done it, and calmly braved the responsibility of taking the law in my own hand, in obedience to the Mosaic law that "whoso sheddeth man's blood, by man shall his blood be shed".

Therefore leaving the wounded wretch to his fate, a fate which I could not have ameliorated even had I had the wish to do so, I turned my horse from him and rapidly made my way to Burrel's hut.

On entering, I saw at once, by the extinguished embers, that Burrel had not been there on the previous night. The shepherd hutter, as one who manages without a hutkeeper is termed, in the morning, ere leaving his hut, to which he is not supposed to return until the evening, carefully covers up his embers with ashes, and they are by this means kept in until his return at night.

This discovery greatly increased my uneasiness about his safety, and with the almost despairing hope that being warned by symptoms of danger from the blacks, Burrel had driven his flock for greater security on to Crawford and Macalister's camp, I urged my now jaded horse thither.

On my approach to this camp, to my great joy I found that what I had regarded as an almost forlorn hope, though the only one that remained of Burrel's safety, proved to be in reality true. Of this I was assured by both eyes and ears on coming in sight of the bushmen's tent, by seeing and hearing Burrel's sheep bleating in the newly-made yard. On approaching nearer, I observed all the men standing close together. At the same moment I noticed Lilly suddenly dismount from his horse, and step towards Crawford. By their gestures they appeared to be in violent altercation with each other.

But before explaining the cause of this altercation, I will first state all that had transpired here on the previous night, when the blacks had made a most determined attack upon Crawford and his companions, by whom however they had been repulsed with some loss.

Evidently, on desisting from their baffled pursuit of Charlie Knight, the blacks had returned to Lampiere's hut; for Lilly, who had stayed there for some time looking about for his body, that however he failed to find, stated that the hut had been looted of its contents and had then been set fire to. Also, from the tokens of freshly picked bones it had been evident that they had been feeding on the carcase of the sheep that Knight mentioned he had gone to cut down when he had noticed them as they made their attack upon Lampiere. Thence the blacks, with the evident purpose of attacking and exterminating the whites piecemeal, proceeded over to where Burrel lived, with whom in his solitary and defenceless condition they evidently doubted not that they should easily accomplish their fiendish purpose.

Now, it had happened in a most providential manner that Burrel, being rather fond of a gun, and observing the number and tameness of the paradise ducks that were on a part of the run where he fed his sheep, had just the day before driven his flock over to the bushmen's encampment, and asked the loan of their gun, promising to return it in a day or two. He then, with this gun in his hands, was rather surprised when late in the afternoon he observed making straight towards him a number of naked black fellows, who, on being confronted with the unlooked-for spectacle of a gun in the hands of the man they had probably imagined to be unarmed, suddenly came to a halt.

Now, Burrel was a shrewd man, and the unusual sight of so many blacks, all armed and naked, to his mind boded no good, so he at once got his gun in readiness for any suspicious indications that the blacks might give of their intentions towards him. Their sudden halt too at the view of his weapon made him still more suspicious of the honesty of their intentions, so he at once brought his piece to the carry. For a short time the blacks appeared as if considering how to get at him. Then after a short pause one of them advanced from the rest towards him, and with what English he was master of endeavoured to lull Burrel's suspicions by the most friendly professions, as, "Budery (good) white fellow, budery black fellow," all the while advancing towards Burrel with a smiling countenance, until the latter at last beginning to imagine that his suspicions might after all have been groundless, suffered himself to be so far thrown off his guard as to let the butt of his gun drop upon the ground. On this the treacherous black fellow, who by this time was close upon him, suddenly bounded forward, and seizing hold of the gun attempted to wrest it out of Burrel's hand.

But Burrel, though only lightly built, was wiry. Holding on with his right hand to the breech of the gun, with his left he directed the point of the muzzle against his antagonist's body, at the same time pulling the trigger, when the black fellow, shot through the abdomen, instantly dropped on the ground. Recovering his weapon, Burrel at once presented it, with the remaining loaded barrel, to the other blacks, who, during this scuffle, that had not lasted over a second, had been rushing towards him, but who, on witnessing their companion's sudden fall, and the white man's terrible weapon again pointed in their direction, turned instantly and fled with great speed towards the thicket from which they had lately emerged.

CHAPTER XXVI.

KNOWING how unsafe it would be for him to remain longer by himself in the hut, in view of further molestation from the blacks, whom he had now more reason to dread than ever, Burrel determined at once to seek protection at Macalister and Crawford's camp. He knew that from his solitary condition he was exposed to the dangers of a midnight attack from his enemies, who would now be incited by a desire for vengeance for the blood of their companion, who was still writhing on the ground, but whether mortally wounded or not, Burrel did not wait to ascertain, for, hastily loading his gun and sending his dog round his sheep, he drove them straight for the bushmen's camp, which he reached just about sundown. The men had just ceased from their work, and were washing themselves as Burrel arrived. On hearing Burrel's account of the manner in which he had been attacked by the blacks they became greatly excited.

"Depend upon it, lads," said Crawford, "we are bound to be attacked by the blacks to-night, in revenge for the man that Burrel either killed or wounded. There wouldn't have been so much danger if Burrel had only managed to frighten them; but now that he has drawn blood, their fury will be roused, for they are very revengeful, and what their want of courage would prevent their doing, their desire of vengeance will incite them to attempt. Now, it is my opinion that they will try and sneak upon us here towards morning, when they think that we are asleep, so we had better sit up all night, and be ready to give them a warm reception if they do come. But, Jehosaphat! we have only one gun among us, and no bullets! As for the bullets, however, we can soon make a good substitute for them by chopping up some nails. That will do far more damage among them than a ball would, for that matter. But, lads! what about the two other chaps, Lampiere and Charlie? They are bound to be murdered if they are left by themselves."

"But what can we do now to help them?" asked Macalister. "They are more than four miles from here, and it will be dark in no time. It would be as much as any man's life is worth to venture out to give them the alarm, with these savages prowling round the place in all directions."

"What do you say, then, to our going along there in a body? If we meet the darkies on the way we can fight them as well there as here."

"Yes," remarked Billy Stack, "that's plain enough, and we may be in time to save the lives of these two chaps yet, and if so, then by joining them there will be all the more of us to show face to the blacks if they do come near us."

Crawford here made some rough comments about my carelessness in leaving so many men unprovided with firearms against the desperate emergency that was now imminent. "What the blazes could he have been thinking about?" he passionately asked, "or does he value the lives of men no more than if they were so many dogs, when he would leave only one double-barrelled gun amongst so many in a wild country like this is, teeming with blacks that have never seen a white man before? And he knew very well, too, that blacks were never to be trusted."

This rebuke, indeed, seemed well merited; for it did appear an act of desperate rashness on my part, with the responsibility of making some provision for the protection of the lives of these men resting upon my shoulders, that the sudden contingency that was now menacing the lives of all should be found to be so utterly unprovided for.

I could, indeed, have retorted, had I been present, that, though so censorious after the event, yet he, too, had been as utterly unsuspicious before, and also that a gun would have been provided for any man who had expressed any feeling of fear in being left there without one; besides the fact that, though hitherto unacquainted by face with the whites, they could scarcely have been so by report. With these blacks, too, both numerically and physically so contemptible, our original idea that in our dealing with them the only precaution necessary was a judicious exercise of firmness and kindness, would have proved perfectly well founded, but for the wholly unlooked for contingency of the advent among them, and consequent demoralisation, of the Warego murderers. Yet, even so, attention on my part to the note of warning as to the treacherous propensities of the blacks, given as a decided maxim by Crawford in his account of these same Warego murders—viz., "Black fellows are never to be trusted"—would have found me prepared.

But to return; in reply to Billy Slack's endorsement of Crawford's plan that all should go at once to Lampiere and Knight's hut, Burrel said, "It is my firm belief that the time is now past for being of any service to either Lampiere or Knight, and that what murdering the blacks may have intended doing there has now either been accomplished or attempted, and that Lampiere and Knight at the present moment are either dead men or skulking somewhere about in the bush. I have been thinking of it all as I came along here. It was from their direction that the blacks were coming when I observed them approaching me, and we know that neither of these chaps had anything with which to defend themselves against the blacks if they had attacked them; and even if poor Bill Lampiere had a gun, he would have been so utterly unsuspicious of them, that they would have knocked him on the head before he imagined that they had any evil intentions against him. He was not like me in that respect, for I am by nature apt to be suspicious, and am quick in reading men's faces. Charlie Knight, poor Self, would be quicker, but without a gun what could he do if he saw the blacks coming, except run? and if he did they would just run him down, unless he managed to dodge or baffle them, which God grant that he did; but I feel certain that it is no use to go over to that hut now."

"Burrel is right," replied Crawford, "I can see it now as plain as a pikestaff. The blacks have been intending to make a clean sweep as they went, and even if they had managed to kill Burrel they would have been upon us to-night, and if they had, every man Jack of us would have been knocked upon the head, for they would have sneaked upon us during our sleep; and you will find they will try and do that, as it is, to-night. And you know, boys, I told you before that Billy the Bull, and Tommy the Turk, who murdered the two whites on the Warego, were supposed to have made their escape down to this back country. Now, mark my word, you will find that these two wretches are among these blacks, and it is they who are putting this devilment into the heads of the others to kill the whites, as they seem bent upon doing. Now, as matters stand, we can do no better, as we have no time to fortify ourselves with logs, than to make them believe that we have no suspicion of their coming, but we will keep on our clothes, and hang a blanket inside in front of the tent door, to prevent the blacks from seeing our light, and be ready when we hear the dog growling. I will take the gun. I know these two black satans by sight, and if they are there I promise that one of them at least shall lose the number of his mess before morning. But now to supper boys, and have a good one whilst we are at it, for fear it should be our last on earth; but make your minds easy for a few hours yet, for the blacks never attack till near morning, when they expect people to be in their soundest sleep."

These words of Crawford's seemed to all the best counsel that could be acted upon in their present emergency. It was now about dusk, and the bullocks were unyoked, otherwise they might have constructed a hasty fortification of logs, thereby preventing their being surrounded by the blacks in their attack.

The soundness of the advice to lull suspicion by an appearance of slumber was evident, whilst keeping on the alert for their secret approach, so as to enable Crawford to give them the benefit of at least one of the barrels of his gun with some certainty of aim, and then to stand to their arms of working tools and do what they could, if the blacks should still come on.

Accordingly, at the usual hour for retiring, the tent was darkened by a blanket, carefully hung in front of the slush lamp, so as to intercept any rays from its light that might have indicated its presence to the watchful eyes of prowling enemies, while all the men listened in silence for their first approach, Burrel keeping his dog beside him at the door of the tent, and occasionally inciting him to watchfulness by a whispered "Look out for them, Tweed".

At length, after the suspense of a few hours, the wisdom of their precautions became patent from the dog's manner. At first sniffing the air, and growling suspiciously, he at length darted out into the scrub barking furiously. This action he several times repeated. At length the suspicions of the watchers as to the cause for this was confirmed to a certainty, for the dog, after one of these furious charges, was suddenly heard to make a howl, as if from the effect of a violent blow that caused him to retreat, yelping back to the tent. All the men were now ready for what was to come, Crawford in front, resting on one knee, with his gun held in readiness. The flap of the tent door was held also in readiness by one of the men, to be drawn aside the moment that the blacks made a rush. The night outside was clear and partly illumined by a waning moon that allowed objects to be easily distinguished for a few yards in front of the tent.

All at once Crawford whispered, "I see them, by ——, they are coming out of the scrub, crawling on all fours. Stand ready boys, I see a lot of them coming; pull the tent flaps aside, and drop the blanket from before the light. Jehosaphat! Billy the Bull!" As he uttered these words, Crawford instantaneously fired his piece. There was a scream of pain that testified that his shot had not been thrown away. Following this yell, there came another, but more defiant one, as the blacks all sprang to their feet and rushed towards the tent, when the report of Crawford's second barrel again rang out, but failed to check them. On the same instant the white men were fighting for their lives against an over-powering force of black fellows. One of these, a powerfully built man, seizing Crawford's gun, endeavoured to drag it out of his hand. "Strike, lads, strike, and keep together for God's sake," shouted the latter while desperately endeavouring to free his weapon from the grasp of his assailant.

The white and black man seemed fairly matched in strength, but not in wit. Finding it imperative to immediately free himself from the struggle with the black from the fear that others might come to his assistance, and being unable to do so by sheer strength, Crawford suddenly released his grasp of the gun, and seized hold of the black fellow by the hair of his head and instantly dragged him to the ground, and almost within the tent, beyond which he had been carried in the struggle;

then, placing his foot upon his neck, he seized hold of the gun, that his now almost paralysed antagonist at once quitted hold of. Instantly clubbing his gun, Crawford lifted it over his head with the intention of dashing out the brains of his prostrate foe, but with a writhe, a bound, and a side spring, the twisty black fellow was on his feet and bounded out of the tent, and into the scrub, all his companions instantly giving way too and fleeing after him.

"Come away out into the open, lads," shouted Crawford on this fortunate issue of this desperate affair, "don't wait here; they may sling their spears at us from the scrub."

As a matter of fact, however, though in his excitement Crawford had forgotten it, the natives of the Darling and its surrounding districts never throw their spears like those of other parts of Australia; but the advice at the time seemed too prudent to be disregarded, and was at once complied with by all retreating for some hundred yards out on to the clear ground, where, by Crawford's directions, they hastily gathered a heap of rubbish and set fire to it, so as to be enabled by the light of the flames to see if the blacks made any show of renewing the attack. But they made none.

Whatever blacks had been wounded had contrived to go off with their companions. For when, shortly afterwards, the white men were able to scan the scene of the late conflict by the light thrown by the fire they had kindled no signs of any body lying there could be seen. Yet that there had been some wounded was evident, for, besides the first shrill yell of pain that had followed immediately upon Crawford's shot, Macalister on the first onset of the blacks had been seen to fell one with a blow of his axe. The blow, indeed, had been partly broken and changed in its direction by the interposition of the black fellow's narrow shield, but that it had taken some effect on his woolly head was certain and borne witness to from the bloody condition of the edge of the tool. And there was blood also on the sheath-knives of both Burrel and Stack, who both fought with these weapons in their left hands whilst provided with a short stout stick to guard their heads against the clubs of their assailants.

With all the excitement consequent on this victory, it may be easily believed that the idea of slumber was but little thought of for the remainder of that night, though in fact there was by this time but little of the night left even if they now had desired to sleep. They did, however, after a while, when assured of the retreat of the blacks, return to their tent. This assurance they obtained by observing the cessation of the barking of the dog, that had been the first, by his keen scent, to admonish them of their danger.

On leaving Lampiere's, Lilly had come on their camp at about nine o'clock, and found them all in the act of setting out on a raid upon a camp about a mile from their own, where some blacks were, who, whether they had been connected with the misdemeanour of the others or not, had appeared all along to be of a most peaceable and friendly disposition.

Great was their satisfaction on Lilly's producing another double-barrelled gun, with ample ammunition for their requirements. And great was their sorrow, especially Burrel's, on hearing Lilly's sad verification of his too well founded fears as to the fate that had befallen his former hutmate, whose loss he could the more appreciate as that of one whose tastes had coincided with his own.

"Come along then, lads," cried Crawford, whose blood was now up at Lilly's narrative. "We have plenty of weapons now; let us go over to that camp and shoot down man, woman and child: it will be the right way to strike a proper terror into these fiends"—a resolution which was at once vetoed by Lilly, who said, "Shoot the men, but curse the idea of shooting women and children".

"What is there worse in shooting them than in killing young snakes? they may be harmless enough now, but the children will be just the same as their fathers when they grow up, their treachery is ingrained; and, as for the women, I reckon it is the only way of striking terror into the men to kill them. You do as you like, I intend to blaze away at whoever I come across."

"No, Tom Crawford, you will do nothing of the sort; wait here till the boss comes, and until he does come I reckon I'm boss."

"You're boss, are you?" Crawford replied excitedly; "then why weren't you here to boss us last night, when four men were fighting with their lives in their hands against twenty murderous savages?"

"I am here now, however," replied the other, "and ready to fight as many murderous savages as you will, without blowing about it, and without playing the part of a coward in shooting helpless women and children either."

"Look, Lilly," replied Crawford, now in a fury, "I have faced singlehanded Warrugul blacks where few men would care to be seen even with help beside them, and I will not be dictated to by either you or any white man breathing as to what I am to do and what I am not to do with these blacks after the manner in which I and these chaps have had to fight for our lives against them, and I am now going over to that camp to do as I said I would do, and the man that interferes with me, 'boss' or no 'boss,' I'll put a bullet into him whether I am a coward or not."

"You will?" answered Lilly, his cheek whitening with gathering passion.

"Yes, I will," replied Crawford, his eyes actually blazing.

"Well, blast you, let us see who can aim the straightest," Lilly fiercely replied, springing from his horse and drawing his revolver.

"For God's sake, chaps," cried Macalister in horror, "mind what you are doing; it will be downright murder if you shoot one another."

The others seconded the speaker with their words of entreaty, but out of fear of their weapons refrained from going between the disputants.

For a moment Crawford looked resolutely at Lilly, as if on the point of at once accepting his challenge, but in Lilly's stern, concentrated gaze, he read the unyieldingness of a rock. Perhaps too, he might have thought, that, in the event of this encounter Lilly's wonted superiority in most things would not be to his own (Crawford's) advantage on the present occasion. A habit of bluster, even when supported by a foundation of natural courage, usually impels a man of this kind beyond the point he intended, and from which a calm undaunted opposition soon makes him retreat. This was the case with Crawford. For a moment, as I have said, he paused, and during this pause his better sense prevailed. "Lilly," he said, dropping his gun stock that he had just before got ready for action, "you are game I know, but so am I game; and I know of no other man besides yourself to whom I would now have budged one step, but on the whole I believe you are the smartest man of the two."

"Why, Tom Crawford," said Lilly, readily softening at the other's concessions, "I know myself that you are as bold as a lion; but ask yourself what sort of manliness there is in shooting women and children. You will have plenty of men to shoot at before it is all over; and now you come along with me, for I swear that I will get on the tracks of the cursed gang that has killed that poor young fellow, and follow it till I can get at them, and then I'll shoot the men down like so many dingoes."

"Well yes, Lilly, that is the best way; but in Queensland, where I have been, people have not been so particular as to what they shot, I can tell you. Many's the time I have been with squatters who have ordered us to shoot right and left, women, and children, just as we came across them," replied Crawford, without the slightest sign of compunction at such a horrid confession.

"Well, Tom," replied Lilly, gravely, "I don't pretend to have much religion about me, but I would sooner that you should have to answer for that work than I, in the next world."

It was at this time that I rode in among them.

"No, no, that would never do," I said hastily, on learning the cause of the dispute, that had ended so happily. "This is a bad business, men, and I am afraid I am to blame for leaving so few guns among you; but, as you are all aware, there had never been the slightest suspicion of any danger from these blacks before. It is useless however to talk about that now. We must try and find out in the first place what has become of Lampiere's body if dead; or, if he is alive, of which I fear now there is but slight hope, where he can be hiding; but as for killing the women and children, why, Crawford, I am surprised at you for thinking about it, if for nothing else, why, then, for fear of the danger you would bring your own neck into. Depend on it, if once a rumour of such a thing were to reach the police, you would be taken up and arraigned for murder; and as for the blacks at that particular camp, I will not hear of their being interfered with till I have clear proofs of their connection with this disturbance. They have been very friendly to us all along, and it is my opinion they have had nothing to do with it, for I have just come across the murderers and had a narrow escape myself, and there wasn't one among those whom I could recognise as having seen before." I then gave them a detailed account of my encounter with the blacks on the scrub, and of shooting one of their number and my subsequent meeting with "Billy the Bull".

"I'll bet my life, Mr. Farquharson," said Crawford, when I concluded my narrative, "that that big black fellow you shot was Tommy the Turk, and if so the murders of that poor German and Captain Bruce are now avenged, from what you say of your certainty of Billy the Bull's death from the dose he got of my chopped-up nails. I am glad that I had something to do with paying up that score anyhow. And now chaps, that these two bloody scoundrels have been wiped out, you will find that this disturbance with the blacks will be at once properly quashed."

I now listened to a detailed account of their proceedings from the men.

I then proposed that Lilly and I should have some breakfast, and give our horses a rest and feed of grass, as both were looking considerably the worse for their long and rapid journey. Burrel I also now directed to let his bleating sheep out of the yard, and with him sent Billy Stack with one of the guns in case of any further attack.

After we had eaten our breakfast and a couple of hours had been allowed to elapse for the sake of our horses, Lilly and I again mounted, and accompanied by Crawford and Macalister on foot, went on our way to the place, where from Knight's description I understood that Lampiere had been struck down, to see what could be there discovered of him dead or alive.

There we expected to be joined by Cabbagetree Jack and Snowball, the latter of whom would at once, through his native instinct and early training, enable us, if baffled, to follow the trail we were seeking.

As it happened, we were able to do this without Snowball's assistance.

On reaching the gate of the sheep-yard, a large, dark red stain on the ground, as if from recent blood, gave only too clear a proof of the tragedy that Knight had described.

But as we were gazing at these sorrowful evidences of poor Lampiere's fate, noticing too that all round about the ground was marked with black fellows' tracks, Lilly suddenly pointed to the footprints of a single black fellow, deeply marked, as if they had been made by some one who had been struggling under the weight of a heavy burden, and they went towards the other side of the yard that was bounded by the water-hole. On following these steps, we found, from the further evidences of blood, that the body had been laid down here and had been temporarily covered over with some brushwood lying there, that showed signs of having been removed from its original position, which was marked by the more tender appearance of the grass on which it had been lying.

Thence the foot tracks of the person who had brought the body there, and who had apparently again removed it, could be easily followed leading towards the end of the water-hole opposite to that on which the hut was situated. Then, crossing the water on the dry channel, they went on through the scrub and bush on the other side. The soil about the place being of a sandy nature admitted of our easily following such heavily weighted steps as we were now eagerly tracing.

We had proceeded thus for nearly a quarter of a mile, following these tracks with but little trouble as they led through the scrub, when with the sudden exclamation of "Look!" Crawford levelled his gun at a black fellow named Charlie, whom he had at that moment sighted coming towards us from a space of rather more open ground in front.

This Charlie, who had been with me at the lambing, tending the young lambs at Mulroy's Station, had come to me under a cloud, having been suspected of having enacted a treacherous part in an affair where a poor cook had been murdered in open daylight, by the blacks in a back block station further up country. On this station Charlie had been retained as a sort of a station hand, his duty being chiefly to run in the horses when any were required. The cook had been murdered in the absence of his employer in broad daylight, and in sight of the people at another station on the other side of the creek, that there formed a broad sheet of water, from which point the people vainly attempted to frighten off the murderers by firing their guns, but the distance was too great for them to have any effect. Charlie, from his position on the station, could, it was supposed, had he so chosen, have easily warned the cook of his impending danger, and this for some cause he had not done. Therefore, for thus having failed to warn the murdered man of his danger, the owner of the station, a ruthless and determined man, immediately on his return swore vengeance against Charlie's life, whilst he forthwith began a war of extermination against all the tribe of blacks, shooting down without compunction men, women, and children. To escape a similar fate Charlie went for his life, and thus arrived at our place, when I took him at once into my service, in spite of this grave charge against him. I had particularly noticed that he was civil and attentive to his work, and, for a black fellow, rather intelligent. It was in view of the suspiciousness of his character from the above circumstance, which in Crawford's hasty judgment laid him open to the same suspicions on the present occasion, that the latter had now levelled his gun at him.

"Hold, hold!" I exclaimed, eagerly restraining his arm. "Let us see what he has to say for himself first."

"Say be d——d. What do you think he would say?" replied Crawford roughly, at the same time impatiently shaking off my hand and readjusting his aim.

"What do you mean? Are you mad, sir, or do you think you are going to act as you like? Put down your gun, sir, and let me see what this black fellow has to say for himself before you attempt to lift it against him again," I said, striking his weapon to one side at the same time. I spoke more sternly to him because of the black fellow's pitiful gesture of supplication at the sight of Crawford's pointed gun, for, falling on his knees and clasping his hands—an instinctive gesture of entreaty common to all humanity when supplicating for life—he was calling out earnestly "Bail, bail (no, no) fire".

On my stern reproof, Crawford sullenly grounded his weapon, and I called out to the black fellow, "Charlie, you come here, me wanten you. Bail shoot 'em you now." The poor black fellow, at once rising to his feet, obeyed my order, encouraged by my assurance of protection against Crawford, who still regarded him with a frowning and sulky countenance.

On his approach I observed that Charlie carried in his hand

a cooleyman—to wit, a piece of bark stripped off a thick branch of a tree, with a sharp bend or knee to it, and used by the blacks as a vessel for holding water in.

"Which way white fellow you know?" said I, pointing to where Lampiere's hut was, to indicate of whom I was speaking.

Thereupon, to my intense relief, Charlie readily replied, "Me know, black fellow no kill him—him there," and he pointed in the direction he had just come from.

"What that you say, Charlie?" here Lilly eagerly broke in; "black fellow no kill 'em Bill, him all right?"

"Bail," replied Charlie, shaking his head decisively in reply to Lilly's latter remark, "him big one sick, black fellow big one cut 'um cobra," and Charlie signified, by drawing his hand over his head, that Lampiere had received a severe wound there.

"Thank God for His gracious mercy," I ejaculated fervently. "Come along, Charlie, you show 'em me white fellow."

Charlie hesitated, glancing at his empty cooleyman, and signifying by a nod of his head in the direction where he indicated that the wounded man was lying, that we should go there by ourselves, whilst he went on for some water.

"Don't let the treacherous dog out of your sight," cried Crawford; "he is trying to work a dodge to get away from you; keep him in front of us till he shows us where Lampiere is, and if he is telling truth he can go for water afterwards, and if not, leave me to deal with him."

Charlie appeared to quite comprehend both the nature of Crawford's speech and of his sentiments towards him. He said nothing, however, only instinctively keeping close to me. On my directing him to lead the way to where Lampiere was, he instantly complied by stepping confidently on in front of us, Crawford following, and keeping his eye steadily upon him in evident expectation of Charlie attempting to play us some trick, which he was as ready to prevent, by instantly shooting him down. However, all this caution happily proved to be needless. After walking for about ten minutes more, on our entrance into another thicket of scrub, Charlie turned round, and pointing with his finger, briefly remarked to me, "White fellow there," and on looking eagerly in the direction in which Charlie had pointed, I joyfully shouted out, "Here he is, lads, sure enough; budgery (good) you Charlie". We then all at once surrounded poor Lampiere, who was lying under a primitive structure of boughs, in the form of a black fellow's camp, and in a couch of leaves and broom. He was perfectly conscious, but very pale and weak, and faint from the loss of blood, caused by a scalp wound that appeared to be so dangerously deep, that it was indeed a marvel how he had survived it at all. His right arm, too, between the elbow and wrist, was also broken, as if he had instinctively interposed it between his head and the weapon that had fallen so heavily there, although its momentum had been doubtless considerably reduced by the action.

I now saw at once what had been Charlie's object with the cooleyman, and at once said to him, "You go along to the creek, Charlie, and fetch water now". Charlie gave an involuntary glance at Crawford, but the latter on this proof of the groundlessness of his suspicions against the poor fellow, instantly replied to that distrustful glance, " Me no sulky, Charlie; me think 'em you b——y rogue, you budgery black fellow—you save white fellow's life". Charlie thereupon immediately proceeded on his mission of charity, that Crawford's hasty weapon had very nearly fatally interrupted.

Lilly and Crawford now vied with one another in their attentions to the wounded man, by at once with their knives cutting timber into splints and binding up his broken arm.

Macalister, by my directions, then took my horse and rode to the camp for a blanket and some tools, a stretcher was soon got ready, and the wounded man borne tenderly along to the camp tent.

CHAPTER XXVII.

MY first care after the departure of the litter with its wounded occupant was to dispatch Snowball, who joined us at about this time, in search of Lampiere's flock of ewes and lambs. Of the escape of this flock from the ravages of the dingoes I now entertained the gravest doubts. As it turned out, however, my fears proved to be groundless. In a short time Snowball returned driving his bleating flock to the yard, having found them feeding very comfortably about a mile from home. Putting them into the yard, and counting them out, I found to my great satisfaction that to all appearance my noble tally of lambs, that I had begun to feel so despondent about, was still unharmed. Such an altogether unhoped for result would have seemed perfectly marvellous, but for one natural explanation, that I thought of at the time, and found afterwards to have been true. This was Lampiere's conscientious scrupulousness in carrying out my strict injunctions to always lay poison about for the wild dogs, to which they were attracted by the drawing of a trail round the sheep-yards at night by means of a scorched sheep's head, and dropping baits on it here and there as they went along. By the diligent observance of this custom the men had bagged—to use a hunter's phrase—so many dingoes, that the others began to grow wary of coming near the yard at all.

That night, in the tent, where Lampiere was laid on the softest bed that by rough hands could be fashioned out of the rudest materials—all hands vieing with one another in their attention to his wants, but Lilly taking upon himself the position of head nurse—we questioned Charlie on the manner in which he had succeeded in effecting Lampiere's deliverance from the murderous attack of so many blacks.

Of the manner in which this was done Lampiere was himself in entire ignorance. He only recovered consciousness, of which he had been deprived by the blow on his head, as he was being carried through the scrub upon Charlie's shoulders.

But Charlie's account of the affair, given in the broken or piebald English that only men like Lilly or Crawford, long accustomed to intercourse with the aborigines, could have fully understood, would be but a tedious task for an unpractised reader to attempt to follow. I will, therefore, give a more connected version of the story. To account for Charlie's action in this affair, however, the reader must first understand that not only had he been well acquainted with Lampiere, both having tended lamb flocks at the same place under Mulroy, but also (for that cause of itself would hardly have weighed much in the mind of a rude savage) that Lampiere, from a desire of studying Charlie's native language, had been in the habit of spending a considerable part of his evenings at Charlie's camp engaged in familiar intercourse with him, besides commending himself to his favour by presents of tobacco and pipes, etc.—attentions that may influence even a rude savage—and consequently a very friendly intimacy subsisted between them.

Charlie, naked and armed like the rest, being among the mob of blacks who went to attack the whites, though why, was not very clear, had determined in his own mind to effect, if possible, Lampiere's rescue. Knowing, however, that, single-handed against so many, this would be hopeless for him to attempt openly, he, to lull their suspicions, affected as great a bluster and eagerness for the destruction of the whites as his companions did. So zealously, in fact, did he appear to wish for the success of their bloody enterprise, that he shouted as in triumph —the shout probably that aroused Knight's attention—as the blacks rushed at Lampiere, who was felled senseless by a club blow from Billy the Bull. At that moment Charlie rudely jostled Billy to one side, as if in ferocious eagerness to be allowed the task of despatching the senseless victim, threw himself upon Lampiere's body, and, seizing him by the throat, as if with the design of strangling him, shouted to the others to go on after the other man, whilst he would throw this one into the waterhole.

As he spoke, the blacks, looking in the direction of the hut, just caught a glimpse of Knight disappearing into the scrub, after whom they, with a yell of ferocious triumph, instantly darted in pursuit.

Charlie's action proved to be the very best that under the circumstances he could have taken ; although perhaps Charlie Knight's appreciation of it might not have been quite as emphatic at the time, had he known that he had been thus made a sort of scapegoat, for his mate's safety.

On the flight of the blacks in pursuit of Knight, Charlie immediately getting Lampiere, who he saw was still breathing, upon his shoulder, went across the yard and dropped him gently across the fence at the other side, that is hidden from view from the hut by an incline in the ground. First concealing Lampiere with a large decayed bough that he threw over his body, he then jumped with a great splash into the water. Emerging from this at once, he with a loud shout dashed after the other blacks who were in pursuit of Knight. These he overtook while still puzzled about the disappearance of Knight's tracks under the box tree.

Whilst moving round this tree in their attempt to again lift the broken trail, Charlie stated that he came across Knight's boots, where he had flung them from the tree. On telling us of this discovery Charlie, by way of expressing his secret desire for Knight's escape, emphatically remarked, in his own piebald English idiom, " me no make a light," which signifies, he did not inform the others of this discovery, but left them to grope their way out of their difficulty by themselves, as best they could. However they soon found their clue. In the course of a wider circle round the tree, one of the blacks fell across the suspiciously deeply indented prints of naked feet, and called the attention of the others to the circumstance, when as the saying is, they smelt a rat, and at once darted forward in swift pursuit —with what result the reader already knows.

Here, however, Charlie contrived to leave them, by a swift double in the scrub, and returned in hot haste to where Lampiere was lying, and again taking him upon his shoulders, went hastily on to the place that has been already described, but which it cost him a tremendous effort to reach with the pressure of Lampiere's eleven stone.

Debating afterwards upon the circumstances of Charlie's

narrative, Crawford, who was by nature suspicious of blacks, was still inclined to doubt the honesty of Charlie's intentions. "What did he want among that lot at all?" he said; "he knew they intended murdering the whites, and even although he did do so well with Lampiere, yet, with any of the rest of us, his hand would have been as red in our blood as any of the others; or what did he want in such bad company at all? I tell you these blacks are never to be trusted."

"Well, no, Tom," replied Lilly after a few moments' thought, "I am inclined to think different. I can see how Charlie might have gone along with the other blacks, knowing well what they were going to be up to, yet he himself be innocent of any intention to do harm."

"Yes! well that does seem a strange contradiction. Going along with a mob, whom he knew intended murdering the whites, and he himself guiltless of any intention of harm; what was he doing there at all if that was so?" was Crawford's reply.

"And you can't see how that might happen for all that? I should have thought that a man like you Tom, that has travelled a bit, and been among a lot of rowdy men so often as you have been, would have known better than that!"

"I don't see what my travelling and being among rowdy men has to do with these blacks, Lilly."

"It should have a lot to do with it," replied Lilly decisively, "for it should have shown you what human nature is. And human nature is always the same wherever you go, under white skins or black; and it is from what I have seen of that myself, Tom, that makes me feel that the Bible is true when it says, that all the people on the earth have the same origin. Now Tom, you have often been in strikes among a lot of hands on a station, haven't you?"

"Certainly I have," said Crawford, "and led them too, for that matter."

"That's it; now, it's no matter whether the hands that struck were in the right or in the wrong, you would expect all hands to go together."

"Yes, that's right. I have been more than once at a back country station, where we knew they could get no shearers, and we put our heads together, and stuck out for a higher price than we could have got, if there had been more shearers to be had."

"Now then, tell me this: after you all, or nearly all of you, had made up your minds to go in for this strike, if there had been a few perhaps among you who refused to go in with you, what would you have said to them?"

"What should I have said? Why, called them b——y

crawlers, and perhaps have kicked them off the place," replied Tom impetuously.

"Just so," said Lilly with all the satisfaction of a successful logician, who, by drawing his opponent into the concession of several inductions, shows him that he has thereby completely taken the wind out of his sails, with reference to the matter under discussion. "Now, Tom, as I said before, human nature is always the same; now just think for yourself. Might not this have been the exact case with Charlie there? The other blacks had resolved upon a strike against all the whites here, which with them meant just knocking you all on the head. Charlie refuses to join them in that strike—and mind you it would be all the worse in this case for Charlie, because he was a stranger here, and on that account more open to suspicion from the others of treachery towards them if he refused joining with them in the murder of the whites. Now, that for which you would only think of calling a man a b——y crawler, or perhaps kick him off a station for, they might think sufficient to smash a head with a club, or split it open with a tomahawk; now, don't you see yourself how for such reasons—and mind you I think they are the true reasons—Charlie might have been against his own wish amongst the other blacks, whom he yet knew meant to stain their hands with the blood of the whites?"

"I see it now, Lilly," replied Crawford, in a tone of conviction, "but I never looked at it before in that light. As you say, human nature is human nature, and after all, I believe it is so in a sort of a way all the world over."

Next morning Billy Stock yoked up the bullocks to the dray, in which, by Macalister's handiwork, a frame was erected, to which a swinging couch for the convenience of the wounded man was slung, and the whole covered in with a tarpaulin. By this means we determined to move Lampiere at once to the home station. His wound, that had been kept carefully fomented, was now bound up by Lilly, who had previously carefully removed all the surrounding hair—for by the time that Lampiere had been discovered the wound had been too much inflamed to admit of its being drawn together by stitches. Though an ugly cut it was not a fatal one, because the weapon that had inflicted it had been partly turned aside in its descent.

On satisfying himself on this point, Lilly had joyfully remarked that "a poet's skull was not such a weak one after all". For in his simplicity in this point, Lilly could scarcely imagine other than that such a purely intellectual cranium must of necessity be made of thinner stuff, and consequently be more easily broken than those of ordinary mortals.

Lilly, tying Coleena on behind, went in the dray with

Lampiere, attending on the sick man all the way, by keeping his wound constantly fomented, by which means alone—aided by the use of some ground blue stone, for the reduction of proud or unhealthy flesh, Lampicre's wound was eventually healed at Lilly's hut, and his broken arm restored again to a perfectly sound condition. With the team Cabbage-tree Jack also returned to his own hut.

To Charlie I gave the charge of Lampiere's flock, until I could get another shepherd, telling Jack to desire Bellamy to send one out at once. Afterwards, on my return, I took Charlie with me to the home station, with the design of finding him in permanent employment there as a station black, fearing lest, after his services in preserving Lampiere's life, he might be exposed to the resentment of the blacks, after we had all left the neighbourhood.

But, accompanied by Snowball, each of us well armed, I scoured the run in all directions for further traces of the gang that had occasioned so much disturbance and danger, but no trace of it could be found. The other quieter blacks in the camp, that Crawford had proposed killing, all declared their belief that "Warrugul (wild) black fellow go away". The black fellow whom I had shot I next day had interred, taking Crawford and Macalister over with me to aid in the work, and Crawford instantly identified him as Tommy the Turk.

On going to where I had seen Billy the Bull however, he had disappeared. Evidently, by the numerous naked footprints around him, he had been borne off somewhere by his companions, but to whatever locality they might have taken him, it could only have been to die, for I felt convinced that he could not survive such a dreadful wound as I had seen. The shepherds, inclusive of those in charge of the other two lamb flocks at the other end of the run, were staying together, in one hut, where there were double yards, and had been happily unconscious of all our troubles, until I rode out to inform myself of their safety.

All being now armed—after, as it proved all necessity for arming had passed away—every one relapsed from their state of lively expectation into the state of monotonous indifference habitual to such men. As for myself, my fears too were quieted—from the conviction that the summary vengeance that had befallen the two desperado ringleaders would have instilled such a wholesome fear into the hearts of the others, as would effectually prevent the recurrence of another such outbreak during the few weeks that would still intervene between then and the time, when, for the purpose of shearing, I should finally evacuate the back country for the summer.

CHAPTER XXVIII.

ON my final return to the station and the monotony of station life there followed a natural reaction from the excitement I had lately passed through, and though I had the shearing operations to look forward to, I began to experience a recurrence of the morbid thoughts due to the total eclipse of my lover's hopes, as well as anxiety as to the fate of her whom I loved. The utter aimlessness of my life, should it be passed year after year in the present routine and in solitude, seemed to impress itself deeply on my mind, and I began to ask myself was it to be always thus? to what purpose did I imagine I ought to live? I could almost hear a voice within me whispering, "Go forth and see the world; why waste your life amid these solitudes?"

Such was my mental state, when one day I received two letters by the mail, whose tidings at once gave a fresh impetus to my feelings, and immediately afterwards a new shape to my destiny.

One was from Mr. Rolleston, briefly announcing his intention of at once disposing of his Darling sheep run, and requesting me to have everything in readiness for handing it over on the conclusion of the shearing; the other was from Mr. M'Elwain, with whom I had occasionally corresponded since my arrival on the Darling. His letter was dated from Melbourne, and ran as follows:—

"DEAR FARQUHARSON, - I have just met Mr. Rolleston, who has informed me of his intention of immediately disposing of his station on the Darling. It seems that he has never recovered from the shock occasioned by his daughter's elopement with that villain Marsden. Marsden is now supposed to be harbouring about the upper Murray, where I believe that the police are on his track. If so, it is to be hoped that the scoundrel will soon be brought to justice. As for the girl, she is supposed to be still with him.

"But it was not of this matter that I meant to write to you. As the station, then, is to be sold, I have been thinking of something that I think should suit you better than attempting to retain your situation under a fresh master, which I doubt not you could do if you so elected. I have just arrived from New Zealand, where I own some property, and in Invercargill I was talking to a Mr. Roscoe, a merchant of that town, who has just taken up some country near —— Lake, in Southland.

"He has no inclination to manage this business himself, and he is also averse to a merely paid substitute, desiring rather to get someone as manager who would be able to go in as a part shareholder in the business. As I have heard Mr. Rolleston speak very warmly in your favour, I instantly thought of you as being very fitted for this opening.

"If you should think so, too, I should recommend you to at once avail yourself of this advantageous offer. As to the money required, I shall have no hesitation in endorsing bills to the extent of Mr. Roscoe's minimum terms for such an engagement.

"The station will carry about 15,000 sheep, but 5000 will be sufficient in the meanwhile for you both to start with.

"Yours sincerely,

"Daniel M'Elwain."

It would be difficult for me to adequately express the satisfaction that this letter of my truly generous kinsman inspired me with. Not only for the agreeable and advantageous prospect of change that it offered was this so, but also because this change would bring me nearer to my kind friends the Campbells. As I knew that it was at some lake in Southland that they had settled, might not this be the same lake where my future home was to be established?

The pleasant excitement consequent on this prospective change in my circumstances, together with the activity that the preparations for the most important operation on a station—the shearing—imposed upon me, again succeeded in diverting my thoughts from the melancholy channel that they had been inclined to move in lately.

Over the particulars of this work I will, however, pass. Suffice it to say, that Crawford was now in his element, and "ringer" of the board, save on such days as Lilly found time to go there, when Crawford had to content himself with playing second fiddle.

Lilly's duties did not indeed admit of his taking the part that he usually did in these operations, and, indeed, I question if but for a mischievous desire of taking the "shine out of Tom Crawford," as he phrased it, he would have found time for being there at all, as in preparing for the coming transfer of both stations he could have been sufficiently occupied among his own stock. But as these were usually, by his excellent management, so well in hand, he was enabled to spare an occasional day from them.

The final transfer of the station into the hands of its purchaser, Mr. Jamieson, was at length accomplished, and I prepared to hand over my charge to my successor, who was likewise to be a manager as I had been. Mr. Jamieson, like Mr. Rolleston, himself residing on another station situated on the river Murray, and not very many miles from Mr. Rolleston's station on that river.

Ere leaving, I thought it but right to direct the attention of the new manager, Mr. Myers, to the claims of the black fellow Charlie; for I considered him entitled to the protection of the station in return for his signal services in preserving the life of one of the station hands. This claim Mr. Myers, who seemed to be very pleasant, readily admitted, and promised that so long as Charlie behaved himself in a reasonable manner, he should always have his protection and encouragement to remain about the home station. I also particularly impressed upon his mind the merits of both Burrel and Lampiere as shepherds, and mentioned the literary tastes of both. In this, however, I was afraid I had said very little to their advantage in Mr. Myers' eyes, who, a practical man himself, had little relish for poetry at any time, and particularly not in a shepherd, his former experience having led him to believe that sheep had been lost by shepherds with poetic and dreamy proclivities, who he said had been musing or reading when they should have been attending to their flocks.

Of Lilly I spoke frankly and warmly, and pointed out his rough independence of spirit and invincible repugnance to any intermeddling in his own domain, and whose prejudices I recommended Mr. Myers to respect, and to endeavour to avoid coming into collision with.

This advice he also civilly promised to observe, but with no very great ardour, for although of affable manners, he seemed to be a person who was more than sufficiently conscious of what he considered to be the duties and requirements of his own position. I therefore scarcely in my own mind anticipated a continuation of the same harmonious relations between Mr. Myers and Lilly as he and I had enjoyed.

Lampiere was still an invalid, though getting rapidly better, when I left, and with the new manager's consent was still staying at Lilly's hut, with the promise of employment either as shepherd or in whatever berth might be open when he was ready for work. His weekly wages, that had been allowed to run on by me, were indeed disallowed under the new management, but as he expected to be at work again in a few weeks time this did not matter so very much.

At length I bade good-bye to all the men and with a cordial invitation to each of them should they chance to come to the neighbourhood of my new home, and a particularly cordial farewell to Lilly, whose few rough and ready words on the occasion I knew came from the brave fellow's heart, I left a place so full of memories of both happiness and sorrow to me.

CHAPTER XXIX.

I CAN hardly explain how it was that I had always associated an idea of dreariness with New Zealand. Whether its unpromising name had anything to do with this or not, I cannot say, though probably it had, but on sighting land, on my voyage thither, and gazing at its wild, precipitous mountains and bare hills, my old prejudices returned in full force.

But had this prejudice been tenfold greater, the prospect of meeting the kind friends whom I had not seen for so long would have made the country still delightful to me.

I had an interview with Mr. Roscoe, a portly, bald-headed, keen-looking man of business. Our arrangements were soon completed. I was to have a third share in the station, with Mr. Roscoe's proportion of the amount requisite for a manager's salary also allowed me for the trouble of working the station.

The picture, as I started on this new venture in my chequered career, was not a very bright one. It was drawing towards the close of what had been a miserably cold drizzling day, one of many during which I had been on the road in company with a shepherd and several other men who were driving a flock of about 5000 sheep in all, almost all of them ewes, and with them a few head of cattle. We were going towards my new run. Along with us too there was a bullock dray, laden with provisions and necessaries for our first encampment and subsequent work in the construction of more substantial station buildings.

My shepherd was a Highlander, with the shrewd, strongly marked features so peculiar to his race, and was of strong build with bushy red brown beard, whiskers, and moustache. Besides his accredited qualifications as a thorough shepherd, I was further prepossessed in his favour by his reputation as a skilful performer on the bagpipes, an instrument to whose wild thrilling notes I am passionately attached, as indeed are most Highlanders deserving of the name. And often have I amused my English friends since my arrival at Invercargill (for I had scarcely ever heard the sound of this instrument in Australia), by my sudden movement in the direction of the music when the wild wailing measure of the bagpipes fell upon my ear.

As I have said, the day was drawing to a close, and we were cold and wet and anxiously looking out for Mr. Campbell's station, to which we were then bound. I had soon ascertained, to my intense satisfaction, that my old friends were actually located on the lower extremity of the same lake, and only about twenty miles distant from my new run, so that I now actually had the delightful anticipation of having as my next neighbours friends whom I at one time thought I might never meet again.

"There is Campbell's place at last," exclaimed Munro, the shepherd, as our party at length debouched from the hills, among which we had been travelling the whole day, and that here terminated in a large plain on the other side of which, at a distance of two miles, the blue smoke was seen curling upward from some station buildings that were snugly sheltered by a dense bush, that could be seen stretching as far as the eye could reach along that side of the plain. As we at length reached the paddock in which the buildings were situated, Mr. Campbell came out to meet us, having been told in advance of our coming, and intended occupation of the neighbouring run, but not of my partnership in the concern.

I could not refrain from smiling as I observed first the puzzled, then astonished expression that overspread Mr. Campbell's quiet, genial countenance, on his first beginning to recognise my identity, as I rode towards him. "Bless my soul, Mr. Farquharson!" he cried, as I reined in my horse, and he reached out his hand to me. "Who would have ever thought of meeting you here? When did you come over to New Zealand? How are you my boy?" I sprang from Selim's back, and warmly shook his hand, and in the fewest possible words informed him of the circumstances that had caused me to leave the Darling Station and enter upon my present situation.

"Come away in! Come away in," cried the hearty old fellow. "Give your horse—is it Selim? and so it is, man! but he is looking grand—give him to the man, he will put him to rights in the stable for you; never you mind about the sheep and cattle, the men here know where they are to be put, and they'll be all sorted properly. Come you away in! the wife and the girls will be right glad to see you. Mary is always talking about you."

"Well, girls," he exclaimed, in opening the door, "here's a friend of yours come to see you." I followed close behind him as he entered, the better to enjoy their expressions of surprise, on seeing me.

"Why, mother! it is Mr. Farquharson! Well, I declare, who would have thought of seeing you here of all people in the world!" It was Miss Campbell who uttered these words of unqualified pleasure on seeing me, words that she followed up by impetuously bounding forward and shaking me warmly by the hand.

"Oh, I am so glad!" cried Mary, following with scarcely less haste and not a whit less warmth; "I have been thinking about you so much ever since we left Australia. I am so glad that you have come to see us at last;" and she kept on shaking my hand as if she would never leave off. Next came Mrs. Campbell, whose genuine welcome I felt in the firm, cordial grasp of her hand as she laid it in mine, and that was a more convincing proof of her regard, than fifty words could have been.

Their pleasure was even greater, when Mr. Campbell announced the nature of my present business, and that henceforth I was to be their own closest neighbour.

"Oh, won't that be nice!" cried Mary, fairly clapping her hands in glee; "but what did you do with Selim, did you bring him over with you? Oh, I hope you did!"

"Selim," I answered with a smile, "is in the stable here now."

"Oh, that's grand; but, of course, you would never think of leaving him behind. Dear old Selim, how often I have thought about him, since I came to New Zealand."

We soon after this sat down to supper; the men at the same time being as hospitably accommodated in the kitchen as I was in the parlour, and, by the sound of their boisterous laughter, I could tell that they were pretty well satisfied with their entertainment.

Our conversation naturally turned upon the scenes amongst which we had met before, but the pleasure of our memories was sadly dashed by the recollections they brought with them of Rachel Rolleston's rash flight with Marsden; and the girls and their mother wept as they spoke of her, and their sorrow was increased on my repeating to them the information contained in Mr. M'Elwain's letter, about Marsden's supposed neighbourhood and harassment by the police.

Leaving this sad theme after a while, Mrs. Campbell said: "You must have had a dull time of it at the station after we all left. We are quiet enough here, I know, but somehow I can never associate the same dreariness and solitude with this place, that seems to be natural to that wild country, where everything looks so dry and parched up, as though nature herself were ready to faint of sheer inanition. It is solitary here too, as far as distance from neighbours can constitute solitude; but to me, the constant companionship of the hills, with their ever-varying tints, overtopped by the mountains in the background, is such a source of pleasure and interest, that it almost compensates me for the absence of neighbours; but in that totally uninteresting Darling country, coupled with the routine of station life, where no interest or excitement occurred to break each day's monotony, I know not how you could keep your blood from stagnating altogether."

"I fear, Mrs. Campbell," I answered with a smile, "that you have conceived a rather unfavourable idea of the Darling scenery, though indeed it is tame enough, we know, compared to these magnificent hills, that meet the view on every side. As for interest and excitement, however. to keep my blood from stagnating, I assure you, I have had no lack of that lately, but rather too much if anything."

"Indeed! and what might have been the nature of the incidents that have been exciting you so much?" asked Mrs. Campbell, with a smile. But the smile soon altered to an expression of alarm, when I began to give an account of my experiences with the blacks in the back country.

My story—that included the narrative of the murder of the white men on the Warego—as may be imagined, excited no small interest among my listeners; and many were the expressions of sympathy, of horror, or of admiration, according as the various phases of my narrative moved them, that escaped from Mrs. Campbell and her daughters, as I told my tale.

Mrs. Campbell, who always traced the over-ruling of Providence in all the occurrences of life, as I concluded my narrative gave utterance to a devout sense of God's goodness in the mercy He had thus so signally vouchsafed to us all in our difficulties; an expression of reverend gratitude, indeed, in which I silently joined.

After a slight discussion of the matter I, by way of changing the subject, enquired after Miss Brydone, whose quiet, lady-like manners had prepossessed me much in her favour during her Christmas visit to the Darling.

"She is in Invercargill, Mr. Farquharson," Miss Campbell replied, in answer to my enquiry. "On hearing of our intention of coming to New Zealand she instantly elected to come over with us; as she had no friends—that is, no relations in Adelaide, and looked upon us as her chief friends, she thought she should like to live in the same colony as we did, so as to have an opportunity of occasionally meeting us."

"Indeed? and whereabouts in Invercargill is she?" I demanded with an air of such sudden though unconscious interest that Miss Campbell smiled mischievously as she answered :

"Ah, you are too late, Mr. Farquharson; Miss Brydone has been engaged some time to a Mr. Ayson, a nice, gentlemanly man, who is a teacher, and to whom she was introduced a few days after our arrival at Invercargill, and I believe their marriage is fixed to come off shortly, so you see you have lost your chance!"

"Not altogether perhaps," I answered in the same jesting tone; "has Miss Campbell forgotten poor Miss Rolleston's very dramatic prophecy that happy Christmas night on the Darling with reference to the probability of my second love?"

Whether from the audacity of my remark, or rather from its very unexpectedness, I know not, but Miss Campbell's eyes fell and her colour slightly rose, a rather unwonted thing with her, whose vivacity was not easily subdued.

But here a more prosaic remark from Mr. Campbell instantly spoiled the effect of this by-play with his daughter.

"Well, Mr. Farquharson," he asked, "what is your opinion of New Zealand?"

"To say the truth, Mr. Campbell," I replied, "when I look at these bare, rugged peaks, where there seems to be scarcely enough soil to admit of a covering of the scantiest vegetation save in the hollows, where everything looks sharp, and angular, and crude, I can not help thinking that civilisation has taken nature by surprise 1000 years before she was ready or expecting to receive it. This idea, I think, is confirmed by the appearance of your native coal—or lignite as they call it here, which I understand will require something like another 1000 years to bring it to maturity."

"That may be," responded Mr. Campbell laughingly, "but I think mother Nature has plenty finished all ready for people to be going on with in the meanwhile. I am told that in the neighbouring province of Otago, on the Taieri Plain, and other places besides, farmers have reaped crops of both oats and wheat, yielding as many as 70, 80, and 90, yes, and over 100 bushels an acre; that's more than Australia could ever do."

"Yes," I replied, "I believe that for the production of cereals New Zealand is a noble land; and no doubt after I have been here for a short time I shall like it better, especially when I get accustomed to these piercing winds, that seem to cut right through me."

Mrs. Campbell here asserted her decided opinion as to the advantages of the bracing climate of New Zealand over that of Australia, remarking, "I never feel that languor here with which in the summer I used to be so oppressed in Australia, and as for the headaches to which I used to be such a martyr there, why, since I came to this country I have scarcely known what such a thing as a headache is, which is a matter of no small blessing in itself".

"I like the country much better than I do the town," here chimed in her eldest daughter, Jessie. "I think Invercargill with its wide streets is such a dreary looking town."

"Yes," I returned with a smile. "A sense of dreariness in connection with the extraordinary breadth of the streets struck me also. To use an Australian phrase, the opposite rows of houses are scarcely within cooey of each other, and the 'stand off' aspect occasioned by this, combined with the cold, bleak wind that came sweeping along them and blowing the dust all over the place, gave me a rather unfriendly impression of this rising town of yours."

"Invercargill certainly has a cold look at present from the width of the streets," replied Mrs. Campbell, "yet I think when the councillors find time to plant the sides of the thoroughfares with trees, as I doubt not they will some day do, what now seems a cause of dreariness will then be found to be the chiefest cause of attraction, and a proof of enlightenment in the founders of the city."

"I see you still have Tiny with you," I remarked, as that rosy-cheeked damsel, with her tall and graceful figure, was clearing away the supper things.

"Oh, yes. We could not do without Tiny, nor would Tiny have cared to have stayed behind in Australia when she understood we were coming over here, as her parents are settled in one of the farming districts near Invercargill."

"And how did you leave Mr. Lilly?" asked Mary, who, during the previous conversation had, as was her habit, sat quietly and attentively without taking much active part in it. "I thought so much of that kind man; he was always so obliging and so ready with a joke. And Tiny says when he was down at the Murray he would never allow her to break a piece of wood or carry a bucket of water. He used to do it all for her."

"Tiny is perhaps more indebted for such attentions to some spell of her own than to any particular spirit of courtesy inherent in Lilly. I can assure Miss Tiny that it is not every young damsel who could have made that boast of her experience with Lilly. On the contrary he is rather inclined to take exceptional views about girls as a rule, probably from the many unlovely specimens of the sex that he has so frequently met in his rough bush life. Tiny has some reason to plume herself on having overcome Lilly's prejudices in that respect, I can assure her. See that she has not made a conquest of my old sarcastic friend's heart as well as his prejudices."

"Oh, no," replied Mary, smiling in her turn, as ladies will smile at such allusions, and bluff them too, "but how did you leave Mr. Lilly? Was he not very sorry at you going away?"

"Well, yes, the brave fellow made no secret of that. You know that Lilly has a rather expressive style of his own, so I will tell you the exact words that he used if you will excuse my doing so. He simply remarked on my shaking hands with him —" Well, Mr. Farquharson, blast me if I ain't jolly sorry that you are leaving here. I never yet was with a boss that I could pull so pleasantly with as I have done with you."

"It is a great pity that Lilly has such a rough way of speaking," remarked Mrs. Campbell, "and I am sure that I often seriously talked to him about the habit, but he would always maintain that there was no harm in people expressing themselves just as they felt, until I on one occasion completely silenced his arguments by asking him if he would have considered it right in me to use the same coarse expressions as he did. However, he got out of that difficulty by the rather ingenious argument that as we women wore different clothes to men, so there was also a quiet style of language to which women ought always to confine themselves, and which he considered it was as unsightly for us to depart from as it would have been to part with our feminine clothes."

"And yet, Mrs. Campbell, you partly wrong him, for although when excited by fierce anger, he will give way to profane expressions, yet apart from such occasions, Lilly's own natural sense induces him usually to eschew the habitual use of wicked and disgusting expressions so common among people of his class, and indeed I might say of a good many other classes in the Colonies. I have even known him to silence others who made themselves conspicuous by the habit of such offensive language.

"I have closely observed Lilly," I continued, "and I believe that what is obnoxious in his habits should be wholly attributed to the circumstances of his early experiences and surroundings. Constitutionally, I believe the man to be as true as steel, and as precious as gold. That his expletives are occasionally coarse I admit, but this coarseness in him merely results from the force of habit and training, and not from the spirit of blackguardism."

"Yes," remarked Mr. Campbell, sententiously, when I concluded my rather warm defence of my old friend, "Lilly is not a bad man at heart, and he was worth something to Mr. Rolleston."

"I am to write to him as soon as I have settled down, and give him my opinion of this country."

"Be sure and give him my very best regards when you do write," said Mary.

At this juncture a sound hideous to English ears, more especially in its premonitory utterances, though even then, from associations, dear to the ears of a Celt, was heard suddenly ringing shrilly from the direction of the kitchen and caused Mr. Campbell to jump to his feet with unwonted alacrity, while he earnestly ejaculated:

"Bless my soul, is that the pipes? Have you a piper with you, Mr. Farquharson?"

It was the pipes sure enough, that Munro, in sign of his exceeding satisfaction at his entertainment, had now, for his own delectation, and with the warm approval of his companions in the kitchen, begun to inflate and to

> "Gar them skirl,
> Till roof and rafters a' did dirl".

We all hastily adjourned to the kitchen to be nearer the performer, though by so doing we rather lessened our pleasure, as the strains from the pipes reached our ears considerably softened of their natural harshness, through the intervening barriers of door, and wall partitions. But with a true highlander the enjoyment of the bagpipes is nothing unless he is alongside, or at least in the same room with the piper.

We remained for about half an hour listening, with spirits in thorough accord, to the animating strains, as rendered by Munro's skilful fingering. Whilst myself always an enthusiastic participator in the pleasure that pipe music gives to a genuine Celt, I was yet amused in observing the extraordinary effect it appeared to exercise upon the spirits of the usually staid and apathetic Mr. Campbell. Beating time vigorously with his feet, I could see by the expression in his face a half-formed purpose of being on the floor to give the exuberance of spirits that he now experienced a freer outlet by the lusty performance of a strathspey or reel of Tulloch. Fortunately his spirit was not moved to give the burning desire within him the effect of a formal proposal for a general reel, or else he was restrained by a warning glance from his good dame who had some feeling for the wearied limbs of her guests, after their toilsome journey. Accordingly we shortly afterwards retired to the sitting-room, and presently, after the performance of family worship, I was shown to my room, where a warm soft bed, with snowy sheets, were a welcome change after the fatigues and discomforts of a long, chilly day in the saddle.

Whether owing to my fatigue or to the novelty of my situation, but probably to both, I tossed for hours in bed without closing my eyes; ruminating chiefly on my present undertaking and surroundings, and in my thoughts the graceful and sprightly form of Jessie Campbell occupied a central position. Again and again did the strange random speech of Rachel Rolleston flash across my mind, till I began to wonder if, after all, they might not have been fraught with a prophetic meaning. At last I fell asleep.

Next morning I was again in the saddle and on the move at

an early hour. Mr. Campbell now accompanied us, to show the way through the bush that, at about twelve miles distant from his place, crossed our path at a point where it narrowed until only about three miles in width, and through which our track was as yet merely indicated by blazed trees. On penetrating this, we found ourselves in a finely grassed country, open, but still undulating. The blue waters of the lake could also be seen in the distance shimmering in the rays of a cloudless sun. Above its waters, on the farther side, the land could be seen towering up in abrupt bluffs and jagged mountains.

Mr. Campbell and I rode on to the place, that by his advice I selected for my future homestead; a spot where the lake lipping into the land formed a calm and sequestered bay.

After seeing the tent for the present convenience of the men duly pitched, I rode back with Mr. Campbell to his house, it having been settled by the family that I was to pass as much of my time there as I could spare from the necessary inspection of my building arrangements, until they were in readiness for my permanent reception. However, after the first day, I availed myself of their invitation only once a week; merely riding over on Saturday afternoons, and spending Sunday with them, which to me was a great boon, for, from my youth up, I had had instilled into me a veneration for the Sabbath, and not all my experience amid Australian back blocks since, could ever wean me from a sense of inward horror in being forced to spend that sacred day among a lot of unruly, godless, and swearing men.

And thus with this kind family I came at last to be on such terms of close intimacy that I began to be regarded by them as one of themselves.

But meanwhile I was not neglectful of my own affairs. At length, my own house being satisfactorily completed, and Mrs. Munro installed as housekeeper therein, a duty that she discharged in a very orderly manner, I ceased my weekly visits to my friends the Campbells, spending my Sundays in my own house. Here I had often cause to felicitate myself in my fortunate acquisition of a shepherd who could play the pipes; of an evening Munro would make the primeval woods ring again with the wild notes of his pipes, and often as I sat listening to some stirring march or wailing pibroch, has my spirit swelled with patriotic ardour, and my mind been stirred by visions of the stern, kilted clansmen of old, whose brave deeds these strains did honour to.

According to promise I took an early opportunity of writing to Lilly, and described to him the particulars of my present situation, as also my opinion of the qualities of the country in general.

At this time (1861), the excitement over the New Zealand diggings was about at its highest pitch, and therefore such flourishing accounts of such matters as Lilly might have seen in the Australian papers, I was now fully able to corroborate ; and I gave it as my candid opinion, that he should endeavour to come over, as there was, for a man of his energy and steadiness, a grand chance of his quickly securing a fair competence for the remainder of his life.

About three months after sending this letter, I received one in return rather illegibly scrawled indeed, but informing me that Lilly had quickly decided on taking my advice, and was coming over—will the reader believe it?—with Colcena, dray, bullocks, and all in the same boat ; for this team Mr. Rolleston, on depositing of his station, had, by way of acknowledgment to Lilly of his valuable services, made him a present of. After receiving my letter he had become exceedingly restless in mind, and, not caring very much for his new overseer, whilst at the same time, with his natural shrewdness, seeing from my accounts that there was a good prospect of making a "rise," as he termed it, by starting as a waggoner to the diggings, where the prices paid for carriage were in those days something enormous, he, after some careful cogitation, determined to take my advice, and take passage for New Zealand forthwith ; and that was not all, for considering rightly that the price of freight for a good team of bullocks to New Zealand would be considerably less than that demanded for an inferior team when there, he determined to take his noble team of bullocks with him, and with them a dray, and also his mare. After this I heard from him occasionally and understood that after losing two of his bullocks, 'Bauldy' and 'Dainty,' he became more vigilant in preventing a like fate from overtaking the rest, and was now doing well. In my replies I kept urging upon him the importance of keeping his money together, and, to prevent what he earned being heedlessly squandered, advised him to deposit it in the bank: an advice that I afterwards had the satisfaction of knowing he had followed.

CHAPTER XXX.

TWO years had now nearly passed away without the occurrence of any incident worth recording here. My intimacy with the Campbells meanwhile continued, and by this intimacy my feelings for Jessie gradually strengthened in character. On her side, however, there seemed to be nothing beyond what might be termed a feeling of frank confidence, or kind friendliness, and I could not feel that her impressions in my favour were so pronounced as to warrant the hope that my attachment was returned.

Lilly, in the meanwhile had found time to pay us a visit, and glad we all were to see him; yet I very much question if this pleasure on our part was equal to that experienced by him, at meeting Tiny again, or perhaps to that of the shyly blushing girl, at seeing him. Lilly was rough and unpolished, but he was none the less a diamond for all that; and although Tiny was shy, she had still discernment enough to appreciate a genuine article when she saw it—though what she thought on this subject, she kept quietly to herself, and beyond the confidential pleasantry of her manner, that now seemed to predominate over her wonted reserve, there was nothing to indicate in her a partiality for Lilly above any other man. And this confidence might after all be merely the result of the long standing acquaintance between them.

In the meantime too, another acquaintance had been added to the list of our friends, who made periodical visits to the Campbells. This was a young man named M'Gilvray—the son of a neighbouring squatter, whose station had hitherto been solely worked by a manager. Mr. M'Gilvray, senior, was a merchant in Invercargill, but his son being naturally fond of a country life, and having therefore relinquished a share in his father's business, had now taken over the control of this station.

John M'Gilvray was an active young man, rather slim in figure, but wiry and agile. His features, that glowed with health, were well formed, his hair and whiskers black, and his eyes of a dark brown. His manners were easy and agreeable, and were made more attractive by a constant flow of humour, and it was with a feeling of anxious jealousy that I observed the perfect harmony between his qualities and Jessie's tastes, and she soon grew to be on a footing of the most confidential intimacy with him. Indeed, had it not been for my own too great personal interest in the matter, I should certainly have admitted that they seemed well suited to one another. So patent did their partiality for one another appear to me, that I resolved to at once smother the flame of passion within my own breast, ere it could master my power to control it. But as my conduct towards Jessie became more guarded, hers increased in friendly demonstrations to me, and this was especially the case when Mr. M'Gilvray was present. Whether this arose from the natural kindness of her disposition, that was distressed in witnessing the tokens of my discomfiture in the presence of my rival, or whether from a wish to make use of me as a foil against M'Gilvray, or from a spirit of feminine vanity, that was gratified by my attention even whilst compelled to discourage an unsuitable lover, I could not tell; I only knew that so it was.

Shortly after my arrival at the lake, I discovered that I had a still nearer neighbour than even the Campbells. This was a solitary digger, named Howden.

On the completion of my buildings, my next care was to provide myself with a boat for the convenience and pleasure of being able to indulge myself occasionally with a row on the lake. When enjoying this pleasure one summer's day, I had rowed right across the lake, that here was not more than two miles broad. On getting up on the bank on the farther side to look about me, I discovered by the smoke the proximity of a digger's hut.

This digger, who happened at the time to be inside, was a man of about fifty years of age, tall, but of a rather spare figure, and with an air of superior intelligence. He had, it seems, prospected to where he then was, all the way through from the west coast. He did not inform me as to how he was prospering—or if he were prospering at all—but as he told me that he had been working where he was for about two years, I drew my own conclusions from the circumstance, and decided that he must have found something there that made it worth his while to stay in such utter solitude for so long.

His hut—that was constructed of sods, and well plastered within and without—was lined inside with prints from *Punch* and the *Illustrated London News*, which gave a most cheerful aspect to the interior, while on a shelf beside the window, several volumes—some of a scientific character and others the works in prose and poetry of the best English and American authors—betrayed in their possessor a mind of no ordinary taste and elevation. Ere this, his only mode of supplying himself with provisions had been by means of two stout packhorses that were feeding not far off, with which he took, when occasion required, long journeys through the bush to the nearest settlement on the west coast.

But now on being acquainted with my settlement on the other side of the lake, together with my means of transit across, these tedious land journeys were discontinued, as all he required in the way of rations I could supply him with.

Yet, though by this means I had a frequent opportunity of conversing with him, I never seemed to get on a familiar footing with him, so that to me there seemed to be a sort of mystery about this man, and his solitary manner of life. On all points relating to the manner of his education or his original status in society he seemed singularly reticent. Yet it was plain that his position in life, judging by his educated manner, must have been at one time a respectable one.

The Christmas of my second year's occupancy of my station was now approaching. Already I had commenced my shearing, and among the men who came to me for this employment were two ill-looking dogs. One of these was a burly, pockmarked man, with a most brutal cast of countenance, whom I instantly recognised as the same that Lilly had handled so roughly some years before at the free fight at Wentworth.

Whether the fellow had noticed me sufficiently then to be able to identify me now of course I could not tell; and, as a matter of course, I treated him as if I had never set eyes upon him before, though, in fact, had it not been that shearers were scarce I should have hunted him at once off the place. I, however, made the best of the matter as things were, and engaged him for the shearing.

His companion was also an ill-looking customer, though of scarcely such a pronounced type of rascality as the other. My original prejudices against the latter, who called himself Michael Hennesy (his companion's name being Elias Jones) was considerably augmented on seeing him at his work, which he performed in a most slovenly manner, cutting and maltreating the sheep so much that I threatened to turn him out of the shed. His conversation, too, betrayed the presence of a mind of the most loathsome cast, his language being of a peculiarly revolting nature. Over his companion, who appeared to be a man of weak character, he seemed to have complete control.

Christmas Eve had arrived when the digger, Howden, in answer to his signal, had been brought across the lake, for he was in want of mutton. While standing beside me and talking quietly, I noticed that Hennesy, who was passing by, suddenly thrust forward his ill-looking countenance with an eager glance into Howden's face, then, hanging his head in his usual fashion, he went carelessly slouching past. Howden was shortly afterwards rowed across the lake. I should mention here that besides my own boat, that was constantly chained to the bank, and padlocked there—the key of which I kept in my own possession—I had another of stouter build that was kept entirely for the station use, and that could be also used by any of the hands who, in their spare hours, had a fancy for a row on the lake. By this means the nuisance was avoided of my boat being used by them at their own pleasure.

CHAPTER XXXI.

ON Christmas morning the sun rose clear and cloudless, revealing a landscape of glorious beauty. The water of the lake glistened and shone like polished steel at the foot of the hills covered with deep woods, and vocal with the cry of the parroquets, the clear ringing notes of the tuis, and the metallic echoes of the bell birds close by my house, and on the farther side the mountains, descending into the water in steep, precipitous bluffs, revealed a prospect such as a landscape painter would have delighted to depict.

The shearers having resolved to keep Christmas after the good old English fashion, the day was observed as a holiday.

As it was but dull work loitering about the place, and I could not lie and read all day long, after breakfast I solaced myself with a row out upon the lake, and I amused myself thus for several hours until it was time, I thought, to return for dinner. A slight breeze had by this time begun to ruffle the surface of the lake, and just as I had entered the bay at the head of which my homestead was situated, happening to glance towards the lower end of the lake, my astonishment was great when I beheld Mr. Campbell's cutter coming rapidly towards me, with sails bellying out in the rising breeze. Instantly divining that the family had resolved to pay me a visit in honour of Christmas, I hurried towards the shore and hastened up to the house to give Mrs. Munro notice of the company she might expect, so that she might set to work to provide suitable entertainment for them when they arrived.

In view of the coming of the ladies, I involuntarily cast an anxious glance round the room, but I might have saved myself the trouble, for, as far as mere housewifery was concerned, Mrs. Munro was method and orderliness personified.

About half-an-hour later the vessel arrived, and was laid alongside the mooring place, where some stakes had been driven in as an apology for a wharf.

On board the vessel there was, indeed, a party one of whom at least I did not expect to meet. Besides the Campbells and Mr. M'Gilvray, whose presence, despite the claims of hospitality, I could have spared, there was another lady with a child in her arms.

"Ah ha!" laughed Mary, clapping her hands gleefully, "Merry Christmas to you, Mr. Farquharson! You did not expect to have such a grand Christmas party as this—now, did you?"

"None the less welcome because unexpected," I answered,

as I walked out along the planks that were laid on the stakes and stood by the side of the yacht. " Merry Christmas to you all, ladies and gentlemen ; you really have given me great pleasure by this unexpected visit."

I then took Mary, who in her eagerness to greet me stood nearest, and lifted her bodily out of the vessel, no other gangway being available, and on to the platform. " Good day to you, Mrs. Campbell, and what is your opinion of the appearance of my habitation now?" I remarked, as that good lady followed suit, by the same primitive mode of conveyance.

"Really," she replied, "it does look pleasant. I had no idea that you had such a nice place as this; the prospect is truly lovely." I then shook hands with both gentlemen, but Mr. M'Gilvray, however, by springing lightly to the platform, obviated the necessity for my landing Jessie in the same mode as I had landed her mother and sister.

It was then for the first time that, in the lady with the child, who now awaited my assistance to reach the platform, I, to my no small surprise and pleasure, recognised my former Christmas guest on the Darling, *i.e.*, the quiet, lady-like Miss Brydone, or rather matron, Mrs. Ayson.

"Truly," I said, while laughing and cordially shaking her by the hand, " we shall soon have our old Christmas party back again. But why, Mrs. Ayson, did you not bring up Mr. Ayson with you? Surely I should be delighted to make his acquaintance."

"Thank you, Mr. Farquharson. I am indeed sorry that Mr. Ayson is not here, but he was detained upon business, and could not come along with me this time. Some other time however he may be able to pay you a visit. He would very much like to do so, for I have often talked to him about you, and the happy Christmas that I spent with you on the Darling."

Lifting her off, I next approached Tiny. " Well, Tiny," I remarked, " and how are you? What a pity that Mr. Lilly is not here to meet you now, eh ! "

"Mr. Lilly cares nothing for seeing me," replied Tiny, smiling shyly. She was a stout, plump girl, and perhaps that might have accounted for the fact that as I lifted her off the boat I found her face very close to my own, her weight pressing heavily upon me, although I was not a weakling either. Be that as it may, her ripe lips so close to mine were rather too great a temptation for one who was by nature no stoic in the matter of woman's charms, and a sly pressure of her lips by mine was the natural result.

Tiny, as if shocked at the liberty she had been unable to prevent, very properly to preclude its repetition, put up her hands, and turning her head aside, laughed blushingly, and ran away.

After seeing the yacht safely moored, I invited the two men who had navigated her to come up to the house with me, where I offered them a cordial, which, by the way they smacked their lips over it, was evidently to their taste.

On entering the house, laughter and good humour prevailed on all sides. Everything was examined with pleasure, and the rude furniture admired by the ladies with laughing good nature, though Mrs. Campbell, with her usual foresight in such matters, had brought an ample supply with her, in the matter of sheets and blankets, for all the party.

For, be it known, that so primitive was my mode of living, and so little in my sequestered situation did I look forward to such a contingency as the present, that my domicile consisted of but three apartments. Of these one was reserved as a sitting and dining-room, and another was my sleeping apartment, whilst the third was the kitchen, in which the Munroes also slept.

My furniture, too, with the exception of three bought chairs and one table, was all bush made, so that to furnish my present party with sitting accommodation, several four-legged stools of very primitive construction had to be requisitioned from the kitchen.

But in spite of these drawbacks, I question if the hilarity and good humour of the party could have been increased had the apartment been luxuriantly furnished with a carpeted floor, spring-bottom chairs, and mahogany surroundings.

But if furniture was scarce, provisions were ample. Plain they were, but still tempting, for Mrs. Munro was an excellent cook, and we were preparing to sit down to a table furnished with a substantial joint, with fitting adjuncts, and an ample plum-pudding was looming in the distance, when the crack of a bullock whip sounded in my ear.

Not expecting such a sound as that, as my wool was to be taken away in Mr. Campbell's boat, and thence loaded on drays from his wharf on the lake, some few miles from his home-station, I went out to see what it could mean.

On gazing intently at an advancing bullock team, the tall figure of the driver, the everlasting cabbage-tree hat, the blue shirt and the leggings, and above all the sonorous shout of "Gee, Dauntless," were not to be mistaken.

"In the name of wonder," I ejaculated, "Whatever has brought Lilly up this way!"

"Hilloa, Lilly! what wind has blown you in this outlandish direction?" I exclaimed, as he halted his team with a loud "woa," near to where I was standing shaking him heartily by the hand.

"Ha, ha!" he answered laughing, "I reckon you didn't expect to see me hereaway; but as carriage to the diggings was getting low, I thought I would take a trial at wool driving, and so thinking that Mr. Campbell would give an old acquaintance a show, I came up to see him about what wool he could let me have to drive down for him. But when I was within five miles of the station yesterday I met the shepherd on the road, and getting into a yarn with him, he happened to mention that all the Campbells' were going away in their boat to spend the next day with you. As I wished very much to spend Christmas with them, I thought that was unfortunate; but on inquiring the way to your place, the shepherd pointed out a way by which I could reach the end of the plain from where I then was, as soon as I could get to Campbell's station, and that was eight miles nearer here. So, determined to be among the fun with so many old faces, I, without going near Campbell's, went along this short cut, and camped out at the end of the plain last night. Then getting the bullocks yoked up by six o'clock this morning, I said I would make them do the distance between there and here inside of one o'clock, and I have kept my word; but I made them travel for it! I don't like forcing any animal out of its proper place, let alone my bullocks, but I thought that for this day I would make it a case of necessity; so here I am, and perhaps you'll give an old acquaintance a load to bring back yourself."

"That I will, Lilly, the whole clip if you like; but how will you get it down? I have always had it sent round by water in Mr. Campbell's boat before. Although on coming through at first I opened a way for the team, it was with great difficulty and by the observance of great precautions, that we managed to get along; as we did not take time to form the road by cuttings, save such trees as we found absolutely required removing from our path."

"Well, the road through the bush is rough enough in all conscience, but I never yet saw the road that I tackled (and I have tackled some rum ones in New Zealand), that I got beat at; for why, in the most desperate pinches my bullocks will do everything I tell them, and hang on like tigers, or 'back,' 'gee off,' or 'come to me,' just as I speak to them, and now that I am here I will take one load with me anyhow."

"Well, turn out your bullocks at once, the poor things are blowing hard, after their long tramp. Wallace and Samson are looking well, and indeed they are all looking well. I wonder how you manage to keep them so, with all the work you give them."

"It is all by care in using them. Whenever they look fagged I spell them, and then, I know how to drive them."

I smiled at this reminder of the well remembered, yet well merited conceit of my old friend; for whom now, in my genuine pleasure at meeting him, I thought to dispense with the forms of society by inviting him to join the assembled company in the sitting-room. But I might have spared myself the trouble of such a request. Lilly was too independent to feel even flattered at such a mark of favour. "No, Mr. Farquharson, thank you," he bluntly replied to my invitation, "though I concern myself about no man's position, as I reckon myself as good as the highest in the land. Yet I know my own place, and I shall feel best pleased with what I have been always accustomed to, and with parlour manners I should be very like a fish out of water, so I'll just away to the kitchen and fare with the hands there."

"Well, Lilly, just as you like; we will see you after dinner, however, and have all your news. There are scarcely any but old friends here, you know."

"Aye, aye, I'll be with you then," he replied, as he proceeded to unyoke his bullocks, and I returned inside.

"Well, ladies, our old Christmas party with but two exceptions (one of whom we can well spare) has come together again; here is our old friend Lilly come to join us; he is outside there with his team."

"Lilly here?" replied Mrs. Campbell, "how strange that he should come here just now; however did he find his way?"

"Lilly here?" chimed in both girls with looks of unaffected pleasure at the announcement. "Oh, that is good news, we shall have some fun now." "Yes," added Jessie, "Lilly was always a favourite with me. I think he is so kind, though he speaks so bluntly, and he can look so gruff, when things don't go according to his mind."

"Yes," I replied; "Lilly is not in the habit of mincing matters with anyone, but he is, for all that, a sterling fellow."

"Yes," remarked Mrs. Ayson, "I have often thought of him, and that wonderful damper that he baked for us in the ashes, and I have often made my friends in town laugh at my description of the scene. What a happy time that was, to be sure!"

Thus did Lilly's arrival occasion another ripple to our already overflowing measure of happiness. The conversation soon afterwards became general, Mr. M'Gilvray sustaining a conspicuous part in it, with a highland twang in his accent, that a colonial rearing from his boyhood had not been able to eradicate. His voice at times mingling with Mr. Campbell's more solid brogue in the discussion of the practical details of sheep breeding, and qualities of wool ; or in lively sallies with Jessie, with whom he seemed to delight to engage in a bout of wit and repartee, commended him to my attention as a quick-quitted, intelligent, and plucky young fellow. Occasionally too, Lilly's loud hearty laugh echoing from the kitchen, informed us of his perfect contentment there, in the society of Tiny, whose softer, but merry tones could also be heard ringing out a happy response.

CHAPTER XXXII.

DINNER being over, we spent the whole afternoon until teatime in wandering about the bush, where that was practicable, for the bush of New Zealand, with its thick damp undergrowth, and jungle creepers, parasites, and lawyer bushes, significantly so termed, from the clinging pertinacity of their nature, when once in contact with either clothes or skin—is not favourable to walking parties. However, some open glades, formed by the clearing of timber for our buildings, admitted of our thus enjoying ourselves with some freedom ; so, yielding ourselves wholly up to the influences of the scene —the melody of the birds, or the view of the bright gleams from the broad waters of the lake, as they flashed in the sunshine through the forest openings—our sense of calm enjoyment was perfect.

Tea being over, and the candles lighted, the next source of enjoyment we proposed to ourselves was Munroe's bagpipes.

Munroe was summoned accordingly ; and forthwith made his appearance in full national costume, by which he appeared suddenly transformed from an ordinary loutish looking working man to one of nature's noblemen ; so well did his brawny limbs accord with his picturesque dress, while his Celtic blood seemed to become instinct with fire and martial ardour in the ancient habiliments of his race.

What man with an ear for strathspey music can listen to it, with all the thrilling ecstacy that the bagpipes in the hands of a skilful piper can give, and keep for long inactive on his seat? With the present party, all Celts, as far as the men were concerned, such a thing was not to be thought of. The tables were instantly cleared, and all the party, including the inmates of the kitchen, were on the floor for a dance. Lilly was among them, and with Tiny for his partner, wheeled with great zest and some skill through the mazes of a country dance. Not all, however, joined in, for Mrs. Campbell refused to take part in such active exercise, and I, too, managed to keep out of it for a while at least, as, never having learned to "trip the light fantastic toe," I was unwilling to make myself ridiculous by taking part in an exercise for which I had no aptitude.

But when at length the simple Scotch reels were proposed I was fairly forced upon my feet by Mr. Campbell, who also up to this point had kept in his seat.

For the first reel I chose Mary for my partner, and Mrs. Ayson for the next, purposely leaving Jessie to Mr. M'Gilvray, with whose devotion I began now to imagine that she seemed quite satisfied. In the evolutions of the reel poor Mrs. Ayson seemed quite bewildered, and had to be piloted in her utter helplessness through the figure eight, when in obedience to a hint, she went through some very demure and staid shuffling with her feet, supposed to be expressive of her exhilarating sympathy with the notes of the music.

But of all the dancers, there was not one who appeared to more thoroughly enjoy himself than Mr. Campbell. The music appeared literally to have put life and mettle into his heels, usually so grave and deliberate in their motions; whilst he whooped and snapped with his fingers, and cut the capers of the steps of the Highland fling with the most marvellous agility, his quiet, orderly wife, smiling meanwhile at the enthusiastic absurdity of his manner.

Among the shearers was a smart young fellow named Turner, who was an accomplished step dancer. He was soon sent for, and he greatly contributed to the amusement of the evening, by his artistic performance of sailors' hornpipes and Irish jigs, rendering the latter with great humour and native character.

Meanwhile my heart began to sink heavily within me. Peculiarly sensitive to the influence of music, my morbid feelings were now vividly impressed by the wild strains of the bagpipes until, steeped in feelings of sentiment, I seemed to endow the object of my affections with all the attributes of a heroine of romance.

The animated expression of her mobile features, her pallor, now brightened with the flush of exercise and excitement, together with the charm of her slight, yet rounded figure, only added to the ardour of my feelings, and moved me the more to covet so much beauty, and to bitterly regret the improbability of my ever winning it. For, as I watched her as she waltzed, her waist partly encircled by the arm of her smart, intelligent-looking partner, I could not help feeling—while thus forced to acknowledge their fitness for each other—how utter was the vanity of my hopes in this second love passage of my life.

And yet, when the dancers returned to their seats, I could not help noticing how Jessie seemed by design to seat herself by me, as if by that to compensate me for the anguish to my feelings that she was conscious of having caused me by her acceptance of the attentions of my rival.

With my feelings in this state I was not sorry when I was suddenly called out by one of the men, who told me that the sheep that had been penned for the morrow's operations were breaking out of the wool shed, one of the gates having been insecurely fastened.

Bidding Munro to keep on playing, as we should, with the assistance of the other shepherd, who had been taken on for the extra work of the shearing, be able to bring back the sheep without his help, I hastily went to the shed.

On reaching there I found that all the sheep, save those in the pens, had managed to escape, as the mischief had not been discovered by the men until they were nearly all out. They were now, however, stopped by the new shepherd's dog, who had them all gathered up in a heap on the hill side, beyond a small arm of the bush, this being the way to their former feeding ground on the hills.

It was now about dusk, and, seeing that the sheep were safely in hand, but that it would take some time yet ere they were housed, although several of the shearers had come to help the shepherd, and knowing that they would all be secured without the necessity of my presence, I turned aside and walked down to the lake.

I did this as an excuse for prolonging my absence from the party indoors, for which, in the depressed condition of my mind, I felt little inclination.

Thus I went along quietly musing until I came to where the boat that I reserved for my own private use was fastened. Mechanically unlocking the padlock, I got into it, and, pushing off, seated myself and plied the oars listlessly for a second or two, when the thought suddenly entered my mind of rousing myself from the state of depression I had fallen into by a little lusty exercise.

The lake at this point was not more than two miles in breadth, and half-an-hour's vigorous bending at the oars would send my light skiff across, whilst the excuse of the trouble with the sheep would be ample excuse for detaining me for another hour at least, when, with spirits refreshed by the healthy stimulus of muscular exercise, I could return to the company in a state of mind more fitted to enjoy it.

Moved by these thoughts I rowed on until the waves curled half way up my bows. I never once rested until I was fairly within a few yards of the opposite shore, where, backing water with one oar, I brought my skiff up parallel with the land that there rose up steeply, allowing the boat to float in deep water at its very edge.

Resting on my oars now, I paused awhile to take breath and wipe my brow, for I was heated with my rapid exercise. After having breathed and allowed myself to cool a little, I again took the oars and gently pushed my boat along the coast, heading down the lake. This I did, I believe, more from an instinctive desire than from a deliberately conscious motive. I ascertained this of myself afterwards, when I found leisure to analyse the thoughts that moved me to take this course rather than to head straight back on my return voyage across the lake.

In the course I had taken, from the situation of the mountains that here approached the side of the lake, their dark shadows were thrown a long way out upon the waters. But a little farther down in the direction that I had now unthinkingly chosen, the land made a sudden bend, and the mountain range was also thrown back further inland, so that the rays of the moon (for it was a lovely moonlight night) fell upon the water almost at its point of contact with the land.

Thus the feeling that constrained me to head my boat towards this point ere shooting out again in a direct line across the lake, I subsequently thought must have been my instinctive sympathy with the moonlight. Man, unless depraved, naturally abhors darkness, and it was probably the darkness of these mountain shadows on the lake that I instinctively desired to shun.

With this object I had been moving my boat towards this point with almost noiseless strokes, as I was keeping my powers in reserve until the moment when I should head my boat for the other side of the lake, intending then to row at the same tremendous rate I had kept up on my outward passage, when, just as I approached the point that I intended to double, and being still in deep shadow, I was suddenly made aware of the sound of approaching oars. On this I ceased pulling, and looking round, saw a boat pulled by two men rapidly approaching, and then disappearing on the other side of this point. "Some of the men bringing over a taste of their Christmas cheer to Howden," I thought, remembering the few bottles of spirits that, by way of making the men share in our own conviviality, I had towards evening sent to their hut.

Howden's hut was over against where I was then resting on my oars, some twenty yards or so inland.

I thought first to hail the men as to their business there at that time of night, but thinking it must be, as I first thought, a visit of pleasure, I had just resumed my oars when the voices of the two boatmen suddenly falling upon my ear involuntarily inspired me with a desire to listen to what they said. Though their words were but few and brief, they almost caused the blood to freeze in my veins with apprehensive horror.

"But are you sure," I heard one of the men ask, "that you are right? it will be a caution if you are all wrong after all!"

"I tell you that it is all right, you —— —— fool. I knowed he was here, his brother told me ——" the rest of the sentence was lost in the distance, but I caught enough of it to at once identify in the speakers the hang-dog looking ruffian Hennesy and his well matched, though less villainous-looking, companion, Jones. And with my recognition of their voices there flashed across my mind Hennesy's lowering and prying glance into Howden's face on the evening before, and the loneliness of the situation that seemed to favour with impunity a scene of violence, perhaps murder, upon an unprotected and unsuspicious man.

"Gracious Heaven!" I ejaculated, "it must be this that they are after. Grant me strength, Lord, to be still in time to frustrate their fiendish purposes!" And with this exclamation, I bent to the oars and caused my boat to dash through the water towards the point now about ten yards ahead of me, where only I could get a landing. To reach it and to spring on shore and to hurriedly twine the painter round a bush at the water's edge, was but the work of a few seconds. Then, seizing an oar, I rushed up the little spur and ran with all my speed towards Howden's hut, whose light I could now see burning but a few yards ahead. A dog, not a very formidable one, kept by Howden, was barking furiously, as if at the sight of strangers whose presence he both resented and feared. I had got within ten yards of the hut when suddenly my blood curdled at the sound of a wild prolonged cry, as of mortal fear, but which seemed to be suddenly stopped as if by violence, for I could still hear a kind of gurgling sob.

Bounding wildly forward (I never thought of calling out and so alarming the murderers), on reaching the hut I dropped the oar and caught up an American axe that was lying gleaming in the moonlight on the wood heap, past which I was in the act of dashing. "Villains and murderers, let go!" I yelled, as I dashed through the door, and beheld the two wretches kneeling over the prostrate form of Howden, whose throat was held in Hennesy's deadly grip, striking wildly at his head with the axe as I spoke.

In my haste, not calculating the direction of the blow, in swing the head of the axe came in contact with the door lintel, and turning in my hand, fell violently on the head of the intending murderer, but with its flat instead of its edge, thus without effecting any serious injury.

Both villains instantly bounded to their feet, and terror-stricken at my sudden attack, sprang out through the door, and instantly fled towards the bush, without attempting to defend themselves against me, whom they could easily have overpowered, had they been sustained by the manly courage that in such emergencies is often found lacking in spirits of mere ruffianism. They both dashed out of the hut, dodging on either side of me as they went, ere I could recover my weapon to deal a second blow.

I made no attempt to follow them, but turning my attention to Howden, who was black in the face, although still breathing, I took him up in my arms, first fastening around me a sheath knife that I saw attached to a belt beside me on the table, and, though the burden of his limp body was great, I went out of the hut determined to carry him down to the beach, where the two villains had left their boat, that was then closer to me than my own.

As I staggered along under the weight of my burden, Howden suddenly revived, and desired to be let down on his feet, and in a few minutes later he was able to walk without any assistance.

Scarcely waiting to exchange a word, we entered the boat, the little dog, which had followed, joyously springing in with us, and, each seizing an oar, we pulled hurriedly across the lake.

CHAPTER XXXIII.

AS we were rowing, I asked Howden, who by this time was perfectly recovered, the particulars of his late encounter with his assailants. He replied that he had been reading when they suddenly entered the hut. On looking up at their entrance he at once recognised Hennesy, whom he had formerly seen, and whom he knew to be a most desperate character. It scarcely needed the latter's fierce demand that he should turn out his money, to acquaint Howden with the object of his sudden intrusion. On Howden's attempting to snatch up a sheath knife that was lying on the table they both flung themselves upon him, while Hennesy seized him by the throat, just as an irrepressible cry of terror burst from his lips; the despairing cry that a strong desire of life will wring from the bravest men at the prospect of violently losing it. He shortly afterwards lost consciousness, and only recovered whilst being borne by me towards the boat. His first impression on reviving was that he was in the arms of one of his murderers, who was carrying him away with the intention of casting his body into the lake, but hearing some words that I chanced to mutter, he recognised my voice, and with it his own security.

"You have met this Hennesy before, then?" I demanded.

"Yes, to my sorrow, though then under the name of Morgan, which I believe to be his right name. I knew him, not from personal acquaintance with him, but from his association with one with whom he was only too intimate."

"And this other person was, I presume, a friend of yours?"

"A friend," he replied, sadly, "well, he should have been a friend, for he was a brother."

"A brother!" I ejaculated in some surprise.

"Ah! that surprises you, Mr. Farquharson; if you knew all you would be still more surprised. He was a man who seemed fated by his own mad folly and unbridled passions to set at naught every prospect of honourable distinction that the possession of undoubted talents and influential friends once seemed to place within his reach, for, as if the path of rectitude were too straight for the demands of his vaulting ambition, gambling and horse racing were the means by which he sought to win life's prizes. The narrow straits of necessity that these hazardous dealings forced him into not agreeing with his high spirits, it being a matter of pride with him to be unfettered in his progress through life by any of the ordinary laws of society, from dice to forgery was with him but a step. Then came the jail, from which with his fierce spirit he soon broke away, and the felon's career—but why should I thus dwell on a brother's infamy? To me it is a theme bitterly degrading, for he was still a brother whom I could have loved. May the Lord remember him!"

As Howden was speaking—we had both considerably remitted our labour at the oars—a flash of memory, first I believe inspired by Howden's incidental reference to Hennesy's true name of Morgan, led up to a train of ideas in my mind, which, setting aside as too fanciful, I did not then give utterance to. There was something in his description of his brother that somehow at once suggested the idea of Marsden to me; a suggestion strengthened by the remembrance of the disguised ruffian who had formerly bailed me up, and whom Marsden had then so sternly addressed by the name of Morgan. So strongly was I impressed by this, that I involuntarily keenly gazed into Howden's face to confirm my suspicions by any

possible brotherly likeness there might be between Howden and
Marsden. But Howden's features, of a calm, thoughtful cast,
positively negatived any such idea, and I at once dismissed it
from my mind as altogether unlikely.

After a short pause, Howden continued: "For years, when
he was in jail, or at large, have I kept my eyes upon him,
though for his own selfish ends, he occasionally made me
acquainted with his movements voluntarily; and I have
even at times met him at his desire, and have then done
my best to move him.

"'Abandon your vile courses and give proofs of your
sincerity of purpose in doing so, and I will still help you
to regain, at least in another land, the paths of virtue and
honour that you have so far departed from.' But he seemed
hardened in his depravity, and the seed sown by a holy mother
appeared to have got corrupted within him. Yet still with me,
however turbulent with others, he has always seemed to be
repentant and humbled. But what can this new thing mean?
That wretch must have had my present address from him, or
he would not have found me out in this out-of-the-way locality."

"Surely you do not imagine your brother to have been fiend
enough to have sent these wretches here to rob and murder
you?" I asked involuntarily.

"Not to murder," he answered quickly, "that might be
a gratuitous addition to his commission on Morgan's part,
from his own natural villainy and pride in his professional
dexterity at such work. But as to rob! Although I have
offered to help him if he would reform, beyond just assisting
him to a very limited extent, on the few occasions when he
has sought me out, I have sternly refused to advance him
money on any other conditions. My last information regarding him was, that he was again in jail, but the presence
of this ruffian here seems to imply that he has again broken
away, and is perhaps even now in this country."

We now resumed our former more vigorous use of the oars,
and in a short time had reached the shore. Instead of
being absent for half an hour or so as I had expected, upwards
of an hour and a half had elapsed ere I had recrossed the lake,
and a good deal of surprise had begun to be felt at home at my
unaccountable absence, as the sheep had been secured for some
considerable time; and the men on their return, on being
interrogated, had denied having seen me at all whilst penning
the sheep in the shed, and among the alarmists, as I afterwards understood, no one appeared to be so palpably uneasy as
Jessie; for calmly, as I imagined, I had veiled my feelings,
yet, with the penetration of her sex she had marked my

distress at her pleasure in Mr. M'Gilvray's attentions during the dancing. Moved by this, she, in her increasing anxiety at my unaccountably prolonged absence, was beginning to conjure up in her excited mind all sorts of desperate fears concerning me, which gained in colour by the close proximity of the lake —a suspicion on her part that did not credit me with much strength of mind, and which, considering that for such hair-brained tragedies I had ever expressed the most profound contempt, did me no small injustice.

"Ho, ho, Mr. Truant," she joyfully exclaimed, as soon as her quick glance detected my entrance in company with Howden, "it is a fine way indeed of showing respect for your company, to run away and leave them to look after themselves! give an account of yourself, sir!"

"Willingly, Miss Campbell," I replied in the same lively tone—so thoroughly had my adventure succeeded in diverting my mind into a more natural channel—"and when I have done so, you may probably acknowledge that I have been guided by a divine impulse which has enabled me to render a great service to this gentleman here, however rude and unceremonious my action may in your eyes at present appear. For this impulse has moved me to no less an act than to row with all my strength across the lake, to the farther side of which I arrived just in time to prevent two murderous villains from robbing this gentleman of his life."

In answer then to the exclamations of startled surprise from all the company, I described the whole scene to them, and the simple cause and intention, that moved me at first to enter the boat, or at least as much of that cause as, for certain reasons, I thought good to state. But Jessie was not to be hoodwinked by this caution. By the keen manner with which she regarded me, when referring to the feeling of dullness that had prompted me at first to enter the boat, it was plain to me that she understood all that was implied in that vague term. She, however, made no comment upon the matter, though perhaps, like Mary in the gospel, "she pondered all these things in her heart".

On concluding, I formally introduced the subject of the narrative to the general company as Mr. Charles Howden, at the same time mentioning his occupation.

Hereupon, on the mention of Howden's name, Mrs. Campbell, usually so orderly and composed in her manner, suddenly exhibited signs of the deepest emotion, by first a violent, nervous start, then by her usually pale countenance assuming a more deathlike hue as she fixed her eyes earnestly upon Howden. Then murmuring "My God, is it possible?" she rose from her seat and advanced towards Howden, and stretching out her right arm towards him, in an eager, importunate, nay, almost threatening manner, she exclaimed: "Tell me, where is my brother? you know, you must know where he is now—tell me, I adjure you, for His sake, whose fatherly mercy you have just experienced so great an instance of!"

On Howden's part the effects of Mrs. Campbell's sudden exclamation seemed to be equally emotional. Shading his eyes with one hand, as he returned her gaze with equal earnestness, he muttered: "Mary Carmichael! who would have thought of meeting with her here!" Then in answer to her question he replied coldly: "Why ask me about your brother, Madam? am I your brother's keeper that you should thus charge me? I know what you would say," he continued bitterly, "it was my action that involved him in that ruin that caused him to abscond from creditors, whose claims he was powerless to satisfy. It was my action, my villainy you would say rather. No, Madam. It was through no conscious villainy of mine that that disaster happened, though it was my mistake, the consequence of which has sent me, too, an exiled wanderer, through the world ever since. It was the mistake of wishing to save from ruin that brother of whom you so often heard me speak, but whom you never saw, that induced me to commit the fatal rashness of putting my name to a bill without consulting my partner. This step I was induced to take as not wishing to bring the disreputable state of my brother's affairs under his notice, and through him to yours. At the time I honestly believed that the transaction would be safe, and its negotiation easy. I knew that though our over-draft at the bank was already serious, yet that, though increasing it by another £1000, within the specified time I could count with certainty upon the returns of our business more than covering this deficit. But the sudden collapse of a business in which we had allowed ourselves to become largely involved, at once turned the tide of these hopes, and exposed my dishonest financing: dishonest only in appearance, but not as regards any benefit to myself; I fell through my too great love for my brother. Any hope that we might have had of surviving our financial ruin and of regaining our credit, vanished through the shadiness of my action at such a critical point in our affairs. From the consequence of this ruin, with a debtor's prison staring us in the face, flight was our only remedy. Your brother fled, a victim of circumstances over which he had absolutely no control, but I, oppressed with the double guilt of having brought reproach on my own hitherto stainless name and of being the unwitting cause of disgrace and ruin to another.

"We did not go together, but by different vessels. We both reached America, I with the stern determination of there regaining, if possible, with the payment of all claims against me, the character I had lost; and then returning and claiming the affianced bride I had left behind.

"Within twelve months, however, I heard that she, doubtless convinced of my villainy, had become the bride of another.

"Yes, sir," said Howden, addressing Mr. Campbell, who was listening to all this in silent surprise. "Your wife was at one time my betrothed bride; and yet with you and her children she has doubtless been happy without me. Whilst I, in return for my zeal for the honour of an unhappy brother, am at this day a friendless and a joyless man.

"Do I know where your brother is now, Madam? well, yes, partly. We met when the Californian mines were at the zenith of their fame. I was doing well then; and there, in conjunction with him, forwarded home to our bank creditors all money due by our late firm; I, however, paying, in addition to my own share of the business responsibilities, the £1000 of the bill that had occasioned sudden ruin to us both. I advanced on Mr. Carmichael's account indeed (who was not then in such flourishing circumstances as I was), a considerable sum of his share of the responsibility, he agreeing, when in a position to do so, to repay me. This he subsequently did rather more than ten years ago, for Malcolm Carmichael was an honourable man.

"He was at that time in flourishing circumstances, he informed me, at Valparaiso, where I understand he still is, and his address then I have with me, and you can now have it." And taking out his pocket book, and tearing a leaf therefrom, Howden handed it to Mrs. Campbell, who received it with a fervent "Thank you, Mr. Howden".

As Howden ceased speaking, I gazed upon him with admiration. There appeared to be something truly noble in his manner, as, with his simple eloquence, he thus vindicated his character before the woman whom he once had so deeply loved, and from whom he was so unhappily separated; revealing as he spoke, a view of his past life, that so well corresponded with my own instinctive and preconceived opinion of him.

To me he had always appeared to be a man of superior parts with his high, intellectual-looking forehead, and thoughtful manner.

Whatever effect on Mrs. Campbell the revulsion of feeling occasioned by this clearance of the long-clouded character of the man she once loved may have had, the presence of her husband and daughters prevented her from showing it. But that she was not unmoved was plain by the softened expression of her countenance. But, that she now felt regret at the circumstances that had for ever parted her from a man whom she once loved, it would be quite unfair to suppose; indeed, her long union with a man, who, however inferior in mental parts both to herself and to Howden, still had treated her with uniform kindness and respect, and was moreover the father of her children, forbade the idea.

"My poor, afflicted friend," she said compassionately; "truly yours has been a suffering and thankless life. But God is wise in all His ways, and what He orders must be for the best. Yet the consequences of error He will not always avert, even though He sees the pure intention with which that error was committed. And you doubtless committed a great error, and a most unwise action, that even your love for your brother could not palliate in the eyes of the law. Thus you have entailed much suffering both upon yourself and me; for we are, however guiltless our intention, forbidden to expect that God will interfere to remedy the consequence of a violated law. Yet, though by this act you involved me in the same suffering with yourself, God has given me a kind husband and these two daughters to comfort my declining days, and doubtless, my friend. He will yet compensate you for all the suffering that, by that one imprudent act of your life you have had to bear so long. Still look up and put your trust in Him.

"But you say my brother is in America? Is he not then in New Zealand? Yet what you state I believe, for of old you were ever sincere and truthful. But it is strange—surely it must have been Malcolm that sent all that money to us when we were in Australia, advising us to come over and settle where we are. How should he in America have known anything about this country, and especially of this very district, that he should have been warranted in advising us as we were advised, to settle here?"

"Mr. Farquharson," said Howden, at this point turning hastily to me as if desirous of escaping from the subject under discussion. "I'll wish you a good night, I think. I'll go out to the men's quarters, and perhaps may even return to my own hut."

"To your own hut!" I repeated in surprise. "You surely, after your narrow escape, cannot think of going back to-night with these murderous ruffians perhaps still hovering about."

"Oh, I don't think they will show their faces there again in a hurry! I guess they are now making tracks for the safety of their necks, and will be miles away in the bush by this time, making their way as best they can towards the coast. But I will go to the men's hut anyhow, and if I can persuade one of them to come across the lake with me to-night I will, for I have reasons for wishing to go. Do not keep me, please; I wish to go," he whispered earnestly.

"But surely you could sleep with Lilly here, in the kitchen, till morning at least? No? Well, if you will persist in going I will see that one of the shepherds goes along with you to keep you company." I said this as I saw that he had some powerful motive for wishing to be off, and that no persuasions of mine would move him. Seeing this, I sent word for one of the shepherds at the hut to take my revolver with him and accompany Howden back to his hut, from which he could return early on the following morning.

There was a good deal to discuss, both in what had just occurred and what had been revealed by his words, after Howden's departure.

"The cut-throat looking villain!" cried Lilly in reference to the blackguard Hennesy. "I knowed him as soon as my eyes lit upon him, and so did he me, for he took precious good care to keep out of my sight ever after."

I here, for the amusement of the party, related the scene of Lilly's exploit with Hennesy, in which that worthy came to such signal grief by Lilly's hands; as also Lilly's other exploits on that memorable occasion, all of which afforded no small amusement to my auditors. Young M‘Gilvray, in especial, was so tickled at my description of Lilly's manner of assisting the police in quelling the riot, in which service he contrived to inflict considerably more damage on the Queen's servants than the rioters did, that he fairly held his sides with laughter.

From this subject to our late scene of operations on the Darling Station was an easy transition, and I asked Lilly when he had last heard from his friend Lampiere and how he was getting on. I have until now omitted stating the information I had had from Lilly on my first meeting with him. It appeared that ere Lilly had left the Darling, Lampiere received sudden word from some well-to-do relative who had just come out from home and had established some wholesale business in the provision trade in Melbourne, and who had kindly invited Lampiere to come down and take a situation as confidential clerk with him, for which office Lampiere had had some previous training ere he left home. On this proposal being submitted by Lampiere to Lilly, as in all his emergencies was Lampiere's invariable custom, Lilly very prudently advised him to at once take advantage of his kinsman's very considerate offer. Although town life was never to Lilly's taste, he still considered that, on the whole, Lampiere was more adapted for that than for a bush life. "Besides which," as he very characteristically informed Lampiere, "you now have had a very fair bush training, and are able, if you like, to do things for yourself, and will not be like these town greenhorns, who are only fit to put on their ironed shirts without knowing however the shirt was either ironed or washed."

Lampiere's last letter, received by Lilly about two months ago, stated that Lampiere's friend Burrel, who till then had remained on the station shepherding, was contemplating a move to Melbourne, and, in conjunction with Lampiere, was thinking of starting a paper, a sort of ordinary news and literary journal combined : Lampiere to be merely a sleeping partner in the concern.

Lilly's opinion now of Burrel's own poetical powers had been very considerably enhanced by the perusal of a poem that Lampiere had enlosed in his last letter, entitled : " My Old Quart Pot ". Unfortunately I am not able to give my readers a specimen of this poem to exercise their judgments upon, as I omitted to take a copy of it from Lilly, and I have not since seen it ; but of the merits of this poem Lilly entertained almost as high an opinion as of Lampiere's " Hutkeeper's Address ".

"I'll not say it is just as good as Bill's poem," said he, " but still it is a really good production for all that. He is a smart, keen customer that. But about this paper—I don't know what to think of it. I hope these two chaps ain't going to drop all their money over the concern. I am glad that Bill is sticking to his situation in his cousin's office. He tells me he gets £5 10s. a week there, and I know he won't spend his money foolishly, so that he is bound to lay some by, if he doesn't lose it in that paper.

" What is the journal to be called, do you know ?" I asked.

" *The Australian News and Literary Journal*," was the reply.

" Rather a long title, I think ; however, I trust they will both do well, as they are both very deserving young men."

It was now time to retire for the night, the ladies all in one room, on ample beds, provided by Mrs. Campbell's foresight, on the floor, and the male part of the company in the kitchen in the same manner.

" It seems to me, Mrs. Campbell," I took occasion to say as, with a more than usually thoughtful countenance, she sat by the fire, while the other ladies were assisting Mrs. Munro with the arrangements for the gentlemen on the kitchen floor, " that what is engaging your mind at the present moment is an attempt to solve the problem as to who forwarded you the money to Australia—a problem whose solution, to my thinking, is not a matter of such very great difficulty after all."

"Indeed, Mr. Farquharson, do you think so? for, to tell you the truth, I have been perplexing my mind about that very matter."

"Then I will take the liberty of affirming that your sex's vaunted instinct is at fault this time at least."

"How, Mr. Farquharson? I scarcely understand you."

"Has not Mr. Howden stated that you and he were once engaged to marry one another?"

"Which statement is perfectly true," she replied sadly; "poor Charles Howden and I were to have been married within a fortnight of the unhappy event to which he referred."

"And is it difficult to imagine, madam, that the love of a person of so highly organised and sensitive a temperament as this Mr. Howden appears to be, would still survive the wreck of those hopes, even after the lapse of all these years, and even take such an extraordinary method of showing his undying affection for her who had once been the cherished idol of his soul?"

As I spoke a sudden light overspread and chased away the thoughtful expression that till then had rested on Mrs. Campbell's countenance.

"Surely you do not think that this was Charles Howden's work? How can he have such means at his command and leading a life, too, of labour in order to earn his own living?"

"I see nothing so very improbable in it for all that. Remember, Mr. Howden has been actively engaged in the goldfields during the very best days of California, Australia, and New Zealand, and he is intelligent and of very sober habits, all points sufficient in themselves to account for the possession of no inconsiderable fortune."

"But why, if this is so, should he prefer to lead the life of privation and isolation that he does? Why, indeed, should he labour at all?"

"Such a choice is not easily to be accounted for, especially in a person of his intellectual tastes, and I cannot conceive him to be actuated by the mere grovelling desire to facilitate the hoarding of wealth. Yet many things might bias him in favour of this life away from society, for which, perhaps, he has lost all taste. Disappointed hopes such as he has been the victim of might sufficiently account for such a preference. What particularly roused my suspicions as to Howden's knowledge of this money was his sudden determination to return to his own hut on your reference to the matter, which at the time seemed most extraordinary to me."

"I see it all," replied Mrs. Campbell; "poor dear Charles, how much disappointment and misapprehension has he had to suffer! Yet he comes out of it after all these long years the same noble-hearted, unselfish man as of old. May God reward him for it all, it is more than I ever can."

With these words we separated for the night.

CHAPTER XXXIV.

> " He either fears his fate too much,
> Or his deserts are small,
> Who dares not put it to the touch
> To win or lose it all."
>
> —*Marquis of Montrose.*

THE next morning, it being a working day, I was early in the shed, when one of the shearers handed me a note, which on opening I found to my surprise to be from Howden. In this note he stated that for urgent reasons he was starting off for Invercargill; therefore he requested me to take charge of his effects, and dispose of his horses, which he expected I could easily manage to get on board Mr. Campbell's large boat, that could at some points be laid almost alongside of the land, for what prices they would fetch, as it was not his intention to return again to that locality. Any money that I should realise from his effects, that, with the exception of the horses, were not likely to be marketable, he also requested me to pay at my convenience into the bank of New Zealand in Invercargill to his account. But the books he requested me to keep for myself, together with such articles as were not saleable, stating also that I should hear from him again.

As I had been in the habit of transacting business for Howden, that, besides occasional purchases, consisted in taking down his gold and depositing it in the bank of New Zealand, returning him receipts of its weight, I was quite able to carry out the wishes expressed in his present note.

After breakfast my visitors prepared for their departure. As I was accompanying them to the place of embarkation, I was walking alongside of Mary, Jessie being in front with Mr. M'Gilvray, when suddenly the latter withdrew from her companion and joined me; when, as if by a preconcerted arrangement, Mary stepped forward and took the place vacated by her sister by Mr. M'Gilvray's side.

Looking up into my face with her full blue eyes, clear as a fountain of water, Jessie remarked:—

" What is it makes you look so dull ? "

"Why do you ask, or what makes you imagine that I am dull?"

"I know very well, and I think you are foolish to give way to such fancies. Why should you not look as cheerful as the others?"

"And why do you give me reason to look dull then, Jessie?"

"Now, why should you talk like that? What occasion have I given you to look dull?"

I glanced significantly at the gentleman in front of me.

"Pooh! Mr. M'Gilvray is a very nice gentleman, and I like him very well as a friend, but nothing else."

By this time we had arrived at the side of the boat, and much as I could have wished the conversation prolonged, I had no further opportunity of saying more then.

But the effects of these few words upon me were simply magical.

I seemed to feel as if a load had been lifted off my heart, and in exact ratio to their previous depression did my spirits now rise to an excess of buoyancy.

As I gleefully shook hands with them all, I called out to Mrs. Ayson to be sure and bring her husband along with her if she felt inclined to pay me a visit again, as I hoped she would, next Christmas.

Next Christmas, forsooth! Little did I imagine that I was standing over a powder mine, whose sudden explosion shortly afterwards would send me far enough away from Lake ———— long ere Christmas would be round again!

That same day (I mean that of the departure of my visitors) Lilly was loaded up, and started away likewise.

On the conclusion of the shearing operations, and when the stir consequent thereon had subsided, I resolved, in pursuance of a determination that had lately been slowly maturing in my mind, since the departure of my Christmas visitors, to bring to a decisive issue a question that had long disturbed my peace of mind.

The reader will easily guess what this question was.

I accordingly one day rode down to the Campbells, and was received by them with their wonted cordiality.

After tea, I found myself in company with both the girls in the neat flower-bordered garden. It was a lovely evening. The setting sun flushed all the western horizon. Loud, clear, and musical rang the notes of the tuis in the neighbouring bush, while hundreds of twittering notes sounded through the glades of the forest, as the birds fluttered through the leaves. The gooseberry bushes were bending down with their profusion of fruity treasures, while, as we strayed along, we occasionally stooped down in attention to each, as its bursting, mellow, tasting fruit seemed to solicit our notice.

At length I found myself alone with Jessie, Mary having lingered behind, and then as if suddenly recollecting something, or desirous of giving us a better opportunity of communicating with each other, went back into the house.

The golden opportunity so long wished for had at last arrived. I felt my heart palpitating violently in view of the momentous issue immediately before me. I believe my very nervousness spoiled all.

I took her hand. I blundered sadly, and stated my case with almost rude bluntness.

"Jessie, I love you very much. Will you marry me?"

She tried to withdraw her hand from mine, but I held it more tenaciously. Again I spoke :—

"I have loved you a long time; I cannot do without you. Jessie, will you not give me some hope?"

At last she replied, and to the point. She was naturally of a blunt, straightforward disposition.

"I am sorry to hear you speak as you do, Duncan," the more formal style of Mister Farquharson had long been set aside in my intercourse with all the members of the family, "I have known you a long time, and I indeed do love and respect you, but only as a friend. Let us go inside." As she spoke she rose up and went towards the house, still looking and waiting for me to accompany her; but, stunned by the sudden overthrow of my late hopes, it was only with difficulty that I could force myself to do this.

If, with one sentence, Jessie had ruthlessly slain all my fond aspirations for her heart and hand, to do her justice, having dealt the blow, she did her utmost for the rest of the evening to mollify the wound.

She took her seat close beside me, and smiling pleasantly, addressed her remarks straight to me, endeavouring continually to anticipate my slightest desires, or if I did anything for her it was instantly acknowledged by a bright smile and a frank "Thank you, Duncan".

Yet towards her I maintained a deep, almost stern reserve, only giving brief, monosyllabic replies when necessary to questions addressed directly to me.

My reticence was not the result of anger. It was not that I felt revengeful towards the girl for slighting my offer. No, I loved her too well for that. But my irreparable loss grieved me bitterly in proportion to my love.

Had I not been ashamed to do so I would instantly have mounted Selim and ridden home; but to my kind friends such a proceeding would have been an act of rudeness, that I could not think of showing to them. So, harassed and dejected, I found it a matter of no small difficulty to rouse myself to the task of sustaining some kind of commonplace conversation with Mr. Campbell—interrupted by passing remarks from his wife—about sheep, wool, and other station matters; all subjects of inexhaustible interest to worthy, common-sense, plain, plodding, Mr. Campbell.

At length the weary evening came to a close, and I retired to bed, but as the reader may well believe, not to sleep; my faculties being kept awake by a hard, dry grief, that refused to be comforted. All the future seemed to be a dreary plain without any line to mark the horizon.

But enough of this. I had hardly returned home, when I received a painful confirmation of the truth of the adage, that misfortunes seldom come singly.

I fear the reader will think me a strange hero, when I record this fresh instance of my heedlessness and trustfulness. After having already tasted the bitter fruits of a too unquestioning reliance on the faith of another in the matter of my business solvency, what will any sensible person now think of my imprudence, when I confess that I found myself convicted of the same fatal blunder?

The facts are briefly told. About six months before, Mr. Roscoe, in view of some heavy liabilities in his own mercantile business, received my consent to strengthen his credit, by the use of the cheques of our joint firm for his own private business, though, as he represented at the time, it would only be for a short period.

To any one reasonably cautious in worldly matters, such a proposal would have been viewed with instant distrust.

But so firmly impressed was I with the assurance of Mr. Roscoe's personal probity, and financial solvency, that such a suspicion never once entered my mind.

Besides, I was sustained in my opinion, by the very fact of Mr. M'Elwain's recommendation, for he had known him very intimately for a long time, and I was therefore the more off my guard.

Consequently when the news of the crash came (for Roscoe's business suddenly did crash), it fell upon me with all the more stunning effect, from being so wholly unlooked for.

Mr. Roscoe, as it turned out, was found to be one of those unprincipled scoundrels, who had for a course of years maintained an appearance of fictitious prosperity, by a systematic system of fraudulent book-keeping.

It was with borrowed money he had at first started the

station business, and although under my management this business had paid well, yet such was the serious nature of his over-drafts, for his commercial transactions, that the station profits were instantly absorbed as they came to hand. My own incautious loan of my name for the requirements of his private engagements, had in consequence rendered me liable for all his debts; and these proved to be so extensive that the proceeds of the station, when sold, were insufficient to meet them.

In a word, all my own private means, besides £1000 of Mr. M'Elwain's, which I had intended immediately to pay off—would have paid ere then in fact, but for this arrangement with Mr. Roscoe—were entirely lost through the unscrupulous selfishness of a detestable villain, the more detestable for the Christian profession and standing as an elder in the church, with which he had contrived so thoroughly to veil his real character, and who had thought fit to decamp, taking no small share of his plunder with him.

Bad as the ruin of my business in itself was, the bitterness of feeling it caused me was intensified by the memory of my late proposal to Jessie Campbell, which would now lay me open to the suspicion of having been influenced in my offer to her, by purely selfish considerations; for, from the ruin that so quickly followed that declaration, both herself and her parents must conclude—a conclusion to me most bitterly humiliating—that I could not have been ignorant of my impending ruin and must have been moved by the most mean and contemptible motives —the endeavour to hamper them with my difficulties.

It was the intolerable idea of the certainty of such a suspicion on the part of the Campbells, that made me on the day following the reception of these cruel tidings, when on my way to Invercargill, ride past their house without calling, an act I was never known to have been guilty of before.

The dismal business connected with the transfer of the station property into the hands of the Bank officials (indeed, properly speaking, there was no form of transfer in the matter, they having taken prompt possession for themselves), I will not enter into, but will dismiss the whole painful matter with the remark, that on interviewing the Bank authorities, I found that it was only due to their kind consideration that I was allowed to retain possession of my horse and £50 in money, and that all my toil for the last five years in New Zealand, had been for naught, and that in addition to having worked hard in the colonies for thirteen years with this result, I was now saddled with the almost hopeless burden of a debt of £1000.

Yet against all this I still resolved bravely to struggle, and to rouse my energies, so as to be able to redeem my honour, now compromised by the serious loss entailed upon Mr. M'Elwain, through his generous action in befriending me; for I had little desire to free myself by a bankrupt's credentials from the encumbrance of this moral responsibility.

CHAPTER XXXV.

"Westward Ho."—*Kingsley.*

AT this time the West Coast of the province of Canterbury was the scene of a brilliant rush. Gold had been lately discovered there, and numerous rich claims proved the richness of the land.

The country there being all covered with dense bush impossible to penetrate on account of the under scrub, save by cleared paths—was devoid of pasture, and consequently difficult of approach with stock from the plains. Therefore such stock as was driven there was being sold at exorbitant prices.

I was a practised stockrider, and in possession of a horse that would follow cattle anywhere. As I left the bank, the idea of returning and seeing if I could not persuade the manager to advance me a few hundred pounds on my own personal security, to invest in cattle, and drive to the West Coast, flashed across my mind.

The practicability of this plan, to enable me to scrape together some funds, had been borne in upon my mind before this. I had heard some men, at the hotel at which I put up—who had just returned from Hokitika, as the goldfield township was named—talking on this subject, and descanting on the excellence of the opportunity that this offered to any enterprising person with a few hundred pounds of capital in hand. Therefore, my idea was, if I could procure such an advance, to proceed at once to Canterbury, and to gradually muster among the farms on the plains there, or at sales, a herd of cattle, as I had means to pay for them, and to drive them to Hokitika, and there sell them; and to keep to this work as long as the present flourishing prospect of the goldfields continued to offer any inducement for its prosecution.

Whilst standing on the pavement outside the bank, and meditating on the probability of my being able to get such an advance, some one suddenly accosted me, and looking up, I saw Howden standing before me.

Whilst I had long suspected that this man's monetary resources were by no means to be gauged by his usual appearance, I knew as a matter of fact, that he had several hundreds of pounds located in that very bank, opposite to which we were now both standing; for I myself had at various times paid in sums in gold ore and money, amounting in all to as much. Therefore, after shaking hands with Howden, I bluntly remarked to him, "Howden, as you are doubtless aware, from the report in the papers, how matters now stand with me, I will just ask you, have you sufficient faith from what you have seen of my honesty of purpose, and business capacity, to intrust me with a loan of say £500? I have an idea that by proceeding to Canterbury, and purchasing cattle there, and driving them overland, and selling them at Hokitika, I should soon recoup myself for my risk and trouble. I would allow you full market value for the use of your money."

He replied, "I am glad, Mr. Farquharson, that you have mentioned the matter. I can assure you I was desirous of asking you if I could be of any assistance to you, but was hesitating under the fear that you might be sensitive at any reference to the unfortunate calamity that has befallen you. Will £500 do you for the present?"

"Yes, it will be quite sufficient, as I should prefer not to involve myself too deeply at my first venture, so as to see first with my own eyes what the nature of the demand and prospect for the continuance of this work really is at Hokitika."

"I will willingly let you have £500 now; beyond that it would be rather inconvenient for me to go, as the bulk of my money is locked up in investments, but to £500 you are welcome."

"Many thanks, Howden, I can assure you that you have conferred a very great obligation upon me, by allowing me the use of this money. What rate of interest do you desire me to pay you for it?"

"The precise rate that the value of my life deducted from it will leave."

"Tut, man! never mind that! that was only a simple action of humanity and duty. How much interest shall we say?"

"No more on that subject at present; we will talk about it when we come to settle up about the principal."

"Well, as you will; by the bye, any further news about that other affair?"

"Some; but rather vague—but excuse me, I would rather not talk about it at present."

"Where," I inquired, "shall I address my letter when I want to write to you?"

"Address to the care of the post office here. If I should not be in Invercargill when it arrives it will be forwarded to me. But in all probability, before you require to do so, I shall see you in Hokitika. In the meanwhile let us adjourn to the bank counter and settle this preliminary business at once."

This dispatched, I was once again ready for action and fighting with fortune.

A hope of inducing Lilly, with his ripe experience in such matters to co-operate with me in my new venture, induced me at once to write to him stating my wish, project, and address in Christchurch, for which place I should leave on the following day, and whither he could write, or come in person in case he purposed joining fortunes with me.

In a few days following my arrival at the A. J. Hotel in Christchurch, to which place I had been recommended by an Invercargill acquaintance, Lilly's answer arrived, testifying, in the first place, grief and consternation at the ruin that had overtaken me, and, in the next, a willing compliance with my proposal of casting in his lot with mine. He stated that on the conclusion of his present trip (for my letter, by my direction, had been forwarded to him where he was on the road, by the person with whom he lodged at Lawrence) he would send up his bullocks by an agent to be turned out at Campbell's station, where he knew they would be permitted to graze and be looked after, whilst he would come on immediately and join me.

Pending Lilly's arrival I went about among the farmers and attended sales, collecting a herd of cattle larger than I had at first intended, because of the larger capital that the knowledge of Lilly's intention enabled me to invest for our mutual enterprise. Lilly, I should add, had made such good use of his time in New Zealand that he was able to put £500 into the business with me.

Thus we were both, while working with a common interest, able to start our new business with the respectable capital of £1000.

Besides Selim, and Lilly's favourite mare Coleena, we each had provided ourselves with an additional horse, strong, sound, and spirited, for each of which we paid a tidy sum of money. Thus, on every trip we started with a fresh horse that was alternately used and rested, so that each animal, though severely worked while on his trip, had ample opportunity to recoup his strength by the time he was again required.

I now propose to pass over the details of this, my new experience in stock driving to Hokitika, and without pausing to give a description of the striking and varied scenery along our route, both of rugged mountain and primeval forest, or to dilate upon their effects on our minds on first witnessing the magnificent spectacle through the Otira Gorge, with the winding road that there rapidly descends, hewn out of the precipitous mountain sides on the Hokitika slope of the range, and the bridges there crossing and recrossing and crossing yet again the rushing torrents that foam down that wild ravine, or staying to describe our difficulties along the road, then in the process of formation on a most substantial basis of broad stones, now winding by the edges of precipices, now a simple track following for miles the pebbly channels of impetuous torrents, or crossing over dangerous streams that intersected our toilsome route, or again, a bridle path, garnished on either side with vine thickets and overhanging networks of supple jacks, winding through the primeval forest amid the deep mud formed under the almost daily droppings of that cloudy atmosphere, when our cattle were only kept back from the bush on either hand by our sagacious dogs. Without wasting time over such matters, however interesting in themselves, I will jump over twelve months that were continually occupied with the claims of this business.

Let it suffice here to say, that at the end of that period both Lilly and I found that, after deducting all expenses, we had rather more than £600 profit each left for our pains, a very happy result, with which we were both entirely satisfied.

Yet the events of this long interval shall not be wholly unrecorded. An account of one evening, as a sample of the times and rough surroundings, when money flowed freely in among reckless spendthrifts, the reader may find interesting as a type of what went on in the Colonies then.

The scene was at a shanty or sort of canvas hotel at the Bealy stream, an affluent of the Waimakariri, and a little way above its junction with the latter. There, as I remember is (along the course of the Bealy) a rather picturesque looking valley, bounded by steep mountain ridges on either side, clothed with thick bush almost to their tops—the tops, however, standing out bare, bleak, and verdureless beyond.

We were on one of our return trips, and consequently disencumbered of all impedimenta. The shanty, where there is now very likely a large substantial hotel and growing township, was then, as I have said, a mere calico affair. It was also, as far as my memory serves me, the only thing in the shape of a building on the ground—outhouses of course excepted. But primitive as was the construction of this hotel, it did not suffer from a lack of customers.

The chief of these customers were prospective miners, on their way to the new goldfield, and others returning from it; some flush with recent rises, others who, in sanguine expectations of similar windfalls, had started with small means to support them while endeavouring to realise such golden dreams, and were now returning with considerably less—that is, nothing. The lucky diggers, who were now rapidly spending in reckless "shouting" the money with which on leaving Hokitika they had decided to secure a more luxurious life in Christchurch, were showing signs of quickly arriving at the same low financial position. The swags of these travellers, mostly carried in oilskin wrappers as a protection from the almost daily rains of this region, were seen piled in a heap at the corner of the tent as we entered.

As this place, at least at that time, was the farthest point that vehicles coming from the plains en route for Hokitika could reach—a gap, between here and the top of the next mountain range, where the road was eventually to go being still unmade—there were several drays and waggons congregated together. The animals that drew them were feeding about in hobbles, and ornamented with tinkling bells—more for use, however, than for ornament—and their drivers made no unimportant item in the crowd who were freely liquoring up at the bar. Nor did their thirst for ardent spirits appear to be in the least moderated by the price of the beverages with which they were endeavouring to quench it—to wit—the good old colonial charge of one shilling per each "nobbler".

On that day, like on most of the days during my stay in that region, it had been raining heavily. From these continuous rains, and the trampling of many animals, the ground outside the shanty was literally a mass of black, fluid mud. People accustomed to these parts, however, soon come to regard rain and mud as trifles.

Yet, if the exterior of the calico hostelry looked unpromising, still, when inside, we found that against the discomforts of the evening, there was most substantial provision made. There was, however, no fireplace, whether from prudential, or necessary considerations, I know not, but, in the cook house—a small sod hut, there was an ample fire for any who chose to make use of its comforting glow. And thither both Lilly and

I adjourned—after having first tested our host's whisky, to give a start to the blood in our half-numbed bodies—for, though neither of us were habitual dram drinkers, neither were we teetotallers.

But as it happened that we had arrived just in time for the supper, of which we stood in great need, our stay by the fire was but a short one. We had previously seen to the feeding of our horses in a most dilapidated building or shed that did duty as a stable.

The signal for supper was now given by the cook, who announced the same at the top of his voice, and the intimation was received by the tippling, rough-mannered guests, with a sort of derisive yell, and an instant rush for places round the ample board. The rude table was soon closely surrounded by guests, some eager for food, but more hilarious with drink. Making our way there, too, with but little loss of time—for a long ride through a cold wet day is a wonderful stimulant to a healthy appetite—both Lilly and I soon became as keenly engrossed in the contents of the bill of fare set before us as anyone there. The supper, though but simple, was at least substantial, and well cooked, and the price for it too, by the way, of an equally substantial nature. Instead of cups, we were given tin pannikins, but this we did not mind, for it only recalled the old habits of our station life in Australia.

However, the keenest appetite must get satisfied in time—though ours, if not inconsiderable, were but contemptible in comparison with those of some that we saw there.

The table was again cleared, and the guests once more betook themselves to the consideration of the great problem of happily disporting themselves for the remainder of the evening. To many this consisted solely in "nobblerising," and to most, in card playing.

As the evening wore on the scene among so many tipplers grew so wild that the climax of a free fight, in which all were likely to join, seemed to be a by no means improbable or remote contingency.

Among the roystering crowd were two undersized but smartly built men, who particularly engaged my attention. There was a sort of family resemblance between them, so to speak—for the idea was merely imaginary—though not in their features. In this regard, indeed, they were decidedly unlike. The resemblance was in their comparative sizes and general get-up. Both had evidently been seafaring men and were dressed in nautical fashion, with white moleskin trousers and blue shirts, arranged with that peculiar shirt-fold overlapping the trousers-band that sailors alone seem to have the knack of making, and their similarity of costume was rendered still more marked by their both wearing a Kilmarnock bonnet. Yet not only were they not related, but they were not even of the same nationality, one being an Englishman and the other Scotch. The coincidence of their similarity of costume was the more odd because from what I gathered from their remarks I could see that they were not even acquaintances, and had not met until that evening.

But whether from some mysterious magnetic influence or not I cannot say, but certain it was that all that evening these two were seldom long apart. Yet whatever the nature of this mutual attraction on their part, it could not have been one of love or admiration, since the principal topic of conversation between them was of a decidedly belligerent character. The Scotchman, indeed, appeared to be most anxious to try conclusions with his English acquaintance, and he was continually swaggering up to him and taking him by the shoulders and shaking and pushing him roughly about.

The Englishman, indeed, seemed gifted with a rather more placid temperament, and bore this rough handling with surprising equanimity. Not that he was always patient, however, for his words at times betrayed considerable irritation, as he kept telling the other not to be a d—— fool. Half-a-dozen times or so, at least, these two men moved towards the door, apparently with the determination of "seeing it out" on the "lovely grass," as the Scot in a flight of high poetic fancy characterised what was in reality for some distance round the tent only a sea of black mud. Invariably, however, on these occasions, ere reaching the door their warlike spirits evaporated, leaving them apparently in a more amicable mood, and then either the one or the other swung his would-be opponent round in an opposite direction with the more friendly invitation, "Come along, old man, and have a drink"; only shortly afterwards to become more belligerent than ever.

At one time, however, this belligerent temperament on the part of the Scotchman, who was in reality the more obstreperous and boastful of the two, seemed to be happily diverted into a more peaceful channel, when he loudly signified his intention of favouring the company with the performance of a Strathspey dance. No one, however, happening at the time to notice this most condescending proposal on his part, it, like many a more important motion, fell through for want of support. This was to my great grief, for although I did not care to draw the attention of the others to a notice of this proposal, I was very keen to see the measure performed.

Up to about this stage of the evening Lilly had kept quiet

—indeed, for him, most singularly so, for in company he was not usually so reserved. But now his attention seemed to be attracted to the movements of both of these heroes. In fact, I rather suspect that Lilly had been in reality taking stock of them all the evening. Now, however, he rose from where he was and reseated himself beside them, his eyes twinkling with some humorous thought, while he appeared to be perfectly delighted with their manners.

I took a sharp glance at my quizzical friend, as I felt persuaded by the deepening lines radiating from the corners of his eyes, always a sign of fun with him, that he was intending to get up to some piece of mischief. What the nature of it was to be, however, I could not imagine, for I observed him simply order drinks for three, the nature of his own being signified by a wink, that the shanty keeper perfectly understood to mean something soft—a temperance cordial—in fact.

Lilly apparently became greatly interested in his valorous new acquaintances, feeling their muscles, and declaring that he would be sorry to run the risk of a pounding by either of them. This flattery inflated them much, especially the Scotchman, and he was in the midst of the recital of a conflict, with a fourteen stone weight antagonist, whom he had gallantly felled with a blow, as I left the room to see to the horses, while I heard Lilly again remark, in tones of the most intense admiration, "Drink up your glasses, boys—my word, but you are the right sort, both of you".

As the night looked wild and dirty, I decided on keeping the horses inside, for they were not always accommodated with a shelter on these trips. I stayed talking with the man who officiated as groom, for about half an hour, when suddenly our ears were startled by the sounds of voices raised as if in fierce anger, issuing from the door of the shanty, that was just then suddenly opened. "There is a row on," remarked the groom, looking out from the door of the stable; "there is two coves going out to fight. My word, they'll fall soft anyhow."

On going to the door in my turn, I beheld several people hurrying out from the shanty, one with a lantern—for the night was pitch dark—and in their midst two men stripped to the waist, each moving with very unsteady steps in the direction indicated by the man with the lantern, who was leading the way.

Now there are few sights more repugnant to my nature than that of a drunken fight. Indeed, to all such spectacles, whether in drink or not—and in my experience among shearers, I have seen many a bloody one—I might almost justly say that I have a sort of cowardly aversion, although I have given proof sufficient on an emergency, that of the grit of true courage, I have my fair average share.

Thus, on the present occasion, on seeing these preparations for this drunken encounter, my instant impulse was to at once adjourn to the house, when, to my intense amazement, accompanying the person who was leading the way to a suitable spot for the encounter, and who had now halted, where, in a hollow depression of the ground, the mud was positively deeper, and more diluted than at any other spot, I recognised Lilly. This recognition on my part at once changed the direction of my steps, and I now—all curiosity and astonishment—walked up to the crowd to see whatever had taken possession of Lilly's senses, to induce him to act in a manner so unwonted.

Lilly was the very reverse of a brawler, and brawling, especially drunken brawling, he detested, and was always ready to put a peremptory stop to.

"In the name of wonder whatever can Lilly mean? he surely can't have got the worse for drink that he is taking part in this drunken row!" I ejaculated, as I hastily strode, as best I could, through the mud to the scene of action, where a ring was being formed, and where inside, and behind one of the would-be combatants, Lilly might be said to be acting the part of master of the ceremonies, as it was he who was giving all the necessary directions.

By his orders, the man with the lantern was directed to hold it so as to throw the light straight in the faces of both combatants, behind one of whom, as I have said, he stood to act as picker-up in the approaching deadly slogging match. Behind the other, to perform a similar office, stood another man, with a remarkably good humoured expression of countenance. In both of the combatants I speedily recognised the two quarrelsome little sailors, who by this time were so blind drunk that they stood see-sawing in a most remarkable manner, preparatory to the signal from Lilly to begin the action.

Lilly, I should say, was backing up the Scotchman, who was now uttering the most frightful imprecations against the other's eyes, and livers, souls and bodies, imprecations which the Englishman paid back in kind.

Of course I now at once penetrated the object of Lilly's presence and previous demeanour with these two champions, with their bombastic wrangling, who, despite their protestations, were evidently afraid of each other; the mischievous fellow had hit upon a method of punishing both, that the sight of the mud, and their drunken condition, suggested as being so exquisitely ludicrous.

For this purpose he had deliberately filled them up with a

succession of nobblers till they had reached their present stage of drunken imbecility, by which they were rendered totally unable to inflict any injury whatever upon each other. The amusement therefore that Lilly now proposed to himself, and the other spectators, who had been put up by him to his trick, was the sight of these pot-valiant champions, rolling and floundering in the mud, in their abortive efforts to damage each other, and which at the same time would be a sufficient punishment for them for their wrangling and noise. Nor was Lilly much, if at all, out in his calculation, as to the absurd effects that would result from this encounter.

Were a person disposed to view it from a merely philosophic standpoint, the sight was simply calculated to sadden and sicken him, as affording in itself such a convincing proof of the inherent depravity of human nature, that two reasonable beings could, on any terms, be brought to roll about in the mud in such a truly swinish fashion.

But taking a lighter view of the matter, I confess that among the others, I fairly roared with laughter at the scene.

On the word being given by Lilly to begin the fight, which he did in these terms, "Now lads, go it a-muck," (and go it a-muck they certainly did), they both at once advanced with the most deadly purpose towards each other. For the better development of the scene expected to follow they had been kept considerably apart, so that on the signal for action being given, they had to traverse several yards of deep mud ere they could meet; and that in their inebriated state they accomplished with some difficulty. Ere they had actually done so, in fact, the Scotch sailor's foot sticking in the mud, he, in trying to release it, pitched head foremost into it, and lay there at full length, to be immediately followed by his English antagonist, who, lunging forwards with all his force as the other's descending head seemed within his reach, missed his aim, and losing his own footing, fell on the top of the Scot.

"Fair play, boys," shouted the impartial seconds, darting forward, their shirt sleeves turned well up for the purpose of dragging up the mud-plastered champions and confronting them again with each other. Standing with difficulty, they now darted their fists at each other's faces, with the like result as before: one fell into the mud and the other atop of him; and there, this time, they were permitted to roll and flounder about as they liked until they were actually heard choking with the mud that had got into their mouths. Again lifted by their seconds, who had thus deliberately left them in the mud for a while, they now presented such a spectacle that their features could not be distinguished. This time the Englishman managed to hit his Scotch antagonist so true that the latter fell flat upon his back, and under the impression apparently that he had got into bed, he instantly fell asleep. His antagonist, who from the effect of his own blow had also fallen back, was, while lying helpless in that position, heard loudly denouncing the deepest vengeance against the other. He was now declared victor, and both he and his companion were pulled up and dragged into the tent. The breeches of both were then pulled off, and they themselves well washed by several pailfuls of water being dashed over their naked bodies—a treatment that, in their drunken state, they seemed to be scarcely conscious of —then after being thoroughly rubbed down, they were consigned to their beds, from which they did not move during our stay at the shanty, as we left early on the following morning.

The greatest hilarity prevailed while this wild practical joke was being played, and the combatants escaped unhurt.

As for their clothes, Lilly himself, aided by the man who had acted as second for the English sailor, rinsed them out in water and left them drying in the cook-house ; and thus ended the joke.

CHAPTER XXXVI.

TO return to more serious matters. At this satisfactory stage of our enterprise, when we discovered what our profits had been, we were in Hokitika, where we had just disposed of our last drove of cattle.

At this time the cattle market was showing signs of glutting, though this was less the result of a more than usual supply than from the fact of the decline in gold returns that was then making itself sensibly felt throughout the community, and thereby occasioning the miners to be less reckless in their outlay. This, of course, resulted in a general decline in the high prices hitherto readily paid for cattle.

For this reason I, in particular, was desirous of resting upon my oars, as, in my peculiar circumstances, I was unwilling to risk by an unfavourable venture the money I had lately been fortunately able to secure, but which I merely regarded as held in trust by me for those to whom I owed it, and to whom it was incumbent upon my honour to faithfully repay it.

Musing upon these matters, I strolled down Revel Street one evening about nine o'clock. Hokitika was then at the height of its prosperity, for whatever of failure might about that time have been beginning to make itself felt in the diggings outside, its effects had not yet made themselves so manifest as to affect the business of the hotels within.

And with houses of this description Revel Street was at that time literally studded on both sides. These appeared to be all driving a roaring trade, and were filled with roystering revellers scattering their money about in reckless profusion in shilling "nobblers". In front, the bars were thronged with tipsy diggers "shouting" for their friends, and as many "loafers" as happened to be loitering about, to the great increase of the prosperity of the landlords, if to the rapid diminution of their own; while from within—and all the hotels in this respect seemed to be equally favoured—where the inspiring strains of the fiddles were heard playing dance tunes, a still greater throng of spendthrifts were persuading themselves that they were having a fair return for all the money that they were spending so freely.

Passing in front of one of these houses, within which the sounds of harmony indicated that "the mirth and fun was fast and furious," I paused to listen; then, out of curiosity, stepped inside to view the scene.

It was in a low ceilinged room, accommodated for the purpose it was then serving with benches at the upper end, that rose in tiers above one another for the benefit of the spectators. The space between the door and the benches was reserved for the dancers, and was barely sufficient—so great was the crowd—to admit of their moving freely. Under other circumstances this apartment, with its tier of benches in the rear and clear space in front, would have been well adapted to the service of God; at the present time, however, with its reeking fumes and reckless assemblage, it appeared to be doing excellent duty in the service of the devil.

The musical part of the entertainment was furnished by a piccolo and fiddle, that were by no means unskilfully played. Besides the musicians there was a third person set apart to act as master of the ceremonies. This gentleman's position was evidently anything but a sinecure, judging from the unremitting nature of his exertions, as one set of revellers was instantly succeeded by another, who had been meanwhile impatiently awaiting their opportunity of a clear space on the crowded floor, and who, on the happy consummation of these wishes, were in their turn as eagerly succeeded by a third set, until, what with calling out the orders of the dances, and introducing partners during the whole evening, he was allowed but few intervals for rest, and thoroughly earned his wages, whatever these might be.

On entering, I at once made my way up to one of the raised seats at the farther end of the room. Beside me, where

I was sitting, I observed and was struck by the appearance of an individual, sometimes sitting, but usually standing, who appeared to be keenly observant of all arrivals as they entered the door.

He was a man of swarthy complexion, with hair and beard that had originally been black, but were now slightly grizzled. He was perhaps under the middle height, but of a very powerful build. He was dressed in plain clothes, but a certain squareness of shoulders and erectness of carriage bore the unmistakable stamp of a military training; and this, taken in conjunction with his appearance of watchfulness, impressed me at once with the conviction of his being a detective officer in disguise.

That he was on the look-out for some one whose presence there that night he had some reasons for expecting, was evident by the quick scrutinising glance with which he seemed to scan the appearance of each fresh comer as he entered the room.

For the diggers and others who felt disposed for dancing there was here no lack of female partners. Of the peculiar characters of the latter, however, their flaunting garments, their loud voices, unabashed looks, and coarse expressions, left no room for doubt—" beautiful daughters of sin," as the sweet Danish author, Hans Christian, so aptly characterises them, and the records of whose lives, if written, would present such a harrowing picture of betrayal and seduction, of neglect and depravity, of lives reared 'mid scenes of infamy, or of seeds of virtue, crushed by degraded association. Such a picture as might well put our boasted Christian civilisation to the blush, and render us objects of astonishment and scorn to those very heathens, for whose spiritual welfare we equip missionaries with such pains and cost, to send them to the ends of the earth with tidings of gospel peace and salvation.

Yet among many of these unlovely representatives of womanhood, I observed one whose appearance, manner, and expression of features, at once proclaimed her to be, though with, not of, that company; this was shown, too, by the sudden flush on her pale cheek, as she shrank back haughtily from the attention of a flashily dressed, half-intoxicated young man, who, swaggering up beside her, attempted to place an arm over her shoulder.

"Hands off, young fellow!" cried out a young woman, good looking, but with a flushed and dissipated look, whose boisterous manners I had already noted, and who was seated on the other side of the girl whom the young man had accosted. "Hands off!" she repeated, coming forward and seating herself between them; "this lady is none of our sort, so you just keep yourself square and behave yourself, if you don't want to get yourself into trouble."

She uttered this caution in a tone that plainly intimated an intention of hitting from the shoulder on a repetition of the offence—a feat that with her vigorous frame she seemed to be well capable of performing. The young man sheepishly withdrew into the crowd—many who had witnessed the scene cheering the spirit of the heroine with a loud laugh and exclamations of "Well done, Kate, that's the way to talk to him".

I now began to scan the appearance of this woman, whose presence in that rude assemblage the remark of the girl Kate had drawn attention to, with more interest. Being seated behind her, my observations could only be made by side looks at her face as she occasionally turned her head. Whatever might have been the motive of her presence there, it was evident at a glance that she had at one time been accustomed to scenes of ease and refinement, and traces of this appeared to hover round her still. She was slight and beautiful in form and features, though her face was perfectly colourless. Her eyes, I observed when on one occasion she turned her head round quickly, attracted by some sound behind her, were dark, as also was the colour of her hair. There was something in the quick bird-like motion of the turning of her head that roused in my mind a vague consciousness of something familiar about this woman, that compelled my eyes as if by some secret fascination to rest continually upon her. I at length asked of the disguised detective if he knew who this woman was, but he simply shook his head and made no reply.

Shortly afterwards the girl Kate, who had before championed her, and who appeared, however inconsistently, to be on terms of intimacy with her and to regard her with some rough sympathy, addressed her.

"Come, Rachel," she said, "what are you moping so much about? It will never do for you to get yourself so low as this, you know; have a drink of summat, it will put some life in you. You know——" Here she whispered something into the girl's ear; and then, addressing some young fellow beside her, she said, "Here, young fellow, ain't you going to 'shout' for me and this young lady here now?"

"Rachel, Rachel," I mused, but even then my thoughts failed to recognise the person in front of me, so far were they removed from the idea of such a meeting.

Meanwhile, to Kate's proposal to accept of a treat from this young man, her companion replied with a smile and a shake of her head, "Thanks, Kate, but I would rather not".

"Tuts! this will never do, you know. You must rouse up;

you know what's before you, so you must brace yourself up for
it. You won't dance, I know, but take something to drink,
and sing a song; that will liven you up, and brighten your
wits for your work."

"A song? I am afraid that my song would be still duller
than myself," I heard the other reply.

"Well, have summat to drink first, and then we'll see.
Come, Jim, you 'shout'. Mine is brandy; won't you try
some of that, Rachel, too? Very well; fetch this lady a glass
of sherry."

As Kate had predicted, her companion's spirits seemed to
be sensibly revived by the glass of wine that she had been
persuaded to take.

"Who can this woman or girl be?" I soliloquised. "She
looks respectable in her appearance; but what sort of respectability can this be, that can thus permit her to associate, and
even drink, if only wine, in such company as this?"

I confess that, from the circumstances of her surroundings,
the utmost extent of my charity in my judgment of this woman
—though her face bore the impress of experience, matured by
sorrow—was but to allow her a slightly higher place in society
than that of her degraded companion. In the meantime these
unfavourable suspicions were confirmed by hearing the girl
Kate strongly urge her to give the company a song, for the
dancing at this juncture had temporarily subsided. To any
respectable woman, who from any circumstance chanced to
be one of that assemblage, Kate's request would have caused
a feeling of disgust, but to the person addressed, there was
evidently nothing offensive in such an idea. She proved this
by her manner of refusing to comply at first, but not as if she
felt offended at it, but only disinclined. At length, however,
overborne by the pressure of her stronger willed companion,
her objection seemed to give way, when, as if rousing herself
up for a determined effort to acquit herself at once of a task
that she saw she could not avoid, she suddenly remarked:

"What shall I sing then, if sing I must? Let me see.
Ah, me," she suddenly sighed, as if with uncontrollable feeling,
"it is on me again, and I can sing only this."

Then, with a voice whose tones of plaintive, wailing sweetness instantly pierced through the recesses of my memory—as
the arrow of a strong bowman pierces the heart of his naked
foe—I instantly saw again a house on the banks of the Murray,
where once before I had, on a moonlight night, listened outside
of a window to that same sweet voice, then, too, wailing as if
in forboding sadness the concluding lines of Burns' exquisite
lyric—

> "And my fause lover stole the rose,
>   But ah! he left the thorn wi' me".

And springing to my feet in almost overpowering astonishment, bewildered and transfixed with horror at the sudden discovery, and the surroundings amongst which it was made—all of which flashed through my mind like lightning—I listened to the long-lost, the ruined, Rachel Rolleston, singing to the same air the following dirge :—

> "The flower that with tender care
>   Sweet fragrance was wont to diffuse,
> The loveliest of the gay parterre
>   Fair blooming, with refreshing dews.
> Till smitten by the summer's blast
>   It's velvet petals droop forlorn,
> When from its bed 'tis vilely cast
>   An object then of loathing scorn.
>
> " Like it, her hapless lot o'er whom,
>   A parents' love once watched and smiled.
> Who now drags out her wearied doom
>   From love and honour far exiled.
> Now to those bright and shining years
>   Undimmed by guilt and clouds obscene,
> How near the prospect still appears,
>   But ah! how vast, the gulf between!"

At the moment when the last word of this song was still ringing faintly from Rachel's lips, a stranger suddenly entered the room. He was tall, erect and very muscularly built, with bushy whiskers of a sandy shade. His features seemed indented, or rather rent with deep lines, traced by violent passions that seemed still to flash from his restless grey eyes, and to quiver on his thin, curling lips.

On entering, he fixed a stern, almost resentful glance on her who had been singing, whose eyes at the same moment met his: she suddenly subsided to her seat and clasped her hands tightly over her breast.

As this stranger entered I observed a quick, peculiar, movement on the part of the disguised detective, who chanced at the time to be standing up. As the door opened, I observed the detective suddenly start, then, stooping his head, he put his pipe into his mouth and struck a match. At the noise caused by the match, the new comer glanced up suddenly in the direction of the sound.

For a moment his stern, inquiring eyes endeavoured to penetrate the detective's disguise, then, as if suspicious of what he had observed, he turned round on his heel and abruptly left the house.

CHAPTER XXXVII.

NOT desiring to make myself known to Rachel where she was, as I thought I should be able to do more for her by quietly interviewing her at her own abode, I now resolved to attempt to follow her there, wherever it might be. With this thought, I quietly remained where I was, with the determination of keeping an eye upon her until she should go out, and then to dog her steps.

For a short time she sat as I have described her, wholly disregarding the unfavourable comments of her friend on the depressing effect of the song she had just sung, for Kate evidently had not observed the stranger whose sudden entrance had caused poor Rachel to collapse so suddenly, Kate having been listening to Rachel's song with drooped head; and the entrance and exit, having all taken place in such a short space of time that she had been unaware of the circumstance until Rachel rose and in a whisper appeared to communicate it to her, and then immediately went out. On her heels I quickly followed, keeping well in her wake as she went along the street. While thus engaged, I was soon made aware that I had a companion with me on the same lay, in the person of the disguised detective; but at first disregarding this circumstance, I kept my eyes steadily fixed upon Rachel's movements.

I watched her enter several shops that were still open, as it was not yet past ten o'clock, and in which she was evidently obtaining supplies of necessary stores, and the various parcels, though all placed inside of a Maori kit, seemed to weigh a considerable amount, for she carried it with some signs of distress, judging by the frequency with which she shifted the load from shoulder to shoulder.

Always keeping well in the shade of the timber, for it was a clear moonlight night, I followed her steadily as she quitted the town, and struck off into the bush by the way of an old timber track. Pursuing this track, she kept on till it struck the bank of the river, along which there was now more difficulty in progressing owing to the tangled nature of the under-scrub, so that at times to avoid it she had occasionally to turn into the bush, where the openings, running parallel with her course, admitted of her progress by that way.

All this time Rachel never once turned her head to see if anyone were following her.

Suddenly turning from one of these openings in the bush, she went directly towards the river, when she almost instantly disappeared among the scrub. As I was cautiously endeavouring to follow the direction in which she had disappeared, I was suddenly joined by the detective, for till then, though both pursuing the same object, we had kept somewhat apart. Neither spoke; yet with the same instinctive thought we here paused and gazed around to take note of the appearance of the spot. This a momentary glance at once revealed. We saw that the level space where we were then standing had become narrowed by the proximity of a mountain spur, that a short way further on appeared to abut directly upon the river's bank. Between that, however, and where we then were, the ground was thickly covered with an undergrowth of tangled scrub, besides other timber. As I have said, it was among this scrub that Rachel had disappeared.

Glancing carefully about, in a little while my companion discovered a path leading into it, and (for the first time speaking) whispering the information to me, we both—first feeling that our revolvers were placed conveniently in case of emergency—proceeded cautiously along this path for about a hundred yards to where further progress along the river appeared to be blocked up by the spur, that there reached down to the bank. We then suddenly observed the gleam of a light coming from the midst of a thick tangle of lawyer scrub and other parasites, in the midst of which a space had been cleared sufficient to allow of the erection of a tolerably large tent.

We now, on hands and knees, approached close enough to the tent to admit of our hearing any conversation that might possibly take place inside it. We had scarcely attained to this desirable position when the sounds of a hasty, heavy stride coming down the spur, which immediately passed within a very few yards of where we were concealed, startled us not a little. But the new comer, without the least suspicion of our proximity, strode on, and at once entered the tent. His form I did not see, but the voice, immediately afterwards sounding clear and distinct within the tent, would have revealed to me at once the identity of the speaker, even had I not been prepared from the events of the evening to make such a discovery.

It was Marsden—the same who about two hours ago had so suddenly entered and so abruptly left the dancing-room. Although, with that passion-worn countenance, had I met him under other circumstances, I should, in all probability have failed to identify in him the gentlemanly yet impenetrable Marsden, who had been such a source of mystery and jealous fear to me about six years before on the Murray and Darling. Yet, on reflection, from the glimpse I had caught of them at the dance-room door, the features were indeed the same, but with this difference—that whereas when I before knew them, they were kept masked by his determination to conceal the storm of passion that ever raged within, the mask was now removed and the passion allowed free vent.

He was now speaking harshly to the woman whose prospects in life he had ruined, and whose spirit he had broken.

"Well, what ails you now, girl, that you assume such a whining look? Why don't you look pleasant and smile cheerfully when a man comes to see you? I might as well go and camp, and root with the pigs on the range, as come here for comfort, I can see."

"Why do you chide me so, Randal? have I not always done my best to please and satisfy you?"

"Done your best? humph! perhaps yes, and perhaps no. You can't say you are doing so very much to please me, when after such a long absence, you can meet me with that reproachful look of yours that you know I hate. Why should you reproach me?"

"I am not aware, Randal, that I do; indeed I do not mean to reproach you. God knows that it is only myself that I reproach!"

"Why should you reproach anyone, or yourself either?" he replied, speaking in a loud harsh tone. "What have you to complain of? Who is hurting you? Have you not plenty to eat and to wear? If not, it is your own silly fault. What have you to complain of, then?"

"Of nothing, if I only had the mind of the brute that perishes; of nothing, if I had not been reared in affluence and love; of nothing, if my mind had not been expanded by education to enable me more fully to appreciate the advantages I have for ever forfeited; of nothing, if my tastes had not been trained to loathe the nature of the associations that I have so long been surrounded with, and to realise all the repulsiveness of the abyss into which I have been dragged. Yet Randal, I do not wish to reproach you, though I cannot forget what I have once been, and the bright position that in my love for you I have for ever forfeited."

"What you have been!" he seemed to retort fiercely, "and what have *I* been, might I ask? have *I* forfeited no position? Or, were your prospects and surroundings ever more promising than mine were? Are thief and robber proper designations for me, or their degraded professions my fitting calling by birth and education? I, who was once looked up to and respected, as my commission of captain in the Queen's army entitled me to be respected. And what am I now? A hunted felon, without a place in which to hide my head, except in a scrub, like a wood-hen. You will tell me that it was my own mad pranks that landed me in this position, that I gambled and forged; true, I did do so, but it was under temptation; it was a wrong that, had I been allowed time, it was my full purpose to have rectified. But, how does a slip such as that compare with the deed of the bankrupt, who by deliberate and systematic fraud cheats his creditors of several thousands of pounds, yet he is acquitted of his villainy, and is—by the same impartial law, that now hunts me as a felon for my comparatively venial transgression, for which I fully resolved to make honourable restitution—permitted to go free, and to set up again in the same nefarious lines as before: aye, is looked up to and respected all the time, as a godly, church-going man by his fellow townsmen. Yet, I must be consigned with other felons to the charge of a devil incarnate, whose brutality drove what feelings of manly pride I had left, out of my bosom: I mean that fiend Price. He was murdered shortly afterwards, and so far it was well for me, for I had sworn to tear his cowardly heart from his body though I should swing for it afterwards!

"Look here, Rachel," said Marsden, here suddenly softening his tone, "I have behaved badly to you I know: I decoyed you away from a luxurious home with false ideas of what my position really was; but it was revenge that made me do so. I was, through ill-treatment, at war with all the world; but, coarsely and brutally as I have used you, I declare to you now, when I see how loyal to me you still are, in sticking to me through my evil fortune, that I feel that there are still some ties common to myself and humanity. There is another, too, but it's useless now referring to him—he was always too stern and uncompromising in his strait-lacedness for me; yet that he has borne much with me also, I must freely admit. But now I will tell you what our present plan of operations must be.

"Morgan and Wilson are already here. I could not get into this neighbourhood until two days ago, and have been skulking about ever since; but even in that time I have seen a chance of robbing the New Zealand Bank that will give me a rise that will enable me to get away to America: and when there I will begin life again with a clean sheet, on which for the future I trust only honourable actions will be recorded. The Bank seems to be quite unguarded, and without any bloodshedding—which, bad as I am, I always avoid—we shall be able to make a clean lift of all its contents. To-morrow, about midnight or a little after, the job is to be done. Now, quick, girl, get me something to eat, for I am hungry enough by this time, I can tell you."

At this point, hearing the hurried tramp of approaching steps, that we divined to be those of the two men referred to by

Marsden, and considering the task of arresting these three desperate men to be more than it would be wise for us to attempt alone, and, being besides furnished with full information as to their plans, we both withdrew in the same stealthy manner in which we had approached the bushranger's camp. I have said we, for from the first sight of him, I had reckoned myself to be hand and glove in whatever measures should be taken for securing Marsden.

"Well, sir," I asked my companion, "what is your opinion of that nice plot, and of the plausible scoundrel who is projecting it? Do you think that with those companions he will have the resolution to undertake it?"

"A nest of villains," he replied, without directly noticing my question; "however, I am glad that I have plumped upon their camp at last. I have been a long time on the look-out for them. Has that man the resolution to undertake that project, do you ask? Aye, faith he has! I should like to know what that man has not nerve enough to undertake! Whatever his crimes may be, want of courage at least is what no man can lay to his charge."

"They have come over from Victoria, or at least Australia," I said. "You are a detective from there I presume?"

"You have guessed rightly, sir; by-the-bye, you seem to have been acquainted with that wench, that you followed her so closely to-night?"

"So well," I answered, "that at one time such a term as that applied to her in my hearing would have been instantly answered by a blow from my fist."

"Rather rough, that way of answering," he replied coolly; "but I have heard that at one time she occupied a good position in society, and even now—although such a term comes sort of natural to me, being a north of England man—I believe though the poor girl has had a hard time of it with him, she keeps pretty straight, although she lives with him, and I don't suppose that they were ever married."

"And of the man Marsden who betrayed and seduced her, what might you know of his career?"

"Rather too much for his good," he answered almost sadly, then continued more abruptly, "let it be sufficient for the present to say of him that a more daring and troublesome bushranger than he, has never been in Australia. At first condemned to three years' penal servitude for forgery, he within two years broke out of jail and escaped, remaining at large for near two years, during which he committed some daring robberies in Victoria and in the neighbouring colonies. He was trapped at last, being pounced upon whilst fast asleep; but some time ago he again effected his escape, since which his vigilance has been such as to baffle every effort to secure him. Fortunately, however, I kept my eyes upon Rachel's movements, for by her continual attachment and loyalty to him I knew that she was either his wife or mistress ; and feeling certain that she maintained some mode of communicating with him, I, as soon as I understood that she had sailed for New Zealand, at once followed on her track to Hokitika, where I have been watching her every night for the last month."

"How did she manage to get that tent fixed up there: do you think she did it all herself?"

"No ; I can't say how she contrived it. I fancy that Morgan and Wilson must have done that, and let her know by letter, not to herself (for I have been watching at the post office for her letters), but by means of some intermediate friend ; and I rather think now that that girl Kate, who took her part so fiercely to-night, and was whispering so closely to her at times, knows something of these matters. I know that Kate has come from the Upper Murray, where this Marsden, as you call him, was hanging about for a long time."

"Were both of these men, Morgan and Wilson, in jail with Marsden?"

"They formerly were, and it was by his means that they managed to make their escape with him. But on the last occasion when their Captain got taken they managed to keep clear ; but my belief is, that they had all along contrived to keep up a kind of correspondence with Miss Rachel, and through her with him."

"That suspicion of yours then makes that poor girl guilty of collusion with the deeds of all these desperadoes."

"Aye," he answered, with professional coolness ; "I guess that, demure as she looks, she has always had a shrewd suspicion of all that was in the wind between them ; and yet it is just possible that for the sake of their own safety, what correspondence they may have had through her, may have been so disguised in its language as to have given her a very obscure idea as to what it was all about—a sort of thieves' cipher maybe. As for Marsden as you call him (though I guess he was never christened under that name), I shouldn't wonder if he has not been quietly planted in Melbourne all the time, to get the police thrown off his scent, till he saw a favourable chance of getting over here."

Conversing thus we reached the town, when, after warning me to observe a strict silence as to what I had seen or heard, he went to the police camp to concoct a plan with the officers there for securing the bushrangers, and setting a watch round the Bank.

CHAPTER XXXVIII.

ON the next morning, urgent business obliged me to start early on a journey as far as the Grey township, about twenty-five miles distant. I rode away without mentioning to Lilly my discovery of Rachel and Marsden on the previous evening. This I intended to do immediately upon my return, when I intended, if I could, to secure his co-operation in the service that I determined upon volunteering to the police, to assist them in the capture of the bushrangers.

I took care to have my revolver carefully loaded and stuck conspicuously in my belt, for I was disquieted at the idea of two such desperadoes as Morgan and Wilson being at large in the neighbourhood; to say nothing of Marsden himself, from whose scruples of generosity I had now, from my late view of his appearance, considerably less hope in case of a second rencontre with him. Besides my fears from Marsden, there were at that time some terrible rumours floating about of some atrocious murders recently committed by another gang of desperadoes.

Whether it was that the nervousness engendered by these rumours tended to make me exercise more watchfulness or no, I know not, but the sudden growl of my dog, in a peculiarly lonely part of the road, caused me to at once be in readiness for action with my hand on my revolver. At the same moment I imagined I heard a rustling among the bushes on my right hand, and glancing quickly round, I just caught a glimpse of two men hurriedly retreating through them.

I had ridden about half a mile further when I espied a horseman riding towards me, in whom on a nearer view I, to my great surprise, recognised Mr. John M'Gilvray.

Equally astonished on his part at this sudden meeting, the young gentleman shook me heartily by the hand, inquiring at the same time what had become of me for such a length of time. But with all his unaffected heartiness of manner, I could not avoid regarding him with some stiffness, as, to his handsome person alone I attributed the miscarriage of my suit with Jessie Campbell. I accordingly replied to his friendly inquiries after my welfare in the dry and commonplace manner of one who desires to shorten an unavoidable conversation. Seeing, my reserve, the spirited young fellow at once took offence, and after a few more remarks bade me good day, and rode away at a canter.

As I rode away from him I felt some qualms of conscience for the silliness of my behaviour. " The girl has given you up fairly," I reflected, " then why grieve further about her? Mr.

M'Gilvray has done you no harm in taking, in all fairness, what you could not get." Reasoning with myself in this way I resolved, on my return to Hokitika—if not that evening, yet to-morrow perhaps—to seek him out and make the *amende honorable* for my boorishness to him now. Thus thinking, I had proceeded at a walking pace for about two hundred yards further, when a spirit seemed suddenly to whisper in my ear— whilst with my mind's eye I beheld the loneliness of the road I had just come along, and into which Mr. M'Gilvray would be at this moment entering—" and these two men". Those four words, and no more, seemed, as I have said, to be just whispered in my ear as if by the voice of some warning spirit; but so sudden and powerful was the magnetic influence that they seemed to exercise upon me, that I at once wheeled Selim round and literally tore back in the direction I had just come; nor, as the result proved, was the impulse on my part either a mistaken one or too soon timed. Just as I had entered the thick, dense scrub that here deeply shaded the road, and through which Selim came tearing in at a stretching gallop, I saw poor John M'Gilvray on the ground, with two fiends on either side of him, each pulling with all his strength at the end of a Chinese sash that was twined at the middle round their victim's throat, who was by this time almost black in the face. What words I uttered as I came upon this fiendish scene, I do not now remember, but a bullet that went singing past one of their heads, as I galloped frantically up, sent both wretches, like hunted wolves, back into the scrub, where they at once disappeared. I did not attempt to follow them. It would have been impracticable for me on horseback to have done so, owing to the impervious nature of the scrub through which they had darted. Yet quickly as they had disappeared, they did not do it soon enough to prevent me recognising the bullet head and flaxen hair of Morgan, as a second shot from my revolver knocked the ruffian's hat off ere he managed to vanish from my sight.

Mr. M'Gilvray, who it seems had been totally unarmed, had been bailed up, and sternly ordered to dismount, with which order, under the threat of two revolvers levelled at his head, he had been obliged to comply. He was then seized and placed in the position from which I had so providentially been enabled to rescue him. His consciousness had not quite forsaken him at the moment when I had come up. Consequently, after a minute or so of spasmodic gasping, during which I could do nothing to assist him save to loosen his collar, he quickly recovered ; and, being naturally of a hardy and courageous temperament, his sole sense of the greatness of the danger he had just escaped was expressed by an exasperated oath at his utter unreadiness for such an encounter, which had caused him to fall so easily into the clutches of the miscreants; in contrast to the lively account he would have given of himself had he been armed with so much as a stick at the time.

Securing his horse, that in the meanwhile had not strayed very far, we both remounted and rode rapidly back to the Grey, and, dispatching my business there, as hastily returned to Hokitika, which we reached at about four o'clock in the afternoon. Immediately on reaching Hokitika, we rode straight to the police camp and informed them of our adventure on the way to the Grey, and at the same time volunteered our services in the event of their attempting to secure the bushrangers that night, John M'Gilvray adding in a determined tone that he would like to get a chance of settling accounts with either of the ruffians who had come so very near strangling him.

On the chance of the bushrangers eluding the police at the place where the latter hoped to surprise them, and of their pouncing upon the bank while the police were absent for this object, it was resolved to divide their force; one half to be retained to watch the bank, whilst the other (under the guidance of Duval, the Victorian detective), proceeded to attempt to surprise and capture the gang in their own tent.

The accession to their divided strength, by the tender of our voluntary assistance, was therefore received with great satisfaction by the police; and this satisfaction was increased, when I informed them of the nature and efficiency of a further ally I could promise to procure them, in the person of Mr. Benjamin Lilly, whose zeal in a cause in which the rescue of the long-lost Rachel Rolleston was concerned, I felt perfectly sure we could firmly rely upon

With these preconcerted arrangements for the capture of the bushrangers at their tent, three policemen well armed were thought amply sufficient, as, by the accession of three volunteers, they would thus be exactly two to one. The remaining police, amounting to two men, were left to guard the bank against a surprise in case the bushrangers had changed their plans. As eleven o'clock had been the hour arranged for surprising the bushrangers at their tent, we laid our plans so as to reach the scene of action at that hour.

We then—that is, Mr. M'Gilvray and I—went back to the hotel, where I was then putting up, and at which Lilly was also staying.

On reaching the hotel and shortly afterwards falling in with Lilly, who happened to be out at the time of our arrival, I called him aside and told him all that I had seen on the previous evening, as also the nature of the duty in which Mr.

M'Gilvray and I were shortly to be engaged, and in which I had ventured to promise for him that he would also take a willing share.

On hearing all that I had to say, which, however, was not done without some strongly worded interruptions on his part, as would have been expected by all who knew Lilly thoroughly, the great-hearted fellow pulled out the beautiful gold watch that, on an occasion already related, Rachel Rolleston had given him, and that ever since he had carefully worn in a fob inside his waistcoat, and putting it to his lips and kissing it, said, "If ever I prayed in my life to God, it was for the purty young lady that gave me this watch, and that I might some day have the chance of paying off the cold-blooded villain who induced her to leave her home to spend a life of shame and misery with him. And now, may God help me this night to keep the vow that I swore to Him then, that if ever I was to come across him, that it should be my life for his. And if I get that satisfaction, I shall part with my life this night content."

Within the hour, we were all at the appointed place of rendezvous, from which we instantly rode forth to the scene of action. About one hundred yards above the place where we had lost sight of Rachel in the scrub, that is, about one hundred yards from the tent, we dismounted and secured our horses at the bottom of the steep ridge.

Some passing fancy while securing Selim to a tree branch, over which I simply threw his bridle, induced me to look up and to mentally gauge the probable height of this ridge; and I thought at the time that, though too steep to attempt to ride up with any ordinary horse, yet it appeared to be not a whit more steep than a ridge up which I had been obliged to force Selim to scramble, with me on his back, a few days before. This I had been compelled to do in order to intercept the flight of a perverse bullock, that seemed determinedly bent on breaking away from the others, and that, unless I had contrived to intercept him on the top of that ridge, would from there have got down into a wild ravine, from which there would have been but little probability of again recovering him.

I also took occasion to observe another distinct feature of this ridge—an observation that was of service to me in the scene that followed very shortly afterwards. This observation was that the spur, at whose foot among the tanglewood scrub the bushrangers' tent was situated, was a continuation of this same ridge. There was no gully between them. I noticed, moreover, that the side of this ridge was tolerably bare of timber. So much for these preliminary observations.

We had all stalked on our stomachs—to use a game phrase—to within a few feet of the tent, where, however, we had to remain for about an hour, ere events began to develop. I at first felt my heart beating loudly with nervous anticipation. Marsden appeared to be within, and stalking rapidly to and fro, as if impatiently awaiting some arrival. He at times addressed a few words to Rachel, stopping whilst doing so, and then abruptly resuming his hurried stride.

"What the —— can be keeping them?" I heard him at one time mutter; "they should have been here long before this."

"Ah, here they are!" he at length exclaimed, as steps were heard hurriedly descending the ridge apparently by a cleared track—for there was no sound that could come from contact with intervening bushes—and in a few moments, Morgan and Wilson, almost brushing by where I was lying, entered the tent, when, immediately afterwards, the stern voice of Marsden was heard demanding, "Well, what has kept you coves so long? I suppose you have been up to some of your old treacherous games, Morgan, and have been trying to do something on your own quiet lay, that you keep such bad time!"

"What the —— is that to you what we've been up to? we ain't your servants, I suppose!"

"You are not my servants, no, certainly you are not; but according to our contract while this lay is on, that I have found for you, you are bound to obey me as your Captain, and I am the man, Morgan, that will make you stand to your word with me, or if you don't, will send a bullet whistling through your thick skull, you common blockhead!" He continued, after a pause, the other answering not, "Can you not see that it is to your own selfish interest to take notice of what I have to say to you. I have the head to plan how to do a thing clean, like what we are about to do this night; and for this half hour I have been kept waiting by you, when we ought to start at once, to take time by the forelock, so as to be able to do our work deliberately, without being hurried over it. All your ideas, on the other hand, are to knock people on the head, and by that original means to get your own worthless neck into a halter. Take time, man, take time, your neck will be there soon enough, I warrant; but now I merely want you to be smart and steady, and to keep by me this night, when, if we succeed—as I doubt not but we shall—you can both go your own silly ways, and get hung up as soon as you like."

"How do you know that there is money in that place at all?" Morgan here asked sullenly.

"How do I know? Because I was inside there getting a cheque cashed, and saw enough to convince me that there is no small amount of gold, to say nothing of notes, within that bank at this very moment."

"Well, that's good news anyhow," remarked Wilson, with villainous cheerfulness.

"You are all ready then," demanded Marsden, "and prepared to obey all my orders to-night?"

"Yes, yes," both worthies eagerly replied, for the news of the certainty of gold being in the building that they were about to break into, had caused even Morgan's sullenness to give way to one of cheerfulness.

This kind of work always afforded him the pleasure common to all experts at the exercise of their own peculiar talents.

"That's right. Now, Rachel, put the bottle on the table, and we will have a mouthful of something to eat, and then start at once."

"Surrender in the Queen's name," sounded at this pleasing juncture of the bushrangers' plans, in the stern voice of Duval, by whose whispered orders we had by this time quietly risen to our feet and approached the tent door, whose flap at the same moment one of the officers drew aside, whilst we all simultaneously presented ourselves, with revolvers levelled at those inside.

"Douse the glim," was that freebooter's reply, as in instant action to our threatening demonstration, he snatched up a loaded revolver that was lying beside him on the table, and fired it at the foremost of the intruders. As his shot exploded, I observed Lilly, who had been impetuously advancing, suddenly reel to one side and fall half stunned against the side of the tent; but, as it proved, his weakness was but temporary, the bullet having only grazed his temple. Another shot from Marsden next stretched one of the policemen with a broken arm on the ground. At this moment, instead of supporting Marsden by a display of equally resolute behaviour, the ruffian Morgan, with a cowardly concern for his own safety only, and trusting amid the confusion and smoke to his bull strength, made a desperate rush to break his way out between us; but meeting his charge with the firmness of a rock, Duval, against whom he came in contact, seized him by the collar with his left hand, and his waist with his right, retaining his own revolver as he did so, and, with a sudden exertion of his powerful muscles assisted by athletic skill, he lifted him clean off his legs, when both men came to the ground with a shock that—even amid that scene of confusion—seemed to make the ground vibrate beneath our feet.

In the meanwhile my position, as we attempted to force our way in a body into the tent, chanced to be on the side where the table with the light stood ; and seeing Rachel, in obedience to Marsden's command, rush forward to extinguish the light, I too darted forward to prevent her object, and seizing hold of her hand—that hand that at one time even to touch had caused the blood to thrill through my veins—I roughly thrust her back from the table.

It was no time then for ceremonious considerations : the whole scene that I am now attempting to describe began and ended in about a minute. As the wounded policeman fell, John M'Gilvray, slightly built but wiry, sinewy, and an excellent light-weight wrestler, sprang in, and grappling with Marsden, attempted to trip him. A blow from the butt end of Marsden's revolver on the head, however, and his spirited young assailant lay senseless on the ground. This occurred as I had thrust Rachel from the table. I then sprang across Marsden's path as he was in the act of bounding over M'Gilvray's senseless body, and striking up the muzzle of his revolver just in time to cause the bullet that was intended for my head to pass through the roof of the tent, I threw my arms around him and endeavoured to detain him. In physique I was at that time a well-built man, and one whom it would have taken a man of more than ordinary strength to overthrow ; but, in the iron grasp of him, against whom I now pitted myself, I felt as if held in a vice. Lifting me from off my feet he fairly dashed me to the ground, and with a wild shout, the bushranger leapt out of the tent and escaped into the scrub. I sprang to my feet shouting excitedly, and fearless of danger instantly rushed in pursuit, followed by Lilly, who, now recovered from his stun, with an imprecation of fury and despair, saw the man against whom he had sworn life for life, now likely to escape his vengeance after having been almost within his reach.

I was also at the same moment followed by Duval, who, with revolver pointed to Morgan's head, had compelled the coward to let the handcuffs be slipped upon his wrists. Wilson too had, in the meanwhile, been secured, and was then in the hands of the remaining policeman.

On rushing out from the tent, I made instinctively for my horse, as the conformation of the spur that I had already noted flashed through my mind and struck me as the probable route that Marsden would pursue. The hundred yards between I must have traversed in a few seconds. Tearing the reins from the branch to which they were hung and flinging them over my horse's head, I vaulted into the saddle.

"Up, Selim, up," I shouted, forcing him straight up the face of the ridge before me. My gallant horse refused nothing that I put him to. Stretching himself out, clambering and clinging, he fairly seemed to drag himself up the face, I standing high in my stirrups and leaning over his wither to help his equilibrium. Snorting, panting, but never hesitating, while my "On, Selim, on," still urged him to more strenuous exertions, on and up he went for about two hundred yards, till he gained and passed the brow of the hill. From there, glancing behind me, I observed Lilly on foot, dragging his horse desperately after him by the bridle, and Duval doing the same, though somewhat further behind.

On reaching the top, to my great joy I just caught a glimpse of Marsden, going at an easy sweeping canter, about twenty yards ahead. This I could easily observe, as it was a beautiful moonlight night, and the ground for some little space about here was comparatively free of timber. He was mounted on a tall chestnut horse (a pure blood) that, as I afterwards understood, he had assisted himself to from some squatter's stable, down on the plains, about a week before. He was evidently unconscious that he was being pursued, or was possibly under the belief, from the nature of the ground, that active pursuit was impracticable, so for some minutes after I had headed Selim upon his track he kept going at the same easy pace. On my part, as a matter of policy, I refrained from attempting to undeceive him on this point by instantly pressing upon him, as I was desirous, by a short continuance of this easy pace, to allow Selim to recover somewhat from the exertion of scrambling up the ridge that had so severely blown him.

This for a few minutes I was enabled to do, till Selim's impatient snort in his desire to press on after the horse he saw he was in pursuit of, betrayed my proximity to the bushranger. I saw him turn his head round suddenly to discover the meaning of the noise he heard behind him, when, with a savage imprecation, he urged his horse to the top of his speed. Selim by this time, however, had somewhat recovered his wind.

That in a flat race he would have been a match with his blood opponent is more than I could deliberately affirm. Selim was not a pure blood, yet his speed was by no means contemptible. But here, though a little winded, the chances were altogether in my horse's favour. Marsden's thoroughbred on plain ground might possibly in a mile race have come in a full neck ahead of Selim, but where he was, the scrub confused and handicapped him, whilst Selim was a scrub horse. He had been bred amongst it, and had subsequently been trained to muster cattle from such ground, and was therefore now in his natural element, and the manner in which he dashed straight through thickets, or under low spreading boughs when he almost had to crouch, or with his fierce snort, darted to either side according as the direction of the animal he was pursuing was changed in the slightest degree, or cut off angles to intercept him when this unpractised courser swerved aside from any obstacle, was such as would have boded ill for the safe seat of a greenhorn or unpractised rider, had such then chanced to have been on his back. The old stock horse knew his duty, and I knew that while his breath lasted he would never lose sight of the object he was following.

Circumstances, however, favoured the pursuit. Whether Marsden knew the country or no, I know not; probably, however, he did not. Heading his horse for the heart of the bush, his career was suddenly checked by a steep gully with precipitous sides. This occasioned him to swerve up along the side of it to the left which brought him in a course trending somewhat in the direction from which he had just come, but which he followed with the probable hope of finding a place of descent down the terrace. This direction was unfortunate for him, for, as it turned out, it brought him in a manner up broadside with his pursuer, whom, by taking the opposite direction along the gully, he might still have kept straight in his rear, and as sailors say, a stern chase is a long chase.

Following Marsden closely, I now cooeyed loudly to apprise Lilly and his companion of this change in our direction. It was, as I have already said, a clear moonlight night. As Marsden advanced, he found that the gully swerved so much in the direction that he had just come from that he would now gladly have gone straight back on a line with his former direction, but the appearance of Lilly and Duval advancing through the trees towards him—Lilly, hatless, his bleeding temple bandaged with his handkerchief, and in a voice, hoarse as a boatswain's call, shouting out to Duval who was behind him, to incline farther off so as the better to hem "the dog" in—showed him that by so doing he would have to run the gauntlet of a fire from the three of us.

Firing his revolver again at Lilly, who fortunately escaped being hit, Marsden now wheeled his horse directly for the terrace which here was not more than twenty feet in height, and without a moment's hesitation instantly leapt his horse over. The poor animal lay at the bottom with her two forelegs smashed, and Marsden, springing off her back, could be seen running off into the bush on foot. Lilly was now on his heels like an avenging fury. Riding furiously to the edge of the terrace, he would have recklessly followed Marsden's example, but Coleena, less blooded or more sensible, baulked and reared. Springing from her back, her fearless rider impetuously took the bank himself. Clinging with nervous hands to projecting roots or branches of trees that grew up from the terrace's side, he let himself down to the bottom with the rapidity of a squirrel.

Marsden could be still plainly discerned running about one hundred and fifty yards ahead.

The report of Lilly's revolver was then suddenly heard, and Marsden was seen to stagger, but he instantly replied with the remaining loaded chamber of his revolver; that also told, though with less effect upon his antagonist. This, however, did not stop the latter from rushing up, and, with a fierce shout, springing on to Marsden, receiving, as he did so, on the barrel of his own revolver that he refrained from using, a desperate blow that the other had aimed at his head with his now empty weapon.

Lilly was a nervous and muscular man, yet I question if in sheer muscular power he would have been equal to Marsden, but that the latter was more seriously wounded than himself, a fact that somewhat equalised their strength as they rolled upon the ground with all the ferocity of tigers, locked in each others' arms.

Meanwhile both Duval and I had also dismounted, and were attempting to descend the terrace, where Lilly and Marsden had fired at each other. Duval accomplished this quickly, but I missed my foothold when near the bottom and rolled down heavily, considerably bruising one of my shoulders. "Yield, Howden; the game is up!" cried Duval, sternly, as soon as he reached the scene of the struggle, whilst he knelt upon the body of the bushranger.

"Ha! hell-hound; what sent you here? Strike, dog, I yield to none."

With a stroke from the butt end of his revolver, Duval instantly reduced the furious man to a state of partial insensibility, during which he bound his hands with his belt, and then still more firmly secured him with the reins that I fetched from the maimed horse, putting the poor groaning brute out of pain at the same time with a shot in the forehead. Duval, then, with Lilly's assistance, led their now securely bound prisoner to the tent, that in reality was not much more than a mile away from where we then were, while I returned up the terrace, and, mounting Selim, drove the other two horses before me down the same ridge, that we had ascended in such haste. Ere starting, however, Duval, who had formerly been in the army, and had there seen something of surgery, did what he could in a rough and ready way to staunch the wounds of both Lilly and Marsden.

Marsden, now conquered and passive, submitted to what conditions were imposed upon him, but maintained a stern silence all the time.

We found Mr. M'Gilvray feeling rather sick, but otherwise unhurt, his consciousness having returned some time before. As for Rachel, she had left the tent, no one objecting to her doing so.

While returning however to Hokitika that night with our prisoners, one idea kept constantly exercising my mind, "Howden," I kept muttering, "what name is this? Is this man Marsden, then, after all, poor Charles Howden's wild brother that he once told me of? Surely it can be none other than he? Yes, now I remember his remark in the tent; while defending his conduct to Rachel, he said then, that there was one other who had been good to him, but that he was always too stern and strait-laced for him. This just agrees with what Howden said of himself, with reference to his views of his brother's conduct. And yet in features there is no family resemblance between these brothers; certainly there is none in their minds! Poor Charles Howden! how grieved he will be to hear of this sad ending of his brother's mad career. For be sure this will be the end of it, as the civil authorities will now make too sure of his custody to give him a third chance of escape."

CHAPTER XXXIX.

LEAVING the wounded policeman in the tent, his companion, who until our return had remained on guard over the two prisoners, having then mounted his horse and ridden hastily to the township to bring medical assistance to him, we all started off with our prisoners, and in about an hour and a half afterwards had the satisfaction of seeing them all lodged in secure custody. Up to that time—ever since Howden's capture—the detective had maintained a stern watch over his charge, as if on the look-out against every possible contingency in the way of an attempt to escape, or an ambushed rescue, and uttering nothing the while beyond brief, necessary directions.

But with the responsibility of this duty off his shoulders, his taciturnity at once gave place to a more considerate manner. Heartily shaking the hands of Lilly and myself, he thanked us for the valuable service that we had rendered. Whilst to the other officials present, he frankly confessed that, but for us, his own efforts in attempting to secure Howden would have been utterly abortive.

"But what sort of a horse is that of yours, Mr. Farquharson?"

he added, " has he claws on these hoofs of his, that he went up that ridge in the cat-like way that he did? I should not have believed it possible for any horse to accomplish such a feat as that, unless I had seen it done with my own eyes, as I saw yours do it; it took me all my time to drag up my horse on foot behind me, and he is not the clumsiest-footed horse either, that I have seen."

"Aye, mate," replied Lilly, patting Selim on the neck with almost as much pride as if the old horse had belonged to himself. "You are not like to see two such horses as this in a lifetime, I fancy; aye, old Selim," he continued, addressing the old horse as seriously as if he reckoned that the animal could actually understand him, while Selim acknowledged his sense of the praise he was receiving in his own way, by vigorously using Lilly's shoulder as a convenient rubbing post for his head. "Aye, Selim, old boy," he said, " we have to thank your pluck for our luck this night; but for your gameness in climbing up that face, and sticking to that fellow afterwards as you did— flash horse and all as he had—that dog might have been far enough away in the bush by this time."

"Yes, that is true," replied Duval, critically examining Selim's various points; for the detective was evidently a judge of horseflesh. That the result of the scrutiny satisfied him was evident, for he added:—

"I would give something now to own a horse like that. And this horse is fully as good as he looks! How he did go crashing through that scrub, to be sure, for although I could not see him I could hear him. I should have enjoyed being on his back during that chase."

"You might have found it more exciting than safe, if you have not been accustomed to stock riding," replied Lilly, bluntly, adding, " it takes practised men like Mr. Farquharson here and me to keep on Selim's back when he is hunting after anything through scrub, I can tell you. I once had a trial of him, and I know what it means to stick to him in such places. My own mare, Coleena, there, isn't bad, but she isn't a patch to him among scrub or perhaps anywhere else."

"Well," replied Duval, " I should imagine that there is a good deal of truth in what you say there. On a level or steeplechase course I would stick to any horse that ever was girthed, but I daresay even a man like me would require some little practice to fit him for keeping his seat with such sudden halts and sharp bolts to the one side or the other as I saw that horse give when I came in sight of you, at the time when Howden attempted to double back. To me at that time your horse appeared to literally wheel on a pivot; his action was so simultaneous with every change of movement on the part of Howden's horse, that I marvelled even then how you managed to retain your seat so easily as you seemed to do."

"My dear sir," I replied, "I just then kept my eyes open for all possible contingencies. Doubtless, however, the secret of the whole matter is practice, by which one's body acquires a kind of instinctive habit of accommodating itself to all sorts of possible emergencies without any conscious effort of the mind. Again, on such occasions, my experience of my horse enables me to leave his conduct in the matter entirely to his own discretion. I content myself then with looking out for the safety of my own head and seat. I know that is all that is required for heading or blocking any animal that he is in pursuit of. My horse understands fully as well as I should myself all that there is to do; and he certainly would perceive any sign of a change in the animal's motions much more quickly. Aye, sir, and what you have lately seen is but a slight sample of my horse's spirit. To see him at his best he should be seen confronted by a wild charging bullock among timber as thick as that of our late scene of action. To keep on his back there, or to prevent yourself from being dashed against a tree branch in such sharp work is what a Yankee might well term 'a caution to snakes,' as, now darting like lightning to either side and again over a log or under a low spreading tree, Selim, reckless of all obstacles, instantly wheels, or darts ahead, or halts, with every corresponding movement of the bullock, or, if at length charged by the infuriated animal, then on the signal of the slightest pressure of my hand on his mane or wither, with a lightning-like wheel he sends both heels into the stubborn brute's forehead, a mode of dealing with him that generally sickens the most contumacious bullock. Aye, sir, he is a noble horse, and such a one as I never intend to part with while he has life in his body and I have life in mine."

Soon after this, the wounded policeman was brought into camp, when the doctor who had accompanied him examined the wounds of both Lilly and Howden.

Both were pronounced by him to be but flesh wounds, there being no broken bones in either case. Lilly was hit on the side of his chest just close up to the arm-pit, and Howden, more deeply, on one side of the loin. The wounds of both, though by this time beginning to feel sore and inflamed, were now properly dressed, and would probably be well in the course of a few weeks.

On the next day, when the prisoners, for the sake of identification, were brought before the magistrates, I observed among the audience, towards the door, the sad pale countenance and lady-like figure of Rachel Rolleston: still lady-like and refined in spite of the misery and degradation of the last few years of her life. Now, under the confirmed conviction of this degradation, I felt a sort of compunction at making myself known to her, not because of any abhorrence on my own part in addressing her, but because of the probable effect that a sudden meeting in her present circumstances with one who had known her so intimately in her happy days might occasion her.

I, at all events, resolved to postpone the necessity of discovering myself to her until this could be more effectively accomplished in the privacy of her own home. My purpose was, if possible, to reclaim her and restore her to her father, and I felt that in return for my services the police would help me to ferret out her present place of abode.

For this end, as my evidence was not required in the examination of the prisoners I during its progress kept my back studiously turned towards her.

As the identification of the prisoners was easy, the examination was soon over, and in virtue of an intercolonial warrant they were formally handed over to the charge of Duval, though kept under lock and key, until the departure of the first vessel to Australia, when he would be able to sail with them to Melbourne. As it happened, this took place the following day.

As I felt no little curiosity about Marsden, or Howden as he was now properly called, I requested and obtained permission to visit him in the lock-up.

From his known character and the insecurity of the structure that did duty as a jail, he was kept manacled and vigilantly guarded. His two companions were confined together in a separate cell, but as I had nothing to say to them I did not disturb them.

Howden's features, altered as I knew they were by my brief glimpse of them in the dancing-room, presented now a still more haggard and wasted appearance—the work of the maddening sense of his loss of liberty, together with the loss of blood from his wound, which seemed to have exhausted even his strong frame. To me, in his unkempt state and with his fierce expression, he suggested the idea of some untamed beast of prey that had suddenly been brought in from its native jungle and there bound, and was glaring with all its native ferocity at the spectators, that its chain alone prevented it from at once springing upon and tearing to pieces.

On my entrance, he at once remarked with a fierce scowl (for he had recognised both Lilly and me at once during the chase and capture): "So, sir, you seem to have known how to show a due sense of gratitude to me for my having once saved your precious life from the hands of those who in a minute more would have made dead meat of you: for that act of weakness on my part, has led to your being a main instrument of my being run to earth, and consigned to the life of a felon".

"For my present action, sir," I calmly replied, "you have only to blame your own unprincipled conduct in seducing to a life of shame and misery one whose virtue and grace so well fitted her to adorn the sphere that her fortune entitled her to move in. It was solely to bring you to justice for that heartless act that my friend Lilly and I were among the most bitter and unrelenting of your pursuers, and that, by God's blessing upon our efforts, we have been the chief means of a period being put to your career of unbridled lawlessness."

"And you have come here, sir, I presume," retorted Howden fiercely, "to torment my spirit!—as if that was not sufficiently maddened by other thoughts, without having to listen to one of your canting lectures on moral propriety."

"No, Howden—for that I now understand is your proper name—I have come here for no such purpose: the gaol chaplain is far more fitted than I to give you such advice. Although, let me say, it would have been to your advantage, even in a worldly sense, if you had paid a little more attention to such lectures than you ever seem to have done. You would not now stand there like a beast of prey restrained by irons, if you had. Indeed, I hardly know why I have taken the trouble to come and see a man whom personally I have only reason to regard with the deepest hatred, unless from a strange sort of curiosity and wish to find out what the motives were that could have impelled you to perpetrate such an act of villainy, as to so degrade that high-spirited, impulsive, but foolishly romantic girl. In distress and danger you went to her father's house, and were treated there not only as a gentleman, but almost as a relative; if you had been one, the attention to your comfort and the studied courtesy of their bearing towards you, could not have been exceeded. Was your nature akin to that of the hyænas, that no kindness can affect, that your flinty heart felt no compunction at the crime you meditated towards those who so cherished you?"

"Sir," Howden answered, "think not that for an instant my conscience will be touched by such high-flown language as that. Yet still I will choose to stoop to your own level of argument. I'll show you, little as you seem to think it, that I too, have a basis of reason for my actions, that I am not altogether the slave of mere passion, or that my own caprice only has been my rule of life in all that I have done. You suppose my nature to be akin to the hyænas. And what pray has reduced me to that state? Because I was born with a nature that has ever been at war with restraint, am I to be reviled for simply following the dictates of my own heart? Because I committed a fault, to enable me to retain my position in society, a fault that I honestly meant to repair had I been allowed time! In rebelling against the undue severity of the punishment meted out to me on that score, have I forfeited the privileges of my manhood? Without a home, and hunted like a wolf, was it an additional crime in me to yield to the instincts of nature, when I met with one whose mind, rising superior to conventional forms, I found in harmony with my own? or was I not to be allowed the privilege of mating like the wolf? You speak of the sphere in which she once moved. Was that sphere superior to that which I once occupied? or was the subsequent descent from that position greater in her case, than it has been in mine? You speak of the treatment that I received at her father's hands, and my subsequent ingratitude. Would such treatment have been accorded me by that wealthy conventionalist, had my true character been suspected by him? If you infer such an understanding on his part, your reproach holds good, not otherwise. I accepted his hospitality at its own value, well knowing, that had a real suspicion of my true character and pursuits dawned upon the mind of the prudent and correct Mr. Rolleston, I should have been at once hunted from his place with no more compunction than he would have shown to the dingoes that worried his sheep."

Thus did Howden seek to justify his own unprincipled life by the plea, that his own peculiar situation was the proper basis to which his rule of action should conform. In reply I said to him. "Your sole argument seems to be, that because you were ruined, you were therefore justified in dragging down this innocent girl in the same ruin with yourself. Why, sir, such an argument as that, can simply apply to the condition of a beast of prey; but you are not a beast, and such sentiments show that you glory in your shame—the desire to obey your hellish instinct to prey upon any thing or person you choose to take a fancy to, no matter how good or unsuspecting that thing or person may be. Surely, sir, your moral understanding is not so utterly crushed, as to be wholly blind to the fact of an original principle of right and wrong? To make my meaning plainer, can you not recognise, that the original offence, that occasioned your first punishment, was in itself a crime against the laws of society? Why then should you pretend to arrogate to yourself the right of immunity from a punishment that you had brought upon yourself? Have I, for instance, the right, for the sake of easing my circumstances, to commit a breach of the law by committing forgery? and if I do this, ought I to expect to escape the consequences of this crime? Or, having committed the wrong, do you think I should be justified and have a claim upon the public sympathy if, instead of bearing my punishment patiently, I persisted in fighting against the law of the land, and continually added to my crime?"

To reason with this man was hopeless. His conscience seemed to be deadened by the pernicious effects of a long habit of false reasoning; and the only effect upon him, of my argument against the root of his so-called justification, was only to add to his fierce impatience. It is only too true, that a long habit of sinful indulgence tends to blunt the moral faculties, until that which appears so reasonable, so logically convincing, to a person whose moral nature has not been warped, to the sin-hardened criminal is utterly incomprehensible.

"Enough of this, sir," he replied, "and of this interview at the same time. We look at the question from different planes, and that which I occupy I shall continue to occupy to the end of the chapter. Yet, ere you go, I would ask one favour of you and only one, and were it not that I know you for a man of spirit and courage, even this favour I would not seek at your hands; but courage is the only virtue that I respect, and in its possession I can always feel that I can repose some sort of confidence. You loved that woman: I know it. She is both spirited and loyal, and for that cause only I plead guilty to some feelings of compunction for the position I have brought her to. But low as has been the life which she has led with me, I believe that she is still honourable. I understand that we are to sail from here to-morrow, and I have a foreboding that I shall see her no more. To-night I would rather not see her, as her pale, patient face would seem to reproach me, and as it is, with all my d——d reminiscences, I suffer quite enough of that kind of thing already. Will you use your influence to restore her to her father? She is now destitute of means. Had you and that other bloodhound not so cleverly circumvented my plans, I had hopes, with the rise I should have made with the booty from that bank, to have taken her with me to America or India, and there to have turned over a new leaf in my life, and to have blotted out the hateful memory of the past. Your zeal has helped to prevent all that, and therefore all the satisfaction that your action in the cause of morality and injured virtue may give you, I wish you joy of. Now, sir, would you be pleased to leave me?"

"Ere I go," I replied, "have you no word for your brother Charles?"

"My brother Charles," he answered, quickly lifting up his head, while an expression of intense astonishment overspread his rugged features, "What the d—— do you know about my brother Charles?"

"That which a two years' residence in his vicinity ought to teach me of a man of whom the more I knew the more I wished to know. It was chiefly to talk to you of him that I came to visit you, for on hearing your true name from Mr. Duval, who, however, has given me no other particulars of your life, I concluded that you must be the same wild brother to whom I, on one momentous occasion, heard Charles Howden so sadly refer."

"And this momentous occasion—what was it, that Charles Howden should find sympathy enough in his heart to refer to me?"

"The occasion, sir," I answered sternly, "of a most foul attempt to murder him, when, through God's providence, I was just in time to prevent your brother being strangled by two cowardly ruffians."

"Ha! do you say so?" he answered, hastily, while a troubled expression passed across his countenance. "And pray who were the would-be assassins? Were they taken, or did you identify them?"

"These same villains from whose clutches you were once man enough to set me free, but whose companionship you still seem to delight in, notwithstanding. Was it by your direction, sir, that these same miscreants attempted to murder and rob the brother who had so often tried to befriend you? It seems so palpable that without information from you they could hardly have found out his place of residence in that far and out-of-the-way corner of the land."

"The cowards," cried Howden, grinding his teeth in passion, "it is well for them (or rather it might have been better, for they are both now bound to be hung) that we are all under lock and key now, for had it been otherwise I would have shot both the poltroons like crows! No, Farquharson, whatever I may be, I trust I am not so bad as to connive at an act of butchery on anyone, let alone on a brother against whom I have no other cause of complaint than that of his continual reproaches against my mode of life. But, being reduced to the most desperate straits whilst in hiding about Melbourne, as a last resource I thought if I could only let him know of my condition he might—as he had often done before—give me as much help as would at least keep the wolf from the door. This and no more than this was my intention in letting Morgan, with whom I contrived to keep up communication, and who was in as desperate circumstances as myself, though not under such vigilant espionage from the police, know of my brother's whereabouts. But the villain! I might have guessed that his avarice could be restrained by no sense of manliness or even of thievish honour."

"That being the case, for your own, and your brother's sake, Howden, I am glad that you have cleared yourself from the horrid suspicion of complicity in your brother's intended murder, until now this has looked black against you. I would again ask ere I go, then, have you no parting message for me to bear to this brother?"

"Parting message!" said he bitterly; "what parting message could I send that he would care to hear from me, whose life can offer such excellent matter for reflection to his virtuous mind? Yet stay; why should I express myself in this way of one who I know had ever my welfare at heart? You can tell him then, Farquharson," he added—whilst for an instant his face softened into a more tender expression than I had ever before seen there —"did I feel assured that I was at this moment on the threshold of eternity, from which it strikes me that I am not very far removed, I would say this, 'had I hearkened to his wise counsels, it would have been for my good this day, and that my last thought of him on earth will be a kind one'. And now, sir, farewell; I desire you to leave me. Remember your promise about my wife."

"Your wife, sir?" I replied in astonishment. "Am I to think then from your words that ——"

"Enough, sir; please leave me now. I am sick of these explanations." As his softened manner had again given place to his usual look of fierce austerity, I saw that further conversation would be useless, and left him.

The word "wife" had arrested my attention, but the idea that his use of such a word suggested was instantly dispelled by the look of stern impatience which followed immediately upon my seeking fuller information on the subject; so, concluding that he had made use of the word in a careless sense, I thought no more about it.

On the ensuing day Howden, with his two companions, was embarked on board a steamer, in Duval's charge. But Melbourne he never reached. Taking advantage of a slight confusion—consequent on their passing between the passengers that crowded the steamer's deck—Howden, finding Duval's vigilance flag for a moment, manacled as he was, made a sudden dart to one side, and in another instant was seen bounding over the side of the steamer, whilst at the same moment a bullet from Duval's revolver entered his side.

Sinking below the water, he never came to the surface again alive; and it was only after a long search that it was found that the determined man had managed to dive completely under the steamer's bottom, where, in spite of his irons, he succeeded in retaining his position, by clutching, with all the tenacity of a death-hold, the off-side paddle-wheel, until life was extinct; showing thus that he had deliberately designed to end a life that, with his fierce intractable spirit, he found insupportable under restraint.

Thus died Randal Howden in his thirty-fifth year, a man endowed with great natural abilities and a fierce energy, that, turned to a right end, might have won for him a glorious name in military annals. Yet, a victim to his own passions, and misled by his unconquerable egotism, and the atheistical principles that loosened his mind from a sense of all moral responsibility for his actions, these talents but enabled him to sustain for a while the part of hero of an inglorious career, whose dark record was at times but faintly relieved by the gleams of a better spirit in the acts of daring generosity that he could occasionally display; and the humanity—still conspicuous amid the fierceness of his disposition—that always restrained him from securing his depredations by the shedding of blood. Morgan and Wilson, on the other hand, for the committal of a most foul murder—the proofs of which were fully brought home to them—were both hung shortly after their arrival at Melbourne.

CHAPTER XL.

I WILL now give a slight outline of Randel Howden's career, as I subsequently learnt it from his brother Charles, whom I shortly afterwards met in Dunedin.

Their family was well connected, but owing to their father's dissipated habits—he died after having led a wild and riotous life—their mother was left in very straitened circumstances. She had had a numerous family, but owing to the prevalence of some epidemic whilst she was with her husband in India, shortly before his death, her family of five sons and four daughters were all swept away save her eldest, Charles, then a boy of eleven years of age. Six months after this sad event the dissipated father died also, and six months after his death the widow, who had meanwhile removed from India to Scotland, gave birth to a posthumous son—Randal—who, in his seventh year had the misfortune to lose his mother also.

The child was then taken in charge by a maternal uncle, who had once been in affluent circumstances, but, owing to reverses of fortune, chiefly incurred on account of his brother-in-law's ruinous extravagance, his income was now considerably reduced. By his influence, several years before his sister's death, a situation had been found for his elder nephew in the counting-house of a friend in Glasgow.

Charles, who even as a lad had shown a disposition of singular nobility and purity, had early made a vow to endeavour to redeem his father's unpaid liabilities and with them the honour of the family, and in his situation, such was the assiduity, the steadiness of his character, and the intelligence of his mind, that at the age of twenty-four he had so entirely won the confidence of his employer, that, on an opening offering for the establishment of a smaller business on similar lines in the neighbouring town of Greenock, this gentleman had of his own accord proffered his assistance to establish Charles Howden in it. This offer was most gratefully accepted, and his patron had indeed no occasion for afterwards regretting this act of generosity.

In three years, such was the success that attended the young merchant's efforts that his business was not only able to dispense with all further support from his patron, but the latter had the pleasure of seeing his young protégé on a sure road towards wealth and honour.

Thus it was with Charles until his younger brother had grown up into a tall, promising youth. With singular inconsistency, however, Charles, while devoting himself unsparingly to his business, and working like a galley-slave for the attainment of his first great object, to wit, the payment of his father's debts, cherished the desire of seeing his family restored to its former standing by his younger brother.

For this object, instead of inciting him to work as he did himself, he rather encouraged him to foster the fatal ambition of one day occupying the same position in society as his father had once done. A person of a like noble disposition with himself such encouragement would have merely incited to a meritorious ardour for properly acquitting himself in such a sphere. But, to a nature like Randal's, this was indeed a fatal mistake, as Charles subsequently found out. He sent his brother to college, however, and afterwards found the means for purchasing a commission in the army, that through the help of family influence was readily granted to him.

Finding himself, in spite of his unremitting industry and personal economy impoverished by these serious expenses, he entered into partnership with a young man of the name of

Carmichael, who was of equally industrious habits as himself, and whose acquaintance he had made some years previously. It was then he made the acquaintance also of his partner's sister, Mary, who came to Greenock occasionally to stay with her brother. Between this young lady and Charles Howden a warm friendship soon sprang up, with the result of which the writer is already familiar.

As for Randal Howden, the object of so many hopes and fears, it would seem as if all the benefits heaped upon him went merely to feed a nature self-willed and selfish from the first.

Whilst his dashing manners made him a general favourite with all, he seemed to be utterly unconscious of his proper position towards the brother who had done so much for him. Instead of gratefully accepting all his brother's kindness, he looked upon it as his right, due to his own superior merits.

Although he rose in his profession, yet in the end his intemperate habits of gambling and drinking compelled him to leave the army, after having already been once guilty of forgery. For this offence he was, however, pardoned by his superior officer, on whom he had forged the cheque, for the sake of Howden's family, with whom this officer was well acquainted. This friend, also, on Howden's promise of reform, furnished him with letters of recommendation to several of the leading merchants in Melbourne. In view, however, of the excesses which Howden had been guilty of in the army, these letters of recommendation were made out in an assumed name so as to give Howden a better opportunity for turning over a new leaf. The name thus given him had been that of Howden's mother.

CHAPTER XLI.

SOME hours afterwards, when the excitement consequent on Howden's daring suicide had somewhat subsided, I seriously took in hand the business of hunting out Rachel's lodging.

This, by the help of a policeman, I was easily enabled to do from her intimacy with Kate Dunovan. After some hesitation, as if she had doubts as to the object of our search for her friend, Kate pointed out to us where the poor girl lived, which was simply in a detached room in a neighbouring hotel.

On entering the house, Lilly, who was with me, thought it wise to keep in the background until I had broken the ground, thinking that this would be more easily accomplished if I were alone, than by both of us suddenly presenting ourselves before her.

The poor fellow, on this occasion, appeared to be almost unmanned by his emotions at the near prospect of again beholding his quondam employer's daughter, whose kind, sprightly, yet winsome ways had formerly so deeply impressed his own rough, yet genuine nature, and whose subsequent betrayal and ruin he had so sincerely mourned.

At my knock the door opened, and Rachel Rolleston once more stood before me. There was but little in the appearance of the apartment (into the interior of which I fear I cast a rather unceremonious glance ere I addressed her) that would associate the idea of moral degradation with its inmate, for though the signs of poverty were many, everything betokened neatness and industry.

Turning my eyes full upon her, I paused for a moment ere speaking, to note what possible effect her recognition of me might have upon her. To my intense astonishment she simply extended her hand to me, with calm dignity, and no other outward expressions of feeling than that shown in the slight touch of sadness in her tone as she quietly remarked: "Well, Mr. Farquharson, is it you? how are you? Come in and sit down." Now, an invitation to sit down in a lady's bedroom seems, at any time, a peculiar one. And although this evidently did duty as a sitting-room too, yet that it was used as a sleeping apartment was plainly evident by the bed which stood at the further end tidily and smoothly made.

Her collected tone so entirely disconcerted me, moreover, that I confess my enthusiasm in her cause seemed to suddenly cool, as if with a dash of cold water. The disagreeable impression of her degradation, confirmed by her seeming entire absence of propriety in thus inviting me into her room, combined with her coolness of manner on meeting with one who had at one time so enthusiastically loved her, induced me, in addressing her, to pitch my voice in a tone of rebuke. But from this high ground I found myself quickly taken down.

"I am grieved," I said gravely, "to find you, of all persons in the world, leading such a life as this. Was it madness? was it frenzy, that could have induced Rachel Rolleston, once so admired, to forsake her father's home and the friends who loved her so well, for a life of shame amid the obscene surroundings in which I lately discovered her?"

A slight blush on Rachel's pale cheek together with an expression of haughty surprise—such as I could have imagined might have well become her of old, if offended—made me suddenly pause, and then she answered quickly—

"I hardly think, Mr. Farquharson, that you can rightly understand to whom, or of what you are talking. You used not to be rude when I formerly knew you. Surely it would become you now to inform yourself better ere you venture to address me in such language as this."

This spirited rebuke, that her indignant manner further emphasised, completely staggered me, while, at the same moment, Howden's use of the word wife, in reference to her, came rushing to my mind, suggesting that after all she might not be the polluted thing I had been conceiving her to be. Though as evidence on the other side, the scene of the dance-room, with the unmistakable character of some there with whom she certainly appeared to be on terms of intimacy, rose vividly before my mind.

For a moment I hesitated, uncertain how to address her; then, nerving myself to ascertain the truth of her position from her own lips, I replied firmly: "Am I then, madam, from your words to understand that instead of being the mere plaything of a man's pleasure, as I have been led to believe you were, that you are in reality the wedded wife of that man whose earthly account has been this day passed in at the bay of eternity?"

Holding up her left hand towards me to exhibit to my view a plain gold ring that encircled her marriage finger, she replied: "I am and never have been other than the late Randal Howden's wedded wife, as the ring on the finger testifies; that was placed there two days after I had, so unhappily for myself, gone off in his company. During those first two days I was with him as a companion, but in no dishonourable position, for I declared to him from the first that only as his lawfully wedded wife would I ever accompany him; and it was only from the necessities of the case, no clergyman being available sooner, that the ceremony was postponed till then. When we reached Euston, where one chanced to be staying at the hotel, from which, under the compulsion of Howden's revolver, he was brought at midnight to where we were encamped in the bush, at that place, in that same hour, were those (for me) unhappy nuptials celebrated, in the presence of those two men whom you saw captured with Howden, and whose names as witnesses to that deed I can still show you on the marriage lines that the clergyman, on the conclusion of the ceremony, committed to my keeping.

"On the evening before, Howden, who indeed would have preferred to have dispensed with such a ceremony altogether but for my determination to quit his company otherwise, had ridden to Wentworth to endeavour to secure the services of a clergyman there, leaving the two men with me encamped among the scrub a few miles off, I occupying, however, a separate tent; but his errand proved fruitless, as there was no resident clergyman there."

I here interrupted her suddenly with the remark: "Good heavens! you really were there that night, then? Do you recollect hearing sounds as of a struggle taking place there?"

"A struggle?" she repeated, looking at me with surprise. "Yes, surely I do, for I was very much frightened about it, thinking it was between the two men, of whom I was always in some dread, and that they were killing each other. So, not knowing what might be the result of the quarrel, I sat still in my tent, keeping the revolver that Howden had given me, with injunctions to at once use it in case of any insolent demonstrations from either of them. But what can you know about the struggle?"

"What do I know, madam?" I answered sadly. "Alas! I have reason now to know too much, though, unhappily for you, I then knew too little."

I then detailed to her the account of my encounter with the two ruffians and my rescue by Howden, and the conversation which followed.

She looked at me with swimming eyes, and replied: "My poor friend, how near you were to delivering me from years of misery! But no; I fear it was ordained that I should have to reap the consequences of my own mad folly! Deliverance indeed would have been impossible. Knowing my husband's desperate nature, I feel little doubt but that you would have paid for your discovery of me with your life. Unarmed as you were, you would only have fallen a certain victim to his fury, had you persisted in your attempt to force me away from him. How strange it all seems now! Of course he never mentioned your name to me but merely confirmed my own suspicion as to the struggle having been but a drunken affray between his two companions, and that he had just arrived in time to prevent one from murdering the other."

I felt at her account of her marriage with Howden, as if a mountain had been suddenly lifted off my heart, yet a still lingering suspicion caused me to question anxiously:

"But, madam, what about the strange society I found you in? Excuse my seeming doubts, but the blunt question will out. How could a respectable lady under any pressure of circumstances, associate with those on whose foreheads were plainly written the signs of shame—yes, even sing and take wine in such company?"

"I see," she said slowly and sadly, "I see. Finding me apparently at my ease in such company, you therefore concluded that I must needs be like it! Oh, Mr. Farquharson, how could you, who of old used to be so kind and considerate, how could you, I say, find it in your heart to imagine that Rachel Rolleston, with all her headstrong foolishness, would have had so little virtue, that such a suspicion, under any conditions, could have obtained such a ready acceptance with you?"

At this rebuke I was deeply moved, and answered: "Our Heavenly Father, who knew how great my devotion to you once was, knows also how deep my gratitude is to find the utter groundlessness of that suspicion; but I could not avoid it, because of the conviction I had entertained from the first, that you had been ruined by the plausible wiles of an unscrupulous villain. Yet, would you mind explaining to me how it was that you were so unconcerned in the presence of such company?"

"To explain that properly, Mr. Farquharson, it would be necessary for me to put you in possession of the whole of my life, since I forsook my father's house on the Murray. I believe, however," she added, with a faint smile, "that I was at no time greatly concerned about the forms and etiquette of society. I was inclined—thoughtlessly perhaps—to set such matters down to the score of worldly pride, a spirit that I was rather too prone to despise. Well for me had it been, had I had a little more of that spirit. But when you come to know all, or even a part of all that I have passed through these last few years, it may cease to surprise you, that what now appears so anomalous to you, should have been scarcely perceptible to me. Will you not be still more surprised, when I tell you that through the exigencies of my husband's wild career, women of that class have been the only sort of female society that I have been permitted to enjoy? That girl Kate, whom you might have noticed was so friendly with me and so ready to protect me against insult in the dance-room the other evening—I did not observe you there, though I saw and recognised you afterwards—has been, with all the dissipation of her life and the loathsomeness of her calling, a friend to whose kindness and attention I believe, that, under God, I owe my life; for she nursed me through the heavy sicknesses succeeding the births of my two children, neither of whom survived above a few weeks, and afterwards when I was in want, supplied me with means for the bare necessaries of life. Then there were the low orgies that I was at times compelled to witness, yes, and at my husband's stern request compelled to minister to, my voice being in request and popular; until, habituated to such scenes, language, and manners, that would at one time have filled me with disgust, I came at length to look at them with indifference."

"This is indeed shocking; and your husband, how did he treat you in other ways? Was he as personally brutal to you as he seems to have been morally callous?"

"I believe that, to a certain extent, he loved me—as much as his fierce nature could love any one ; for, whilst he never tried to keep me from the society of his rude companions or to prevent my ears from being contaminated with their filthy talk, yet, further than that, no one ventured to annoy me. For my husband was a man who was feared by the others, both on account of his courage and the strength of his arm ; and more than once have I been forced to witness the sickening exhibition of his prowess with some of the more fractious of his companions, and always with the same result—that is, his antagonist was left senseless and bleeding on the ground."

"And how could you, Mrs. Howden," said I, for the first time addressing her by her married name, " still remain among such scenes as these, with your father's door always open to you ? Did you never think of that, that you never tried to make your escape from such a life as that ? "

"Before you asked that question, Mr. Farquharson, you should first have asked yourself if you could gauge the depth of a woman's love for a man to whom she had fully surrendered her heart ? If you can answer that, the answer to your other question will not be far to seek. I had deliberately cast in my lot with his, and for his sake I had chosen to forgo my father's love and the position of wealth that, as his daughter, I was entitled to ; and, unworthy and desperado as I soon found my husband was, the same passionate love that at first made me to go with him, has still, through all these cruel years, constrained me to cling to him, with the constant and yearning hope that I might yet persuade him to flee with me to America, and in that country under fresh auspices, to lay the foundations of a new life.

"For a time, indeed, he appeared to be considerate of me, and to endeavour to keep from my sight the degraded scenes connected with his calling. But gradually his carefulness on this point lessened, until he at length concealed nothing. After that his manner towards me varied with his circumstances, and although his hand was never raised against me, still his fierceness at times was such as to make me tremble at the thought of what might some day follow. This was when he was harassed by the police, when his temper increased in violence and moroseness, and he would drink until what little of the man that had been left in him before, seemed to be swamped in the passions of the brute. On these occasions his harshness towards me was great, for he was disappointed at reaping no benefit through me from my father. For, though by means of an agent he endeavoured to obtain the money that would have been my marriage portion had I married otherwise, my father, knowing well how it would be used, sternly refused to part with a shilling of it to him ; nor would I ever, even under pressure, consent to write to my father myself with the object of obtaining it. I had disgraced him sufficiently, without seeking to annoy him further by any such request, even had I not felt persuaded that he would not listen even to my suit. Then came the sickness I have referred to, and the destitution that after my husband's sudden capture I was exposed to, when I was nursed and assisted, as I have said already, by Kate. Up to that time we had lived in the country about the Upper Murray amid a people who all appeared to be marked with the brand of infamy—convicts and the children of convicts.

"On Howden's capture, I found my way to an obscure lodging in Melbourne. Thither also Kate and one or two of her companions moved. I supported myself meanwhile as best I could by my needle, whilst through Kate I kept up some communication with Morgan and Wilson, who were also then skulking about Melbourne, until Howden managed to elude the vigilance of his keepers. For a long time he lay concealed in a house that Kate knew of, though I myself, for fear of the police, was not permitted to see him. Neither did I again see him until the night when you saw him suddenly enter the dance-room. But, informed by Kate of his wishes, I came over here with her. Through her I was informed of the whereabouts of the tent that had been constructed by Morgan and Wilson, as also of Howden's expected arrival in Hokitika. The exact position of this tent I then, by Kate's direction, made myself acquainted with by an afternoon's walk that I took in secret shortly after my arrival at Hokitika."

"You mentioned that you had seen me before now?" I remarked, as Rachel paused in her interesting narrative. "I had no suspicion of this. Was it in the court-house? I saw you there, but as I did not think that a suitable place in which to discover myself to you, I tried to screen myself from your view."

"I noticed you all the same," she replied. "Indeed, as I knew that you had discovered me, I was surprised that you had not sought me out before, knowing of old what sympathy you had for me. But I recognised you first, Mr. Farquharson, during that dreadful scene of struggle in the tent, when you seized me so hard by the arm, and flung me so roughly from you. See there how you have left your marks behind you," she added, baring her wrist as she spoke, and displaying just above it some blue marks. Singularly enough now, as she spoke, I remembered what I had forgotten till then, but what I had at the time noted, and that was her pale face with her full expressive eyes being suddenly turned up to mine, as I seized her hand to prevent her extinguishing the light in obedience to Howden's command.

And in the swift glance of that terrible moment, what a train of disconnected thoughts must have flashed through that poor girl's mind!

I rejoined: "I had no idea that you recognised me then, and I feel grieved that even unintentionally I should have been harsh and rough with you, but you know that things were then at a most desperate crisis, and my only idea was the immediate prevention of Howden's escape".

"Nay, indeed, you need not apologise : for that you had then no intention of hurting me, I am now well assured of. Yet I could not help wondering afterwards if you had really known it was me at the time, or if that repulse had been an indication of the contempt with which you had now come to regard me in your heart. You who had once been so kind, and had so valiantly risked your life to defend mine. You of whom I have so often thought with such sad regret. Alas! how great my folly has been, and how bitter the price I have paid for it!"

These words, spoken in the tone of one whose bright spirit had been subdued and broken by sorrow and hardship, went to my very heart, but beyond a glance of compassionate sympathy, I gave no further expression to my thoughts, but after a while remarked: "And now that death has loosened your bonds to this man, am I to understand that this is a grief to you, and that you do not regard it in the light of a happy release?"

She answered in a subdued tone: "The suddenness of his end, unprepared as he was for his Maker's awful presence, though only what his unbridled career gave reason to expect, is a grief to me; but now that he has been taken away, do I count it as a relief? Why, yes; because now freed from my responsibility to him, and the influence of the strange life that his presence could still inspire me with, I can breathe freely and choose my own mode of life for the future. For with him, from the very nature of his circumstances and the desperation of his calling, life was but a mockery and a burden."

"What about the black detective in the scrub?" I inquired doubtfully; "it is evident that it was none other than Howden who shot him."

"It was," she replied, "he himself informed me of the fact, but he did it in defence of his liberty, which he declared he would surrender to neither black nor white; a threat whose stern significance, the determination of his manner when he attempted to shoot you all as you entered his tent to capture him, so well justified. Yet, though I believe he could do things like that without the least remorse, still he would suffer none of his companions to gain their ends by murder, nor would he permit any one to wantonly insult a woman. He was indeed a most strange mixture of good and evil."

I then asked her why Howden hated Duval so. Her explanation was that this Duval had at one time been an officer in the same regiment with himself, and like him had lapsed into habits of gambling and drinking, for which he too had been eventually forced to leave his regiment. In the course of one of these gambling bouts, Duval had accused Howden of cheating, and they had drawn their swords, but were separated by their companions. Duval then, having joined the police force in Melbourne, made himself particularly obnoxious to Howden by the unremitting activity with which he pursued him, and which was in fact the means of Howden's former capture. This, coupled with their former enmity, seems to have peculiarly exasperated Howden against him. Yet it was strange, that during Howden's trial in Melbourne, Duval mentioned nothing in court about his former character, or that he had even known him before that time.

I now asked her if, on the night of the capture or since, she had observed anyone else in my company that she recognised. On her replying in the negative, I told her I would introduce another old friend to her. Then leaving her, I at once went out to find Lilly, and meeting him, told him hastily of Rachel's true position, and the entirely erroneous nature of our suspicions about her; we then at once hurried back to her room. Advancing towards her, with his hat deferentially held in his hand—a mark of respect, that I never knew Benjamin Lilly take the trouble to show, either to Miss Rolleston in her prosperity nor to any other lady, before—the gallant old fellow, extending his arm, seized Mrs. Howden's hand in his rough, cordial grasp saying, "God bless you, Miss Rachel, for I must still call you by your old name, I am mighty proud to see you safe in the hands of your friends again. You must have had a sorrowful time of it, since you went away with that wretch. But he was a bold villain, we will give him his due. And it is myself that is glad to hear, that bad and all as he treated you, he had still some sense of honour left in him, to save your good name. Shure, I thought that he had ruined you entirely, but even if he had, you would be still all the same in my eyes; for good as gold I ever knew you, and gold is gold," added he sententiously, "though buried in a pigsty."

"Lilly, my dear old faithful friend," cried Rachel, her eyes overflowing with tears, while warmly returning the pressure of his hand. "How can you show me respect and consideration now? I am sure I but little deserve such kindness from either Mr. Farquharson or you."

"Deserve such kindness, Miss?" said Lilly stoutly, "and why shouldn't you deserve it, shure? Who is it that ever knew you, with your winning ways, and kindness, and want of stinking pride to a poor man, but would have laid down his life for you? Haven't I still the beautiful watch you gave me," said he, suddenly producing it, "that ever since then, I have worn agin my heart, and on which I swore an oath before my Maker, that I would have revenge in the life's blood of the villain who, I thought then, had ruined you? and I thank God now, this day, that he gave me the chance of as good as keeping that oath. Yet he was a bold and a brave man, that same Marsden, or Howden, or whatever his name was. What a pity that a man who had such good points in him should be such a rascal as he was!"

To Mrs. Howden I now gave a detailed account of her husband's capture, and the leading part that Lilly had played in it. Wiping her eyes at the recital of the close of the desperate struggle with the man whom she had loved, and whose tragic end now sincerely moved her, she said :—

"And my father, my poor forsaken father, can you tell me anything of him? Is he still alive and well?"

I told her that though I knew no particulars about Mr. Rolleston, with whom I had exchanged no correspondence since I had left him, yet that I believed he was still living in Melbourne, as the death of a man of such wealth and position would have been doubtlessly chronicled in the papers, in which event I should certainly have either read or heard of it. I added :—" But as we are about to return as far as Dunedin, our business in this place being now finished, you will of course come along with us, and we will immediately communicate with Mr. Rolleston."

Rachel hesitated a little.

"My father's favour," she replied, "I have but little right to expect after the disgrace I have brought upon him by my disreputable manner of leaving him. Yet he still may forgive me when he learns that I am not quite so vile as he has had reason perhaps to think I was, and when he knows too what I have suffered. But the girl Kate—how can I forsake her, who has been such a friend to me, yet whom I could not think of taking with me to my father's house?"

"Mrs. Howden," I replied firmly, "you must not let your feelings of gratitude overpower your sense of what is due to your own character as a lady, for such, in spite of all that has befallen you, you still are. Looking at the matter in this light, you must see that Kate can be no sort of associate for you now. You lately hinted to me that your close association with her, and companions like her, for the last few years, seems to have dulled your sense of the fitness of things. It must now be your study to regain your old point of view in these matters. When you return to your father, you can find means of befriending Kate, according as the circumstances of her position may make it desirable. At present, however, to attempt to do this by a gift in money I should strongly deprecate, for it would only be dissipated at once among her companions. I will see the girl myself and explain to her your altered plans, and that this renders an alteration in your position towards one another necessary."

"You speak wisely, Mr. Farquharson, and your suggestion I will endeavour to follow, but pray explain to the poor girl that it is not pride but necessity that occasions this seeming alteration in my feelings towards her, for that I shall never, never forget her kindness to me, or lose sight of her if she is ever in want of a friend."

"Yes," replied Lilly, who had so far listened to these arrangements in silence, "that is the right way to do things, and if there is any money wanted for the payment of debts, shure we have both plenty of money to pay them for ye, Miss Rachel"—he could not bring himself yet to address her under her married name—"and if your father will not take you home with him, Mr. Farquharson and I will find a home for you. But your father will take you fast enough, and right glad will he be to see your purty face again, white as it is now; for Mr. Rolleston was always a good 'boss,' and a good 'boss' is always a good man."

Leaving Mrs. Howden shortly after this, I sought out Kate's abode. "Humph," I growled, struck suddenly with the awkward imputation on my character, the sight of me visiting this place might occasion in the eyes of any one who might notice and recognise me. "Humph! a pretty account of me this would make, to find its way to the Campbells. What would the excellent Mrs. Campbell think of me after that, let alone the girls? Bother the girls! I am not likely to see them again, for some years anyhow. Yet I should be sorry to give them occasion for thinking badly of me for all that." I was by this time inside the house, and inquired for Kate, who, on being summoned from an inner room, at once made her appearance. I desired her to come outside with me, as I wished to talk with her about her friend Rachel. This communication at once called up a serious manner, and she instantly left the house with me, when I, in as few words as possible, put her in possession of the fact of the alteration in Mrs. Howden's circumstances, and the necessary change in the terms of their intimacy, that such an alteration involved, delivering Rachel's message to her at the same time.

The poor girl was neither destitute of good feeling nor of good sense. At Rachel's message, the tears flowed from her eyes, but she replied:

"Shure and Kate Dunovan is not the girl to think herself the aiqual of such a lady as Rachel. What for should I be wanting to disgrace her wid my rough company, I would like to know? For among us, the poir girl always seemed like a suffering angel among so many divils. And my heart often pitied her lone condition, and I did what I could to help the poir thing: and she always so maik and gintle too, while I knew she was pining at her heart, tied to that divil's own son of a captain, rest his sowl. For he was not a bad sort at bottim, wid all his roughness. And it's myself that's downright glad to hear that she's now going home to her father, and to live as she should live in her own proper station, and it's nothing that Kate Dunovan wishes her but good luck and joy in the same. But I will go and bid her good-bye, anyhow."

Saying this, Kate hastily went towards the hotel where Rachel was staying, and rushing up to her room, flung her arms around her neck, blessing her after her own rough fashion with all the sincerity and emotion of her warm, impulsive, Irish heart; while, with equal emotion on her part, Rachel promised her that she would never lose sight of her again, but that when she got finally settled at her father's house, she would do what she could to befriend her.

"I'll not send you money, Kate," she remarked, "for it would do you but little good, you would only spend it in drink, or give it away to your companions; but I will do better for you than that, if you will let me."

After this they bade each other farewell.

I then went to the landlord to discharge what debts might be owing to him for Mrs. Howden's board for the past month. This might have been a serious item—for hotel board was at that time an expensive matter in Hokitika—but for the hotel keeper's consideration for Mrs. Howden's evident poverty, and his interest in the lady-like gentleness of her manner, that induced him to supply her with such needlework as she was competent to undertake.

On understanding this, as also that the landlord — who appeared on the whole to be a good sort of fellow—disclaimed any intention of pressing Rachel for any arrears she might have been unable to meet, Lilly testified his high approbation of his conduct by at once "shouting for the whole house". As the popular interpretation of this phrase meant every one who was inside the house at the time, guests, servants, and tipplers at the bar, besides the landlord and his wife—this order represented drinks at one shilling per glass for about fifty people.

On the following day we sailed for Dunedin.

CHAPTER XLII.

> "Oh! never till this breast grows cold,
> Can I forget that hour,
> As standing on the vessel's deck,
> I watched the golden show'r
> Of yellow beams that darted
> From the sinking king of day,
> And bathed in a mellow light
> Dunedin from the bay."
> —Thomas Bracken.

MY sensations on sailing up the bay from Port Chalmers to Dunedin were certainly in harmony with those so effectively recorded in the above lines by the genial and talented poet of Dunedin, as he first saw that picturesque city from the harbour. The city's airy situation on the undulating slopes that rise close to the bay, the sound beyond, and the mighty ocean yet further off, its vast blue expanse and white-crested billows in constant motion, rolling with hoarse murmur on the shore, and, on the opposite side, the dense foliage of the pine hills and other heights broken at intervals by the bright fresh green of early summer—give to Dunedin an appearance of picturesqueness and salubrity that I had not hitherto seen equalled by any place I had been in since I left Scotland.

Towards evening the town appeared very animated. It was New Year's Eve, or Hogmanay, as I heard it more generally called, a term that, after my long severance from the land from whence it came, was strongly suggestive of my early home.

On this day the coaches had been coming into the town laden with passengers from all the districts round. Some had come on visits to their town friends, but the majority to witness the celebration of the Caledonian Games, the chief annual festival of this and the neighbouring Scotch province of Southland, that, beginning on the morrow, were to be continued for two days. From this cause, on this important eve of our arrival, all the hotels and respectable houses of accommodation were so crowded with guests that it was with no small difficulty that Lilly and I succeeded in finding a respectable house where we all could be accommodated. Such a place, however, we happily did succeed in hunting out at last in the "Highland Home"—a well known hostelry in Dunedin, that our ignorance of the town had alone prevented our finding out at once, and that, fortunately for us, was not then so full of lodgers as the others were. This house was kept then, and for some years afterwards, by Mrs. Sutherland, a respectable and motherly widow, at whose hands I felt confident at the first glance that Mrs. Howden would receive such kindly attention as she, poor girl, had of late years been but little accustomed to.

After tea, I took a quiet stroll down the street. At intervals as I passed along, the wild ringing notes of the martial music of the Gael issued from the hotels, and I had already heard similar sounds proceeding from one of the sitting-rooms at the "Highland Home"—that, as its name implied, was a regular gathering place for Highlanders.

At any other time all this pipe music would have sent the blood tingling through my veins, for I have already owned to my degraded taste for this class of music, that in English ears sounds so barbarous; but just then I felt too much occupied with soberer thoughts to spare more than a passing thought to what, at other times, would have so delighted me.

Now that all the excitement through which I had recently passed had subsided, the thought uppermost in my mind was as to how I should in future live. This unromantic question persisted in intruding upon my mind. Ere leaving Hokitika I had lessened the amount of my indebtedness to Mr. M'Elwain by £500, retaining a little over £100 for such emergencies as might turn up whilst I was looking out for fresh employment.

In his reply to my communication concerning the disaster that had befallen my station business, Mr. M'Elwain had shown distinct annoyance at what he termed my inexcusable carelessness, although he made no reference to the loss that my ruin had occasioned to himself. This I believe he would have borne in silence rather than have harassed me for the money. It was evident that his confidence in my business capacities was entirely gone, and that I need look for no further help from him in any future venture.

Doubtless, however, the fact of his having suffered himself to be so completely hoodwinked by the specious professions of the hypocritical villain who had ruined me, prevented Mr. M'Elwain from being more openly censorious than he otherwise might have been.

But though I had reason for thankfulness that my cattle

business had so speedily enabled me to wipe out such a large part of my debt I was still £500 in arrears, a serious, nay, in my circumstances, an almost hopeless sum for me to expect to be able to defray for years. This my present musings showed me all too plainly. "Mighty fine, these late chivalrous bush-ranging exploits of yours, I daresay, Duncan Farquharson," I soliloquised in some disgust, "but you will now please devote your thoughts to the more prosaic adventure of earning bread and butter for yourself. And pray how, besides this first necessity, are you to pay your respected kinsman this £500? This I'm sure it passes my imagination to conceive; yet that mountain must be first levelled ere you can take one step towards your own future welfare."

In the midst of all this depression I was not however without some gleam of hope that the renewal of my acquaintance with Mr. Rolleston might possibly end in something good for me after all. I had already telegraphed to him the tidings of his daughter's discovery and rescue, and I had received a telegram in reply, informing me that he intended taking passage for Dunedin by the first boat that should leave Melbourne for that town.

However much I might be disposed to disclaim the idea of any sort of monetary recompense for the services I had been able to render his daughter, still, it was evident that I could not prevent the fact of these services becoming known, and when known they would probably create an impression in my favour, as to them would be due the recovery of his daughter. Besides, Mr. Rolleston had in the past had some personal experience of my capabilities for station management. Hence, the logical deduction was, that had Mr. Rolleston such a situation vacant, he might bestow it upon me. This reflection revived my spirits upon the whole, and enabled me to get rid of the depression that was beginning to prey upon me.

Next day Lilly and I went to witness the Caledonian Games. We should have taken Mrs. Howden with us if we could have induced her to come, as we both thought that the animating sight of the sports, and the vast crowd of people would have tended to lighten her spirits, but she preferred to remain where she was.

It was the first time since my arrival in New Zealand that I had had the pleasure of witnessing these games; consequently, on entering the large area of ground that has been so liberally endowed for the celebration of these pastimes, I was not a little interested in the animated scene. The ground and grand stand crowded with thousands of spectators—everything gay with the flags of various orders and nationalities fluttering in the breeze—refreshment booths, and, in the centre of the large circular space, round which a deep ring of spectators stood densely packed watching the struggling competitors, the same old lion flag of Scotia that has waved in past ages over so many hard-stricken historic fields : all these things together presented a show that, combined with the influence of a bright day, was truly exhilarating.

CHAPTER XLIII.

I WAS intently watching an exhibition of Highland dancing. This dance was the reel, and the manner of its execution, by four stalwart Highlanders, was exciting my admiration. I thought it a most gallant entertainment, and it gave me a vivid impression of the prowess of old Caledonia's sons. For such exercises as this, methought, could only have been the expression of the genius of a stalwart and active race.

I was thus absorbed when I suddenly felt a hand upon my shoulder. On turning round, in obedience to this claim upon my attentions, to my equal surprise and pleasure, I saw the tall form, with the usually pale, pensive countenance, of Charles Howden—now lightened up with a smile of pleasant recognition. We shook hands with that cordial grasp that only mutual esteem can give.

Turning away in instant forgetfulness of the scene that had been engrossing my attention, I walked on with my friend till we got free of the crowd, when we slackened our pace for the sake of greater conversational convenience. What the topic of our conversation was the reader can doubtless guess. With all the events of his brother's capture and death, and the active parts, taken in connection with them, by Lilly and myself, Charles was already acquainted. This much from the newspaper's reports he had already learned.

For the news of my interview with his brother in the gaol, he was, however, wholly unprepared. His brow, whose wonted gravity had deepened into sadness when we began to talk, now looked sadder still as I faithfully detailed to him my conversation with his brother, and that brother's last message to himself. Yet he thanked God, in a tone of inexpressible fervour, when he understood that he had hitherto been wronging him by the suspicions of his complicity with Morgan and Wilson's attack upon himself.

That evening I took Charles to the "Highland Home" to tea,

and introduced him to Mrs. Howden. On learning his name, Rachel shook hands with him with great feeling. With his name and disposition, she was, indeed, well acquainted already, as at times, when in his better moods, Howden had spoken to her about this brother, in terms that, coming from her husband, had given her a very exalted idea of what her brother-in-law's character must really be.

On Charles' part, Rachel's appearance, with features so worn, and yet so sweet, and still refined, in spite of all the terrible experience of the last few years, appeared to make the most favourable impression. During all that evening, what remarks he made to her were addressed in a tone of most marked respect, whilst he appeared to give the most studious heed to her slightest word, his fine, intellectual countenance being strongly marked with an air of compassionate tenderness as he seemed to realise the full depth of the misery that a person of her disposition and accomplishments must have suffered in the life that her romantic attachment to his brother had entailed upon her. He was, indeed, a man of a rare excellence of spirit, and, as I watched him then, with his attitude of chivalrous respect towards her who had been so degraded and crushed, I felt as if I could have taken him to my heart.

At length the pleasant evening came to a close, and bidding Rachel a kind farewell, Charles took his leave. Lilly retired to his own room, but I went out with Charles to accompany him a part of his way home. On my signifying my intention of doing this, he remarked that he wished to speak to me about something, anyhow.

This something turned out to be a very liberal offer on his part to assist me with means that would enable me to start again in business for myself. For, with my present fortunes and prospects, he had, of course, been made pretty well acquainted by the necessity I had been placed under of applying to him for assistance before; and, although I had repaid him in full all that I had then received from him, he shrewdly conjectured from his knowledge of my character, and his acquaintance with the extent of my monetary obligations to my kinsman, what the probable state of my finances were.

He now informed me that he had several thousands of pounds saved, and in safe keeping. That it was his intention, within a week or two at the furthest, to proceed to Melbourne, and thence to sail for Britain. In the home country, it was his intention to invest his money safely, and to spend the remainder of his days quietly. Said he:—

"I have no ambition for the further heaping up of money. I am the last of the family, and my connections are but few, and these few are personally strangers to me. This £1000 I can easily spare to you until such time as you are able to refund it to me. Interest from you I neither require, nor will accept. You will, by this means, moreover, be able to pay off your debt to your cousin, for, from what I have seen of you, my opinion, Mr. Farquharson, is, that the thought of that debt presses as much upon your sense of independence, as in a similar case, it would, I know, upon mine.

"Were I not convinced that my poor sister-in-law will shortly be placed under the protection of her wealthy father, whom you tell me is even now on his way from Melbourne, I would have entrusted you with another £1000 for her future settlement in life—for in this matter at least I should look upon myself as being heir to my poor brother's moral obligations, and should desire to make her some amends for all the bitter wrong she has endured through her devoted attachment to him. But you will, however, accept of this sum from me."

That I was deeply moved at such generosity I need not say, but strong as the temptation was, in the prospect that such a sum at my command opened up, I hesitated to accept it. I had had drilled into me from childhood, by my mother, a perfect horror of debt, and the lesson had been impressed upon me by the example of my father, whose views on this subject had been only too lax. The difficulties that had been brought upon the family by his weakness, had caused my mother to be peculiarly careful to impress all her children with the necessity of scrupulous honesty in money matters.

On the former occasion when I had accepted a loan from Charles, the case had been different, for then I had had a well defined plan in my mind that I knew would, if properly carried out, realise a handsome profit. Apart from this view of the matter, moreover, I was not without hope of being able to dispense with the necessity for any more capital to enable me to make a fresh start in life. I felt morally convinced that if it lay within his power, Mr. Rolleston would, through his influence, find me some such situation as I had formerly occupied under him. This would be to me a much more agreeable, if slower mode of freeing myself from the difficulties of my present situation, and perhaps yet enable me to occupy a respectable position in life.

With this thought I deemed it best to decline Charles Howden's kind and noble offer till such time as Mr. Rolleston would arrive in Dunedin, when, failing Mr. Rolleston's ability (for I made no doubt of his will) to find such a post for me, I frankly agreed to accept it most gratefully, with this understanding, and wringing his hands with the warmth that such disinterested kindness would naturally inspire, I parted with him, and retraced my steps rapidly to the "Highland Home," with a spirit considerably lightened.

CHAPTER XLIV.

I ATTENDED the exhibition of games on the second day also, though for but a short time only, for, feeling by that time pretty well satiated with the spectacle, I left early.

It was well on in the afternoon, and I was strolling along the pavement in Princes Street, in a fit of deep meditation, and the central figure of all my thoughts was the same that had at one time held supreme control there, to wit—Rachel Rolleston, now Mrs. Howden. The advantageous position that fortune had now placed me in, towards her who had at one time been the object of my devoted love, and whom I still deeply respected, was also very apparent to my mind. And with it the magnificent prospect of worldly advantages, that the possession of her hand as sole heiress of her father's immense wealth opened up. What need indeed for further toil, or chafing over future prospects, with such a means to fortune, almost within my grasp? These were thoughts that I should have been more than human to have pretended to ignore, but when it came to deliberately acting upon them, there were two distinct bars to the vision.

Firstly, as regarded Mrs. Howden herself, I felt convinced that, though all that respect could give and gratitude bestow were already freely mine, yet there was that in her present condition that would have made the idea of an offer for her hand a matter of simple indelicacy, and especially so in face of the advantages that I should gain, which would at once stamp my action as wholly mercenary.

Secondly, there was an even greater bar to such a proposition on my part, in my still smouldering passion for Jessie Campbell.

But Jessie had deliberately refused me already. Then why take further account of her? It was true that she had done so. Yet it was none the less true that her image still reigned supreme within my heart, to the exclusion of her once potent rival.

My thoughts reverted to the time when Rachel Rolleston was all in all to me; Jessie nowhere, by comparison. How little then did her manners please me, when held captive by Rachel's charms. To what, then, could the change be due?

It was not that I thought that Rachel had degenerated, either through her sorrows or, especially, her marriage with another. I could still imagine my old love surviving the last, whilst my compassion would have been enlisted for the first. Nor could it be that Rachel, by her preference for another, had deliberately abdicated her throne in my heart, for Jessie had done the same.

Thus engrossed in my own thoughts, I slowly sauntered along, when my eyes were attracted by a carriage and pair of peculiarly showy horses that were prancing by. Whilst my eyes were directed towards it, it was suddenly stopped and the window let down, and a lady's hand extended eagerly towards me.

Wondering who in fortune's name it could be who could be claiming acquaintanceship with me, I stepped across to the side of the carriage, and, to my intense astonishment, recognised the laughing countenance of Mary Campbell, although she was so richly and tastefully attired and adorned with jewellery that I almost doubted the evidence of my own senses.

"Ah, ha! you naughty man, we have found you at last!" she laughed, as I stood gazing in a sort of bewildered way at the unexpected apparition. "Now, don't stand staring there as if you could not believe it was me, for it is me and no one else. Jump in, and you shall soon learn all about what is evidently now such a source of surprise to you."

On complying with her request, I was rather disconcerted on finding myself suddenly in the presence of her sister, of whom, little dreaming of her whereabouts, I had just been so intently thinking. Jessie's cheeks coloured slightly as our eyes met, while she shook me kindly and warmly by the hand. I had never before seen her so reserved. I, too, felt inclined to be reticent towards her—not, of course, from annoyance, but from confusion—so I continued to address my remarks to Mary.

I should mention that Jessie was dressed with the same elegance as her sister. Both ladies looked superb. Mary was now in the full bloom of womanhood; and, as I gazed on her dimpled cheeks, her graceful proportions, and her bright, blue eyes, I marvelled at the blindness of all her bachelor acquaintances in letting her go free so long. "But bide a wee," as kind, garrulous old Mansie Waugh would have said. I very soon learned that Mary was not quite so free as I had imagined.

Mary, who did all the talking—for Jessie said never a word —now let me into the secret of all their grandeur. It was owing to the sudden reappearance of their long-lost uncle, an immensely wealthy man. Their mother, on receiving his address from Charles Howden, had lost no time in writing to him, when at once determining on putting a period to his career of money-making, he, as soon as he was able to wind up his business, had obeyed his sister's earnest injunction, to set sail for Dunedin, at which place, on his arrival, he at once took steps for settling down for life.

"And nothing would do for him," said Mary, "but we must come and stay with him, and mother has been down since Christmas. And you must never leave us again, for my uncle has more money than he knows what to do with, and I am sure that he will help you to do something better than going wandering about the country. We often wondered where you had got to—Jessie and I were always talking about you and Selim and Lilly. But we were on the look-out for you to-day, for, you see, Mr. M'Gilvray called upon us last night, and told us you were in town, and that he had seen you at the games: and he told us all about your terrible bushranging battles. Why, Duncan, but you are quite a hero now!"

Here Mary stopped, not so much from want of something to say, as for want of breath.

"It is strange," I said, "that Mr. M'Gilvray made no mention of your being in town, when I saw him yesterday, but then I was not in his company more than a minute or so, as he turned round to speak to some one else, and we got separated in the crowd, so I saw no more of him."

"Just like the idle fellow," replied Mary, with a pretty pout. "He never seems to think of anyone but himself, or he would have told you where we were at once. I have a good mind not to speak to him again for that."

In reality, as I very soon learnt, Mary had a good mind to do nothing of the sort. By this time I saw the carriage turning into the drive, and made an effort to get out, but here Jessie joined her sister in resolutely preventing my doing so, and, constrained by their wishes, I accompanied them into the house.

"But why had I made such a silly pretence of reluctance to do that which one would have thought I would have done out of common politeness?"

The truth was, reader, that my feelings in Jessie's presence were so painful that the tumultuous surging of my blood through my veins had made a coward of me and, with or without reason, I was desirous of beating a speedy retreat from her presence.

The thought of my rejection, when asking for her hand, on the last occasion of our meeting, was still bitter to my memory, and the humiliation of it was the more intense, because of the unabated loyalty of my heart for her still.

With these feelings strong upon me, I accompanied the

ladies into the house, resolved firmly not to let my feelings master me.

But when ushered into the handsome sitting-room, the sudden exclamation, in a tone of genuine gladness and surprise, of—"Duncan Farquharson, is it possible! how glad I am to see you again"—as Mrs. Campbell took my extended hand, almost put all my prudent resolutions to the rout. As a sort of compromise with myself, however, I took care, when seating myself, to select a chair on the side of the room opposite to where Jessie was seated.

But to maintain even the semblance of reserve in the presence of Mrs. Campbell and her two daughters, who showed such a warm, eager interest in my welfare, seemed to be so hopeless, that I foresaw at once that an inglorious surrender would be inevitable.

I had also here the pleasure of shaking hands with another acquaintance, in the person of Mrs. Ayson, who had been one of my Christmas guests the year before, and on whom I had laid such strict injunctions to bring her husband with her, when she repeated the visit next year. Here, too, was her husband, a mild, blue eyed, cherry-cheeked, gentlemanly looking man; and thus meeting within such a few days of the time fixed by my invitation, though in such a manner, and in a place so little dreamt of by me when I gave it, occasioned no little merriment at the time. But to be sure, we were feeling so generally happy, that it took very little to arouse our laughter. The first greetings over, the conversation turned upon the events in which I had so lately taken part, and then at once I thought of Rachel, who, till then, through my confusion, occasioned by my sudden meeting of Jessie, had been forgotten. With this thought came also the reflection of the absurdity of my boggling at entering the house of my newly-found friends, when I had such joyful news to give them, of the recovery of her who had once been so dear to them all.

I had given some particulars about the manner of Howden's capture, at the ladies' request, though they had already from Mr. M'Gilvray received a circumstantial account of the whole thing. When Mrs. Campbell remarked, "But Duncan, whatever could have induced you and Lilly and Mr. M'Gilvray to join in this desperate affair? or what possible interest could you have had about this man Butler's capture?"

I stared at Mrs. Campbell in some surprise at this question, when I suddenly recollected how much she had yet to learn about the matter. I forgot to state before that Howden was only known to the police by the name of Butler, and that when John M'Gilvray had agreed to make one of our party in the attempt to capture him, I had not informed him of my motives for my strong interest in this man, beyond the remark that I had the most powerful reasons for desiring his capture, and I had not seen John again until I met him at the Dunedin games.

Consequently from his ignorance of Butler's real name, in his narrative to the Campbells, of the circumstances attending the capture, it was only by the name of Butler that he had referred to the prisoner. All this I suddenly recollected now, and said with a smile: "You consider my action quixotic I see, Mrs. Campbell, but I have only to pronounce one word to cause you to change your views on that point. Then, instead of calling Lilly and me rash for our conduct in venturing into that affair, you will praise us, nay, I venture to say, that you will thank God for having permitted us to render such assistance as we did in effecting that man's capture."

"Indeed," said Mrs. Campbell, looking at me intently, "and what word is it pray, that is to have such a wonderful effect upon me?"

"This man Butler, *alias* Howden, might possibly have had another *alias* that would sound more familiar to you," I answered.

"Whatever can you mean, Duncan?" she replied, turning slightly pale, as if thinking that something horrible was about to be revealed; "what *alias* can that be?"

"Marsden!"

"Merciful heaven," cried Mrs. Campbell, springing to her feet, as that name aroused a train of most painful thoughts. "And that girl in the tent ——"

"None other than Rachel Rolleston."

"My poor, lost darling!" exclaimed Mrs. Campbell, in a tone and manner of the keenest anguish, as she clasped her hands together at my answer, and both girls joining in an exclamation of sorrow.

"No; not lost, Mrs. Campbell," I said, hastily, springing at the same time to my feet, in the eagerness of my feeling to disabuse the minds of these friends of such unpleasant suspicions concerning Rachel's character. "Not lost, though recovered without one stain of shame upon her brow, save that of the folly of her first step in believing the plausible representations of a thorough villain. Though, heaven knows, a deeper stain than that might well have been effaced by the sea of suffering through which she has since passed."

I then gave them the whole history of Rachel's experience. The silent tears that flowed from the eyes of her friends at the story bore ample testimony to the depth of their sympathy with Rachel in this sudden eclipse of her happiness. Even Mr. Ayson wiped his eyes, and blew his nose with extraordinary vigour, to the danger of the healthful condition of that organ. On learning where Rachel then was staying, it was at once decided to go and fetch her in the carriage. The mother and both daughters were unanimous in their desire to proceed with me to the "Highland Home," but it was at length settled that Jessie should wait at home and see to the preparation of tea, as the hour for that meal was now at hand, whilst Mary and her mother should accompany me in the carriage.

"Oh! Duncan," Mrs. Campbell feelingly remarked, "both you and Lilly have indeed acted bravely in what you did, and deep is my gratitude to the Almighty Being who made you both the providential means of this great deliverance."

I involuntarily here glanced towards Jessie. Our eyes met, but, applying her handkerchief to her face, she hastily left the room.

On reaching the "Highland Home," I sprang out of the carriage, and, preceding the ladies, bounded up the stairs and knocked at Rachel's door. On her opening it, I seized her hand, saying, in a low, earnest tone:

"Courage, dear friend. There are those coming upstairs whom you love well, and who love you. They are Mary Campbell and her mother. I accidentally discovered this afternoon that they were in town. Keep up your courage now, as you always do, and do not let this shock of happiness overpower you altogether."

By this time the ladies, who had been shown the way by the landlady, were on the landing. For a moment Mrs. Campbell and Rachel and Mary looked earnestly at each other, as if curious to note the changes that had taken place in the time that had elapsed since they had parted. Then, extending her arms, Mrs. Campbell exclaimed, in a broken voice:

"My own darling, have I found you at last?" when, bounding forward, Rachel threw her arms round the neck, and laid her head on the bosom, of her who had been a second mother to her. Mary also rushed forward, and the three were locked in one another's arms. Feeling that this was no place for me, I turned and hastily descended the stairs.

As we were entering the carriage to return, Lilly, who had been absent from the house when we arrived, just then chanced to come up; and great was his astonishment when, in the occupants of the carriage, he recognised his old friends.

Regarding the handsome equipage with astonishment, Lilly bluntly remarked: "Why, Mrs. Campbell, you seem to have got on in the world since I last saw you?"

"Not much up, Lilly," replied Mrs. Campbell, with a smile,

"although well enough, and thankful for many things. This carriage belongs to the brother you may have heard me speak about sometimes. He has turned up at last: and turned up, I am also happy to say, a wealthy man. Jessie and Mary have been staying with him for some time, but I have only been down since the holidays began. But jump in; you must come with us too. My brother has often heard us speak of you, and I know he will be very glad to see you. I have heard how nobly you acted in concert with Mr. Farquharson towards our former mistress here; but jump in."

"Tiny is with us, Lilly," Mary put in, archly, on seeing him hesitate. The hint about Tiny had the effect, however, of instantly allaying his fears of compromising his independence by taking his place among company with whom he could not feel at home; and he accordingly took a seat in the carriage, when we drove swiftly to Mr. Carmichael's house. To avoid witnessing another scene when Jessie and Rachel met, I allowed the ladies to enter the house, whilst I stood watching the groom take the horses to the stable.

As for Lilly: on leaving the carriage, he was for striding straight off to the kitchen, when I called after him that he had better come into the house with me, as I knew that he would be expected there.

"No, thank'ee, Mr. Farquharson," he answered, bluntly; "what should I do among the parlour mob, I should like to know?"

I smiled to myself as I shrewdly conjectured it was less his aversion to the "parlour mob" than the magnetic attraction of Tiny's presence in the kitchen that influenced him to move so promptly in that direction.

On entering the house, I met Mr. Carmichael, Mr. John M'Gilvray, and a precise-looking, elderly man, to whom I was introduced as Mr. M'Gilvray, senior. Mr. Carmichael was a bluff-looking gentleman of frank manners and intelligent countenance. He looked considerably older than his sister, although they were twins; but this was no doubt due to the harassing cares of a business life, as well as to the effects of long residence in a sultry climate.

The evening that followed was a most pleasant one. We all engaged in a well-sustained conversation, and "fought our battles o'er again," during which I could not help thinking that I looked very much like a hero, as I realised the fact that there were three people then present whose lives I had at one time saved.

There was Mary, whom, with the help of Selim, I had some years before saved from a watery grave in the Murray, and

Rachel, who, on the Darling, I had probably been the means of preserving from the horns of an infuriated bullock, besides the prominent part I had so lately been enabled to play in the scene by which she was ultimately rescued from a miserable state of existence; and, lastly, there was John M'Gilvray, whom I had preserved from strangulation.

I observed, with some surprise, that, contrary to what I had expected, John M'Gilvray attached himself to Mary during the evening, and well did this charming young lady now look with her vivacity and dimpling cheeks as she parried some thrust from her companion, with a quiet counter sally of wit, that generally had the effect of unexpectedly turning the tables upon him, when the merry-hearted fellow signalled his defeat by a hearty laugh.

Jessie was consequently seated by me at the tea-table, and afterwards seemed to prefer to have her chair by mine, though she spoke but little. As for Rachel, she kept her place close by Mrs. Campbell during the whole evening, and I observed that the elder lady's eyes, as if moved by a feeling of maternal solicitude, scarcely ever left the face of her long-mourned friend.

The evening was considerably advanced, when some reference chanced to be made to Lilly as being in the kitchen.

"Eh, what's that? what's that?" cried Mr. Carmichael, quickly, and starting from his seat as he spoke. "Do you mean to say that your friend, Lilly, has been in the kitchen all the evening? Why let him remain there? Bring him in here instantly."

"I am afraid, Malcolm," replied Mrs. Campbell with a smile. "You will find that Lilly has his own peculiar notions about what he thinks to be his proper place, and these notions you will find it a vain task to try to knock out of his head. Knowing this peculiarity of his, I took no pains to disturb him where I knew he was perfectly contented to be."

"Then," said Mr. Carmichael impetuously, "if he won't come to where we are, I will go to where he is. Come with me, Mary, and introduce me to him. Come along, M'Gilvray, and get yourself introduced to this man: after what I have so often heard my sister and niece say of him, and what his conduct has lately been in the bushranger exploit, I am sure he is worth knowing."

We all, that is, all save Mrs. Howden and Jessie, rose and adjourned to the kitchen, where we found Lilly, apparently happy as a king, "spinning," as he himself termed it, "cuffers" to Tiny about old times, and late times, too, and all the adventures he had had since he had last seen her.

"Lilly," said Mrs. Campbell, as we all entered the kitchen,

"this is my dear brother, Malcolm; he wishes to be introduced to you."

"Wishes to be introduced to me?" replied Lilly, bluntly; "that's easily done, for here I am at his service, and I am glad to know your brother, Missis; for if he is anything like you, I know he is worth knowing."

"Thanks, Mr. Lilly," said Mr. Carmichael, laughing. "I hope I deserve some part of this compliment at least. But why, man, should you be content to stay in the kitchen? My sister's friends are surely worth inviting to sit where she and I sit."

"I thank you for thinking that much of me, sir, but Mrs. Campbell knows my way, and it is my way to think that a working man should keep among working men, and a 'swell' among 'swells'. Now, I am no 'swell,' and I don't think I am ever likely to be one. Not but that I think myself as good as any 'swell' for all that, for, on that point, I am like Bobbie Burns, who said that—

> 'The rank is but the guinea stamp,
> The man's the gowd for a' that'."

"Right, Lilly! I honour you the more for sticking to such sentiments as these, and there is no man that I more truly respect than an independent, upright, and down straight working man; I have risen in the world myself, but I have known what it is to turn my hand to a hard day's work with pick and shovel, and it is for that reason that I have learned to respect a man as a man only, in whatever sphere I find him."

With a few more remarks, Mr. Carmichael shook hands heartily with Lilly—the elder Mr. M'Gilvray following suit in an equally frank and pleasant way—and Mr. Ayson in a more genteel though none the less sincere fashion. On our return to the sitting-room, spirits and some supper were laid on the table. Having taken of the former sparingly, I rose to depart, to Mr. Carmichael's great surprise, who had taken it for granted that I was to pass the night under his hospitable roof. But although Mrs. Campbell and Mary joined vigorously with Mr. Carmichael in trying to overrule my decision, Jessie said nothing, though she looked as if she would have liked to very much, so I thought best to be firm on this point. For although I had ingloriously surrendered my position of composed reserve, in front of Jessie, and had, I fear, become as chatty and frankly communicative as if I had never suffered the heroic grief of slighted love, there was still that much of the spirit of pride left in me as to cause me to preserve at least some appearance of consistency towards what I conceived should have been my line of action.

Both Mr. Carmichael and Mr. M'Gilvray, senior, accompanied me to the door, after I had shaken hands warmly with all in the room. When at the door, Mr. M'Gilvray remarked, " You are not thinking of going away from town for a few days, at least, I hope. I have a particular reason for desiring you not to leave Dunedin at present."

I replied that I should most likely be in town for a week anyhow, and perhaps longer—as it depended entirely on the nature of any opening that might offer for my future employment.

"Just so," replied Mr. M'Gilvray. "But anyhow, whatever crops up, be sure and let us—that is, Mr. Carmichael and myself—see you before you take any steps for the future."

"Surely Mr. Farquharson will do so, for I expect nothing less than that he will, at anyrate, pass his evenings with us, even if he is determined not to take up his quarters in my house whilst in town. You will certainly do this, won't you, Mr. Farquharson?"

"I'll not actually promise that I will do so, Mr. Carmichael; so much must depend on other things."

"Mr. Rolleston, on his arrival, will, of course, stay with us, and I believe that the 'Tararue' is expected to-morrow evening at Port Chalmers; you must certainly come up when he does arrive."

"I most certainly will, Mr. Carmichael, not only because of my natural pleasure in seeing him, but for other reasons as well."

"That is right, we will expect you to tea then to-morrow evening."

"If the 'Tararue' should have arrived in time to allow Mr. Rolleston to reach Dunedin by that time, I will be there."

"Then that is settled. I will send the carriage to the wharf, to meet Mr. Rolleston when he lands, and have him driven to my house at once."

With this understanding and a cordial exchange of handshakes we parted. Lilly, who had been warned of my departure, I found waiting outside at the gate, and together we strode off in high feather to the "Highland Home".

CHAPTER XLV.

THE "Tararue" arrived in Port Chalmers several hours earlier than had been anticipated. Consequently, her passengers for Dunedin were landed at the jetty about four o'clock on the same afternoon. Apprised of this in time, I was on the wharf when the steamer arrived, and among her passengers I had the felicity of at once recognising my quondam employer, though now sadly altered by sorrow, against which his immense wealth had proved no specific. Grasping my hand with almost nervous eagerness, he betrayed what thoughts were uppermost in his mind, by the hurried ejaculation, "My daughter, is she well?"

"Well, and as happy as the society of her best friends can make her, for she is at present with the Campbells," I replied with a smile.

"May the Lord reward you, Mr. Farquharson, for your goodness to my unhappy child, and for lifting off such a load of sorrow from my heart. But are the Campbells here? I always thought that they lived in Southland."

I put him in possession of the circumstances that occasioned the presence of Mrs. Campbell and her daughters in Dunedin, as also of the happy accident of my meeting them as I had done.

"But see," I continued, "here is Mr. Carmichael's carriage that he has sent to meet you, as he promised to do; so we had better get in at once, and you must prepare your mind, my dear sir, for meeting with your daughter within half-an-hour."

In much less than that time Mr. Carmichael's spirited horses had borne us up to his house. Mr. Carmichael himself was on the verandah as we drew up. Hastily descending the steps, he met us as we stepped out of the carriage.

"Mr. Rolleston, I presume?" he remarked, as, without further ceremony, he took that gentleman's hand within his own cordial grasp; while, by way of introduction, I said: "This is Mr. Carmichael, Mr. Rolleston".

"Compose yourself, my dear sir," the former now rejoined; "you will find your daughter awaiting you inside; come away." They entered the house, but, not thinking it quite in taste to make myself a spectator of the meeting that would follow between the long-divided father and daughter, I lingered on the verandah for some time.

When I eventually entered, Mr. Rolleston was seated in an easy chair by the fire, showing, by the way he bent towards it, that his thin, Australian blood was particularly sensitive to the chill breeze that so frequently rises of an evening in New Zealand. Beyond, however, a quietly happy expression on his face, there was no indication of any unusual emotion—no expression of delight on the romance of the hour—for the simple reason that there was not one particle of romance in Mr. Rolleston's prosaic nature. Save in the first rush of his natural feelings as a father (and of these he had his full share) on his meeting with his long-lost daughter, when he was overcome for a few moments, there was nothing to show that his thoughts had been moved from their wonted prosaic channel. And, though not without gratitude of the sincerest kind towards me, as his subsequent conduct showed, he had by this time assumed his wonted plain, practical manner of discourse; and business matters and topics of the day, ever paramount in his mind, were now being as calmly passed under review by him as if nothing extraordinary had so lately occurred to displace the habitual current of his thoughts. By his side sat his daughter, with a sweet expression of grateful love that gave a flush to her face and a softer light to her eye.

The conversation I will not attempt to describe. Suffice it to say that that evening was one of unmixed happiness.

As the evening drew on, other visitors, in the persons of the M'Gilvrays (father and son), arrived. Lilly also found his way into the kitchen again, where he appeared to be perfectly at home, chatting with Tiny, as that comely damsel busied herself with the duties of her office.

That Tiny's interest in Lilly's arrival was of the liveliest kind, appeared to be amply demonstrated by the cordiality of her welcome and the liveliness of her speech, so contrary to her wonted demureness and almost staidness of manner.

On hearing that Lilly was in the kitchen, Mr. Rolleston instantly proceeded thither and, greeting him most cordially, took a seat and entered into conversation with him, as in former times had been his frequent habit. For Mr. Rolleston was always plain in his manners, and utterly wanting in that pride of place that causes most people in his position to affect an air of superiority when conversing with their dependents. Long, on the present occasion, did he stay talking with his former stockman, of whose worth and integrity he had always entertained such a high opinion. As for Lilly, he conversed in the same blunt manliness of tone to his old employer that he was ever wont to observe when under him, and for whom, as a master, his regard had been as great as had Mr. Rolleston's of him as a servant. Perhaps, on the present occasion, the wonted bluntness of his manners might have been slightly softened into a tone of more marked respect, for the pathos attending Mr. Rolleston's present circumstances and appearance in meeting with his daughter, to whose recovery Lilly's own chivalrous exertions had so largely contributed.

While Mr. Rolleston was with Lilly, Mr. Carmichael and the elder Mr. M'Gilvray had also risen and withdrawn into another room where they remained a considerable time as if engaged in the discussion of matters of business. In due time I rose to take my departure. As I shook hands with Mr. M'Gilvray senior, he glanced inquiringly at Mr. Carmichael.

"Yes, I know," the latter replied, as if in answer to a mute inquiry, and going to the door with me requested me particularly to come to tea on the morrow, as there was a matter of business that he and Mr. M'Gilvray wished to discuss with me. "Tell Lilly to come also," he continued.

Of course this I readily promised to do, and shaking hands with him very heartily went away, wondering what the nature of this business might be.

On the following evening I duly fulfilled my promise, while, as usual, Lilly made for the kitchen. After tea I was summoned into another apartment, whither Mr. Carmichael, the elder Mr. M'Gilvray, and Mr. Rolleston had already repaired. As this apartment was evidently used by Mr. Carmichael as an office, I immediately surmised that the subject for which I was now summoned must be something official. I was not permitted to remain long in suspense as to the nature of this business, for almost upon seating myself, the precise mannered Mr. M'Gilvray delivered himself thus—

"We have sent to you, Mr. Farquharson, to speak to you on a simple matter of business. I am, myself, a practical man, who seldom wastes time in beating about the bush where business is concerned; so on the present occasion I will go to the point of what I desire to speak to you about at once. From your own remarks, and what we have otherwise learned of your late history, Mr. Farquharson, I think I am right in assuming that you are not very well off in money matters. This being apparent, both these gentlemen and myself, who have all such occasion for gratitude towards you—and, speaking for myself, for your courageous behaviour in preserving my son from a cruel and shocking death"—Mr. M'Gilvray's voice here momentarily quavered—"for which words fail me to express my thankfulness—I say," he continued, clearing his throat and regaining mastery of his voice, "that both those gentlemen and myself have been consulting together, how we can in some measure express our sense of these obligations to you. Now, one way presents itself of doing this in a substantial way, which I conceive should not be painful to you to accept of, nor unreasonable in us to offer you. In my hands lately, as estate agent, there has been placed for disposal a valuable run on the Windaway river. The run carries at present twenty thousand sheep, and its grazing capacities are such that two or even three thousand more added to that number would not overstock it. Now as the occasion of this sale is simply compulsory—being pressure of the mortgagees, who have no confidence in the present owner, who appears to have been a most injudicious manager—all this fine property can be obtained on the payment of the mortgagees' claims—that is £1 a head for the sheep. So that £20,000 will purchase it all, including improvements, horses and cattle, of which latter, however, there is no great stock on the place. Now the case stands simply thus : This property we three have decided to purchase for you, and you can repay us by yearly instalments.

"According to the most prudent calculations made by Mr. Rolleston and myself—who you will admit ought to be good judges of such matters—with such an advantageous start, there should be no obstacle in the way of an intelligent and energetic man making all this magnificent property his own in ten years' time, at the very furthest. Of your own fitness for such an undertaking, Mr. Rolleston's past experience of your efficiency and integrity when in his service, would be ample proof, even if none other were forthcoming; but I have also the testimony of my own son. It may be pleasing to you also to know that Mr. Rolleston himself has offered to supplement his own share of the cost of this purchase, by his personal security to both Mr. Carmichael and myself, against the possible contingency of loss to us in the undertaking. This security we have, however, declined, for, being equally obliged with him, we have thought it only right to be equal partakers in the risks that may be involved in this manner of expressing our sense of these obligations. Interest on this money of course you will be charged, but it will be at the lowest market rate, as our desire is to benefit you and not to burden you. Now, what do you say to this proposal? I should think it is one that is in every way suited to your taste and spirit."

I, at first, felt quite stupefied with the very magnificence of the offer thus suddenly held out for my acceptance. But gradually, as I deliberately contemplated the full bearings of this proposal, so congenial to my taste and habits, my natural hardihood of spirit reasserted itself, and in thought, I could already feel my knees pressing Selim's sides, as I made the rapid circuit of this new field for my energies ; and, with alert eyes, marked the details, whose attention was to crown my undertaking with the stamp of efficient management. Nor in the present instance did my natural horror of debt appal me as it had done at the idea of a loan from Charles Howden, when the prospect of a profitable investment had appeared so vague and shadowy. It happened also that I knew something of this same station property, for a few days before I had heard all its advantages canvassed by a station manager from the same district, with whom I had by chance foregathered at the Caledonian games. Taking all things together, therefore, I saw my way clear to the end, with as good a chance as ever man had, of coming out of this undertaking a winner.

These were the palmy days of wool and mutton, when a man with any practical sense and energy could not possibly have failed of success under such advantageous conditions as were then submitted to me. I, however, thought of Lilly, and determined on securing his co-operation with me in this enterprise. For while the thought pleased me of enabling him to share in my good fortune, I knew also that with his intelligence and energy in co-operation with my own, the risks of ultimate failure were proportionately reduced. To Mr. M'Gilvray's question I accordingly made answer:—

"Gentlemen, that my thanks to you all are of the deepest and sincerest nature, for such unheard of generosity in making me such a magnificent offer as this, I need not say; and though the responsibility of incurring such a monetary obligation seems such a serious one, yet in view of the advantages I can see in this prospect, I have enough faith in my own sense and energy to induce me to accept your noble offer, as I feel convinced that you will never have cause to repent putting such trust in me as you have now shown. Yet, although certain that I shall be able to go through with this undertaking successfully by myself, I should be just as pleased if my friend Lilly could go with me into it. He has some money of his own to begin with, and with him to help me, any risk that there may be in this would be lessened."

"Good! I approve of your idea, Mr. Farquharson," here spoke Mr. Rolleston. "To Lilly's integrity and abilities, my own long experience enables me to bear the very highest testimony, and I should be glad to see him go with you as a partner in this run."

"Let Lilly be sent for then at once, that we may see what he has to say himself about the matter," said Mr. Carmichael.

Accordingly, Lilly was sent for. He came, and on the nature of the proposal being explained to him, he, for a minute or so, considered it squarely, and then gave his opinion, with his accustomed straightforward bluntness. "There is no man I know of that I would sooner go mates with into any speculation than with Mr. Farquharson. And if there is no other way for it, I am willing to go along with him into this run; but, left to my own choice, I would sooner be my own master, with what I have, than go partners with anyone; and no man can be his own master while he is a partner. I have £1500 of my own money, and this money I will put into this business with Mr. Farquharson, if there is no other way for him to get the run ; but, if he can get on without me, I would sooner myself buy a small place, and work it after my own fashion. This is my mind on the matter."

"Then," replied Mr. M'Gilvray, "there is no cause why you should not follow your own wish, the more so, as, bordering this run, I have also for disposal a small place, with a freehold of good arable land, of about 700 acres, and a government lease of mountain country, that, in all, is capable of carrying about 3000 sheep."

"This is the place for me," said Lilly, emphatically, "and let Mr. Farquharson take up the big run ; and when he wants any help for mustering, or anything else, he will know where to find me."

Thus this great matter was settled, and thus once more the clouds lifted from my life's prospect, and another stage in my chequered career began.

I then informed these true friends of Charles Howden's generous offer a few days before, and my motives for hesitating, at the time, to accept it.

"How like Charles Howden!" Mr. Carmichael warmly exclaimed, on hearing this. "My kind old friend, how true he appears to have always proved to that ideal that he seemed so early to have set up for his own guidance! I trust I shall see him to-morrow. I will look him up at his lodgings, if you will come with me, Mr. Farquharson."

I replied that I would do so willingly.

Next day, accordingly, I took Mr. Carmichael to Howden's lodgings, where we found him at home. Long did these old friends converse together of days gone by, and the strange vicissitudes of life encountered since. I told him, too, of the bright changes in my prospects, and to whom I was indebted for it. Mr. Carmichael strongly advised Charles against his idea of settling in the old country, where he would simply lead a melancholy life in scenes from which his tastes and sympathies had been long weaned by his colonial experience.

"Go home, by all means," was his practical friend's advice, "and solace your spirit with a sight of your childhood's haunts, but leave the bulk of your money, at interest, where you are. You will find yourself sated with a very few months of residence at home, with all its formalities and old-world notions, from which your long intimacy with the rough independence of colonial ways has utterly alienated all your sympathies."

To this shrewd advice Charles was eventually persuaded to listen, and the event amply justified its wisdom. Within twelve months of his departure from Britain, Charles returned again to Dunedin quite content to settle there for life.

Impatient to visit our future homes, Lilly and I left Dunedin for the Windaway as soon as the necessary legal business in connection with them was fairly put through. Nor was our gratification at our new possessions a whit lessened on finding that our homes were not more than four miles apart, because my home station had been built near the end of the property marching with Lilly's ground.

Our manner of parting with our kind friends in Dunedin I will not particularise, but I do believe that Mary's regret on that occasion was not a little increased on Selim's account, on whose back, ever since the happy accident of our meeting, she was accustomed to take daily exercises; and for which animal, ever since her adventures with him on the Darling, when on the red bullock's broad forehead he gave the full benefit of his clattering heels, her regard had amounted to positive affection.

CHAPTER XLVI.

"Be thou diligent to know the state of thy flocks, and look well to thy herds."—Proverbs xxvii. 23.

THE end of my narrative is now at hand; and the readers who have found sufficient interest in these pages, to accompany me so far, will now be in a fair position to form an estimate of my individual character. There is little of interest that remains to be told, and what there is, I will tell as briefly as I can.

To begin with, then, I will say that, in my new undertaking, matters prospered so well with me, that, thanks to my unremitting attention and diligent economy, in six years I had little more than £300 remaining unpaid of my monetary obligations to those generous friends who had started me in this run; moreover, I had repaid the £500 still owing to my kinsman.

At this time, however, a change in the conditions of the land tenure of runholders necessitated my making—in order to preserve the best of my station from being cut up into farm allotments—freehold of all this land, comprising 15,000 acres—excellent arable land the whole of it. All this I was enabled to secure only by the use of such influence in high quarters as, I confess, had the effect of raising a howl of indignation from the popular party at such unprincipled "toadeying" to the interests of the "bloated squatter," for I had secured it all at the upset price of £1 per acre. As I, however, felt quite justified in an action that was quite legal, this howl I paid but little attention to.

The necessity for this purchase however threw me back in my monetary obligations almost to my starting point six years before, though of course with property much increased in value, as being freehold. Though a sudden fall in the price of wool (that, however, regained its former value in the market in a couple of seasons) a few years after making this purchase gave me an anxious time while it lasted, yet, steadily pursuing the same method and economy of expenditure as before, I contrived to avoid the rocks on which so many runholders struck, to wit, extravagant expenditure of money on improvements, and wasteful farming.

At the time of my writing this, my place is almost free from debt and in excellent working order. All this, I imagine, is saying a great deal in face of the sharp experience of so many years of depression, that runholders, as well as other business men, have had to face. And now the near prospect of a move from the time-worn primitive residence in which we have hitherto resided, to one of a more pretentious class—a long deferred hope—seems to mark the culmination of happiness to Jessie and myself.

"Jessie and yourself? What! You really married Jessie after all, and not a word to say how it came about?"

Yes! Of further romance in this matter there remains no more to be told. The cause of the sudden change in my love prospects can be related in a very few words.

It was on a hint that I received from Mary—mind you, ere I had been aware of the sudden brilliant change that was dawning upon my prospects—that she had often seen Jessie crying over my photograph, on my sudden disappearance after the collapse of my lake station, that I conceived that my prospects with her were not quite so desperate as I had been led to believe. After this I took the very earliest opportunity I could for an interview with Jessie, and after a very few words of passionate entreaty, she let the mask of indifference fall from her sweet loving face, and with an irrepressible burst of tears, leant her head upon my shoulder, while I pressed her in a fond embrace to the heart that had so long and loyally treasured her. Shortly afterwards we were married by the Rev. Dr. Stuart. Now I have six healthy boys and one girl, who is the youngest and the pet.

It appeared that John M'Gilvray had really entertained

thoughts of asking Jessie's hand; but he was a clever, practical man who would be the very last to break his heart for a woman's love. His common sense soon showed him that with Jessie he had not the ghost of a chance. Yet, by a singular coincidence, at the time that he made this discovery, he also made another, and that was the superior charms, in his eyes at least, of Jessie's sister Mary; and they were married at the same time, and by the same kindly hands as Jessie and I. They immediately settled down on the station owned by John's father that was given to the son by him. They, too, have now several blooming children. Thus was Mrs. Campbell in one day deprived of both her daughters. But she bore her loss the more resignedly, that one of them was afterwards to be her own next neighbour. Mr. Carmichael, however, made a sort of good-humoured protest at being then suddenly deprived of what he termed his two bonnie housekeepers, but he was laughed and kissed into acquiescence with the necessities of the case.

Shortly after my own marriage and settlement in life were thus accomplished, I became concerned about Lilly. We were frequently at one another's places, and I observed that he was becoming silent and moody. It was not that he was dissatisfied with his new home, for he had expressed himself as being delighted with it, and upon his arrival there had set himself to work with his accustomed energy, upon his improvements and management, when all of a sudden he began to grow listless and careworn. The fact was, that Lilly, who never had more than one heart for a woman's love, had offered that to Tiny, who, from what seemed to be mere feminine caprice, had declined the gift. Tiny was now staying with her parents, who owned a farm not many miles away, in the same district in which Lilly and I had settled.

Tiny's motives, however, for refusing Lilly, were not those of mere feminine caprice, for she in reality entertained such a genuine regard for him that her own heart smote her sorely in being obliged to give him the pain of this refusal. It was simply a matter of religious principle that had caused this conscientious girl to act in the manner she did. Tiny was a strictly Bible-taught Presbyterian, whilst, by profession at least, Lilly was a Roman Catholic. Now, whatever effect love's all potent influence might yet possibly have in softening down Tiny's own conscientious objections on this head, by Tiny's mother, a rigid Presbyterian of inflexible principles and old world prejudices, popish views were looked upon with horror, so that this was deemed an insuperable bar to her daughter's union with Lilly.

With all this I was fully acquainted, as Tiny had detailed it all to Jessie, hence my information on the subject. Much therefore did I cogitate on the possibility of being able to reconcile these jarring elements, that threatened to sap the happiness of my old friend's hitherto contented life.

Of old I had known that Lilly's religion was simply traditional, and that real attachment to the leading Romish doctrines he had none; or rather, I should say, that to some such—as the celibacy of the clergy—their claims to the power of forgiveness of sins, and of papal infallibility, he was roundly antagonistic. I knew, moreover, that beyond what information he had been able to gather from the wordy disputes of polemical shepherds, and shearers, in the station huts in Australia, that doctrinal knowledge of the rival merits of Popery and Protestantism he actually had none. For among the elements that went to make up Lilly's early scanty education, religious teaching was an item that had been left out of the question altogether.

Love is a potent factor in helping most people to overcome difficulties that stand in the way of its consummation. In the present instance, at my suggestion, it caused Lilly, in order, as he said, to see to the bottom of this doctrinal difficulty, to resort to the local clergyman for information. A better step he could not have taken, for this Mr. Davidson was a well read person. He was particularly well versed in all the points at issue between Catholics and Protestants, and, being as pious as he was erudite, it was a matter of delight to himself to help Lilly through his difficulties, by faithfully explaining the different renderings of these doctrinal texts. He did not stop there, for besides mere doctrinal knowledge, he revealed in his teaching the Christ who was behind all these doctrines, until such scales fell from my old friend's darkened eyes, that at length, instead of a blind traditional adherent to a Church, the chief of whose doctrines he had ever repudiated, he was able to say with humble earnestness, "One thing I know, that whereas I was blind, now I see".

To Tiny Sutherland, with her rigid, Bible-loving mother's consent, he was shortly afterwards wedded, and the little group of fair, curly-haired and dark-eyed toddlers—according as they reflect the images of either parent—may be seen now about the door, to prove that their union has not been fruitless. And a fond father Lilly makes; for, when in the house, he is seldom to be seen without one of his children on his knee. He has prospered also; and although his bluntness of speech, impatient spirit, and sarcastic wit are, as of old, still outward manifestations of the irrepressible Old Adam, yet behind these is a spirit of earnest, Christian consistency. This, combined with his strong force of character and keen sagacity, has caused Lilly's opinion to be deferred to on all matters of public interest that crop up in the neighbouring borough of Risington. As of old, when the influence of these traits gave him such prominence among the rough and ready and frequently fractious bushmen in Australia, they still carry the same marked influence with all classes here, and all this influence is always on the side of good. His old mechanical and manual ingenuity were also here strikingly and amusingly exemplified. Although Lilly had probably hardly ever seen a plough at work until after purchasing his own farm, yet, determining to obtain a thorough insight into the working of this implement, he undertook the holding of it himself; nor did he desist, until at ploughing matches, in two successive years, he, first with the single, and next with the double furrow plough, carried off the first prize, defeating on both occasions competitors renowned over all the districts round. After this he resigned the further working of the plough to his hired man.

Of William Lampiere, too, there remains something to be told. It was several years after my settlement in the Windaway district, that one evening, among the contents of the mail's budget, there came a parcel, which, on being opened, turned out to be a slim volume of poems, on the gilt lettering of the green cloth covering of which I saw printed, to my great surprise and pleasure,

BUSH POEMS,

BY

WILLIAM LAMPIERE.

My first impulse was to send a messenger post haste to summon Lilly to view this permanent memento of his friend's poetical talent, but a second thought caused me to pause in my purpose. Lampiere had, since parting on the Darling, kept up a regular correspondence with Lilly; and although Lilly's distaste to the use of a pen amounted almost to a positive aversion, yet his genuine affection for Lampiere constrained him to write in acknowledgment of his friend's monthly epistle about twice in the year. I, therefore, judged that the same post that brought this book to me had brought another to Lilly also. Indeed I thought it would be strange if Lilly did not shortly appear to tell me of the very news that I was on the point of sending to him.

It turned out as I had thought it would. In little more than an hour afterwards, Lilly came into the parlour, with an expression of exultation on his countenance, at what he considered such a convincing proof of Lampiere's genius that was almost laughable in its very childlike earnestness, the more so as it was in such singular contrast to the keen and almost sarcastic expression on Lilly's features at other times.

The book consisted of rather more than 100 pages. The poems were short, some of a meditative character, and others descriptive of bush scenes and scenery, and they certainly, in my judgment, showed merit in the earnestness of their tone and truthfulness of their colouring. But best of all in Lilly's eyes there was his favourite poem the "Hutkeeper's Address".

I observed afterwards some critical notices of this book in the Australian and New Zealand papers. Most of these were kindly and encouraging in their tone, save the aristocratic *Australasian*, that only deigned to notice this humble yet genuine contribution to colonial literature with a supercilious snub, "that proved the man who wrote it to be only a common snob," as Lilly remarked in wrath when he read it.

About twelve months afterwards we had a visit from Lampiere himself. I was very glad to see him, and to know that he was doing well in his situation, and that his friend Burrel had succeeded so well with his literary journal, that it was now quite an established and widely-read periodical.

While he remained, he resided chiefly with Lilly. Although he always found a welcome in our house which he occasionally availed himself of. He seemed to be greatly entertained with the children, and a short poem that he wrote descriptive of our house and social amenities, Jessie declared to be prettier than any in the book. It was entitled "On the appearance of a home of Friends".

"Their children numbered seven,
  And beautiful were they,
With eyes like stars of heaven ;
  Brows open as the day.
There were Jim and Jack and Tom and Dick,
  And Rab, and little Sandy,
And saucy Kate, her mother's pet,
  Nicknamed Sugar Candy."

Mrs. Howden and her father returned to Melbourne a few days after the business in connection with the purchase of my run was concluded. Athough in her own quiet way supremely happy, that elasticity of spirit that had once imparted such a ringing joyousness to her manner Rachel never regained. In its place, her manners became pervaded by a deep, unostentatious piety. Of suitable offers of marriage—some of them representing positions in society that might well have tempted her, had she been influenced by feelings of personal vanity—she had several, but declined them all. True to her promise to Kate, her first care when she found herself settled at home was to invite the poor girl to place herself under her protection. This Kate, after some preliminary difficulty, at length consented to do, when by Rachel's unremitting kindness and earnest conversation, she soon became completely reformed.

The happiest times in our home life are when it is brightened with Rachel's presence during the occasional visits to New Zealand, when the principal part of her time is distributed equally among her former friends, the Campbells, M'Gilvrays and ourselves. If her wealth, as her father's sole heir, is great, the sums that she annually disburses on behalf of various charities are also great, and to such as work in the direction of providing for the reclamation of such as Kate Donovan, her sympathy is ever ready and her purse ever open.

As for Charles Howden, the calm brightness of his life's middle age seems to have amply compensated for the troubled clouds that had so long obscured its spring. He breaks the monotony of his town life by visits to his country friends. To me his face is ever welcome, and his society ever a boon. But it is chiefly with the Campbells that his time is spent.

Of the manner of Selim's end, it now but remains to say a few words ere I close this narrative.

About seven years after my settlement on my station, on going out one morning to the paddock, to my sorrow, I found Selim—who, save in the severest weather, was always suffered to go in the paddock, as with his natural hardy constitution he took badly to the stable—lying down, with his legs bent under him, in a paroxysm of cramp. I tried what remedies I could think of to recover him, but in vain. The sands of life with my brave old steed were evidently running out. The effects of former toil had told at last, and his once powerful constitution had suddenly collapsed. I was not able to get him on his legs again. And the most trying action that I was ever called upon to perform in my life, was when, after tying my handkerchief round his eyes (thereby sparing myself the additional pang of observing the intelligent animal's possible consciousness of the action that cruel necessity impelled me to), I put the muzzle of my revolver almost against his forehead, and pulling the trigger, thus put him beyond the reach of pain that I could not otherwise relieve. Digging a deep pit beside him, I there interred him, and at his head, I had shortly afterwards a large square post erected to commemorate the spot where lay the remains of the brave companion of my wanderings.

THE END.

www.ingramcontent.com/pod-product-compliance
Lightning Source LLC
Chambersburg PA
CBHW021159230426

43667CB00006B/474